PROBLEMS IN
QUANTUM CHEMISTRY

PROBLEMS IN QUANTUM CHEMISTRY

Poul Jørgensen
Department of Chemistry
Aarhus University
Aarhus, Denmark

Jens Oddershede
Department of Chemistry
Odense University
Odense, Denmark

1983

ADDISON-WESLEY PUBLISHING COMPANY
Advanced Book Program/World Science Division
Reading, Massachusetts
London • Amsterdam • Don Mills, Ontario • Sydney • Tokyo

Library of Congress Cataloging in Publication Data

Jørgensen, Poul, 1944-
    Problems in quantum chemistry.

    Includes bibliographical references.
    1. Quantum chemistry—Problems, exercises, etc.
I. Oddershede, Jens, 1945-   . II. Title.
QD462.7.J67 1982   541.2'8'076   82-11326
ISBN 0-201-05486-8

Manufactured in the United States of America

ABCDEFGHIJ—HA—898765432

TABLE OF CONTENTS

PREFACE

This problem set grew out of an undergraduate course in quantum chemistry. In the course we have used various textbooks which often had appended to each chapter problems illustrating the physical principles treated in the chapter. These problems did not, however, elucidate real chemical or physical events using the principles discussed in the text. We found that this aspect was important for motivating the students and for giving them a deeper understanding of the text. We therefore developed this set of problems, which represents an effort to bridge the gap between learning and applying quantum chemistry. In a typical problem of this set we calculate a physical quantity and then try to interpret the related experimental data, which may be of, for example, a spectral or conformational nature.

The topics of the problem set are the ones generally treated in undergraduate and graduate courses in quantum chemistry. The main emphasis is on molecular structure problems, but we have included two small sections on topics that are often treated in quantum mechanics courses, namely, simple quantum mechanical systems, and qualitative atomic orbital theory. In the molecular structure problems we have considered applications of both *ab initio* and semiempirical methods. In the sections on qualitative molecular orbital theory, ligand field theory, and application of symmetry in chemical reactions, extensive use has been made of group theory. The topics considered and the treatment given to these topics mean that there is very little overlap between this and other collections of problem sets in quantum mechanics and quantum chemistry.

To make it possible for the student to use these problems for a self-tutorial study of quantum chemistry, we have included solutions to all the problems. Furthermore, we have included a short introduction to each chapter and an appendix that in a telegraphic style defines and explains some of the key concepts needed to solve the problems.

For the student who is familiar with the fundamental concepts of quantum chemistry, the problem text and solutions are self-explanatory. For the inexperienced student, however, this problem set should be used in connection with one of the standard texts of quantum chemistry. The problems are formulated such that most texts can be used, and in the introduction to each chapter we have given examples of textbooks that contain sufficient material to enable the student to solve the problems of that chapter.

This problem set or earlier editions of it have been used in quantum chemistry courses at Aarhus and Odense Universities for about 15 years. Many of our present and previous colleagues have thus contributed significantly to the development of the problem set. We are particularly indebted to N.H.F. Beebe, E. Dalgaard, H.J.Aa. Jensen, J. Linderberg, L. Seamans, J. Simons, J. Spanget-Larsen, E.N. Svendsen, and P. Thulstrup, all of whom have either suggested some of the problems or greatly improved the presentation of them.

The typing of the manuscript was competently carried out by Hanne Kirkegaard and Birgitte Buus. We also thank Arne Lindahl for preparing all the drawings for the book.

Jens Oddershede
Poul Jørgensen

STANDARD CONVENTIONS AND FORMATS

The following standard conventions, notations, and formats have been used throughout the problem sets.

1. Each chapter starts out with a short *introduction*, which explains very briefly the contents of the problems in that section. In a few cases where we feel that the solution of the problems requires knowledge of topics that are not generally included in standard texts of quantum chemistry, we have given a short summary of the theory in the introductory sections. When we have found it more convenient to do so, we have occasionally instead presented the theory directly, as an integrated part of a problem.

2. The use of molecular symmetry or *group theory* plays an important role in the problem set. Unfortunately, many different conventions exist for choosing coordinate axes and the use of character tables. Throughout the problem set we have used the conventions defined in the tables of Atkins *et al.* (1970). This means, for example, that the principal axis (if it can be defined unambiguously) is the z-axis of the coordinate system. We have also often included a figure illustrating the choice of coordinate system for a particular problem.

3. Many of the key concepts have been defined in a telegraphic style in the *list of concepts* that constitutes the Appendix to the book.

4. *Atomic units* have been used exclusively unless otherwise specified. A definition of atomic units and their relations to other units is given at the end of the Appendix.

5. We have tried to maintain a common *nomenclature for orbitals, wave functions*, and the like, throughout the book:
   (a) ϕ_i labels an atomic orbital.
   (b) χ_i labels a symmetry orbital, that is, an orbital that transforms as one of the irreducible representations of the molecular point group.
   (c) ψ_i labels a molecular orbital.
   (d) spin-orbitals of types (a)-(c) are labeled as the orbital itself if it has α spin and with a bar (e.g., $\bar{\psi}_i$) if it has β spin.

(e) Ψ_i labels a state wave function that can be either a single determinant or a linear combination of determinants.

(f) u_i labels an unspecified wave function of types (a)-(e).

6. The *Dirac bra–ket* notation is used in most of the book. A state wave function is thus labeled as, for example, $|\Psi_{LSJ}\rangle$ or, in a shorthand notation, $|LSJ\rangle$. The scalar product is defined as

$$\langle \Psi_i | \Psi_j \rangle = \int \Psi_i^* \, \Psi_j \, d\tau$$

and the average value of an operator $\hat{H}$ is

$$\langle \Psi_i | \hat{H} | \Psi_j \rangle = \int \Psi_i^* \, \hat{H} \Psi_j \, d\tau$$

7. *Operators* have a caret on top—for example, $\hat{H}$.

8. *Equations, figures, and tables* are labeled with the section letter and problem number. Within a particular problem the equations, figures, and tables are numbered with Arabic numerals in the problem text and with Roman numerals in the solutions. For example, Eq. (F.3,2) is the second equation in problem F.3, and Fig. F.8, II is the second figure in the solution to problem F.8. Equation numbers are in parentheses—Eq. (E.11,1)—whereas figure and table numbers are not—Fig. H.3,I.

9. We have tried to limit the use of *acronyms*. However, some of the more standard abbreviations will be found in the text. These include the following:

AO	atomic orbital
CI	configuration interaction
EWMO	energy-weighted maximum overlap
HF	Hartree-Fock
HOMO	highest occupied molecular orbital
IR	irreducible representation
LCAO	linear combination of atomic orbitals
LUMO	lowest unoccupied molecular orbital
MO	molecular orbital
PPP	Pariser-Parr-Pople
SCF	self-consistent field
STO	Slater-type orbital
ZDO	zero differential overlap

SIMPLE QUANTUM MECHANICAL SYSTEMS

INTRODUCTION

In this introductory problem set we use the simple quantum mechanical systems (the particle in a box and the hydrogen atom) to illustrate some of the fundamental principles of quantum mechanics. We consider the quantum mechanical probability interpretation (A.1, A.2) and we discuss the time development of the wave function, the average position, and the average energy for a nonstationary solution to the Schrödinger equation (A.3). The "aufbau principle" for particles in a two-dimensional box is the topic of A.4.

The last two problems discuss the use of approximate hydrogenic orbitals in the calculation of ground-state total energy (A.5), kinetic and potential energy, and dipole oscillator strengths (A.6) for the hydrogen atom. The approximate orbitals encompass Gaussian and Slater-type orbitals, which are the standard expansion functions used in most quantum chemical calculations.

The theory needed to solve the problems in this set is given in any text of quantum mechanics and in the introductory chapters of most quantum chemistry books.

Textbooks that give sufficient background for solving the problems in this section

Chandra (1974). Dekock and Gray (1980). Dewar (1969). George (1972). Levine (1974). Lowe (1978). McWeeny (1979). Murrell, Kettle and Tedder (1970, 1978). Pilar (1968). Slater (1968).

Textbooks that give a more elaborate treatment of the topics

Atkins (1970). Dicke and Wittke (1960). Eyring, Walter, and Kimball (1944). Landau and Lifshitz (1965). Merzbacher (1970). Pauling and Wilson (1935). Schiff (1968).

Problem Description

A.1 One-dimensional box: probability distribution
A.2 One-dimensional box: momentum distribution
A.3 One-dimensional box: time-dependent nonstationary state
A.4 Two-dimensional box: distortion with constant area of the box
A.5 Trial wave functions for the ground state of the hydrogen atom
A.6 Virial theorem and oscillator strengths using the exact and the Slater-
      type orbitals for the hydrogen atom

PROBLEMS

A.1 Consider a particle in a one-dimensional box of length l. Let us assume that the particle is in an energy eigenstate $\psi_n(x)$.
1.   Calculate the probability $P(\frac{1}{4})$ of finding the particle in the left quarter $(0 \leq x \leq l/4)$ of the box.
2.   Determine an explicit expression for $P(\frac{1}{4})$ for n even and odd.
3.   Which value of n gives maximum probability?
4.   Determine the limiting value of $P(\frac{1}{4})$ for $n \to \infty$. How does this value compare with the one obtained in the classical limit?

A.2 Consider a particle in a one-dimensional box of length l. Assume that the particle is in an energy eigenstate $\psi_n(x)$.
1.   Show that the probability for finding the particle with momentum p is

$$P(p) = \frac{4p_n^2}{l^2(p_n^2 - p^2)^2}[1 - (-1)^n \cos(pl)]$$

where $p_n = n\pi l^{-1}$ and the eigenfunctions of the momentum operator are normalized over the length of the box.
2.   Determine the classical momentum of the particle in the eigenstate ψ_n.
We will now show that the quantum mechanical momentum distribution has maximum amplitude for the classical momentum. Consider the momenta $p = p_n + \alpha$.
3.   What is the limiting value of $P(p)$ for $\alpha \to 0$?
4.   Sketch $P(p)$ as a function of α for $n = 1$ and $n = \infty$.

A.3 Consider a particle in a one-dimensional box of length l. Let us assume that the particle is in the nonstationary state

$$\psi(x, t) = \frac{1}{\sqrt{2}}[\psi_1(x, t) + \psi_2(x, t)]$$

where ψ_1 and ψ_2 are the two lowest stationary states for the particle in the box. Let us first investigate the time development of the probability amplitude of $\psi(x, t)$.
1.   Plot $|\psi(x, t)|^2$, $|\psi_1(x, t)|^2$, and $|\psi_2(x, t)|^2$ versus x for $t = 0$, and explain the differences in the shape of $|\psi(x, 0)|^2$ relative to those of $|\psi_1(x, 0)|^2$ and $|\psi_2(x, 0)|^2$.
2.   At which times t are $|\psi(x, t)|^2 = |\psi(x, 0)|^2$?

3.   Plot $|\psi(x, t_0)|^2$ for $t_0 = 0$, $l^2/3\pi$, and $2l^2/3\pi$.

Let us now consider the time development of the average of the position and energy operators.

4.   Determine the average value of x for the system in the nonstationary state $\psi(x, t)$.

5.   Determine the total energy for the system in the nonstationary state $\psi(x, t)$.

A.4 Electrons that are bound to a surface will be restricted in their movement to local parts of the surface if certain conditions are fulfilled for the surface structure. As a model for the movement of the bound electrons we will assume that the electrons are confined to move in a two-dimensional box with the widths a and b. We limit ourselves to considering width lengths in the intervals $\sqrt{3/8} \leq a/b \leq \sqrt{8/3}$, and assume further that the electrons in the box are noninteracting.

1.   Determine the energy levels and eigenfunctions for the electrons.

Assume that each two-dimensional box on the surface on average contains five noninteracting electrons. The total energy of a state is obtained by adding the orbital energies of the noninteracting electrons, which obey the Pauli exclusion principle.

2.   Determine for various values of a and b the occupation of the orbital energy levels when the five electrons are in the state of lowest total energy (the ground state) and discuss the degeneracy of this state.

We will now distort the box, assuming that the surface area that is covered by the box is kept constant; that is, $ab = c^2$ (=constant).

3.   Plot the energies of the orbitals that are occupied in the ground state as a function of the variable a for constant c with indications of extrema and orbital crossings.

4.   Determine the total energy of the ground state (question 2) as a function of the variable a and the constant c. Indicate the values of a for which the ground-state total energy will be minimum. What is the energy lowering of the ground state compared to the energy of the square box ($a = b = c$) ? Plot the ground-state total energy as a function of a for constant c.

5.   Can the minimum of the total energy be determined by considering the change in only the highest occupied energy level (the frontier orbital)?

A.5 To describe the ground state of the hydrogen atom we choose the trial functions

$$\psi_1(r, \alpha) = e^{-\alpha r}, \quad \psi_2(r, \alpha) = e^{-\alpha r^2}, \quad \text{and} \quad \psi_3(r, \alpha) = (1 + \alpha r)^{-2}$$

where α is a parameter to be determined from a minimization of the ground-state total energy.

1.   Show that

$$E_1(\alpha) = \frac{\alpha^2}{2} - \alpha$$

$$E_2(\alpha) = \frac{3}{2}\alpha - 2\sqrt{\frac{2\alpha}{\pi}}$$

and

$$E_3(\alpha) = \frac{\alpha^2}{5} - \frac{\alpha}{2}$$

where $E_i(\alpha)$ is the energy function for the trial wave function $\psi_i(r, \alpha)$.

2. Determine the values of $\alpha(= \alpha_i^{(0)})$ that correspond to a minimum for $E_i(\alpha)$, $i = 1$, 2, and 3.
3. Determine the minimum ground-state total energy for the three trial functions.
4   Plot the radially normalized functions $\psi_i(r, \alpha_i^{(0)})$, $i = 1$, 2, and 3, as a function of r. Comment on the accuracy of the energies determined in question 3 by considering the behavior of $\psi_i(r, \alpha_i^{(0)})$ for various regions of r.
5. Could α in $\psi_2(r)$ and $\psi_3(r)$ have been chosen to reproduce the correct behavior of the hydrogenic orbitals for small or for large values of r?

A.6 In this problem we analyze how theorems that hold for exact wave functions are fulfilled if approximate wave functions are used. We consider the virial theorem and the equivalence between the dipole length and velocity forms for the oscillator strength. We will first prove that the equivalence of the two forms for the oscillator strengths is valid for exact wave functions.

Consider the Hamiltonian for a one-electron system

$$\hat{H} = -\frac{1}{2}\left(\frac{\partial^2}{\partial x^2} + \frac{\partial^2}{\partial y^2} + \frac{\partial^2}{\partial z^2}\right) + \hat{V}(x, y, z) \tag{A.6,1}$$

where the potential energy $\hat{V}(x, y, z)$ is a simple multiplicative operator that commutes with x, y, and z.

1. Show that

$$[\hat{H}, x] = -\frac{\partial}{\partial x} \tag{A.6,2}$$

2. Use Eq. (A.6,2) to prove that

$$\left\langle \psi_i \left| \frac{\partial}{\partial x} \right| \psi_f \right\rangle = (E_f - E_i)\langle \psi_i | x | \psi_f \rangle \tag{A.6,3}$$

where ψ_i and ψ_f are eigenfunctions of $\hat{H}$ with the eigenvalues E_i and E_f, respectively.

The intensity of light absorption for a dipole allowed transition $\psi_i \rightarrow \psi_f$ is proportional to the oscillator strength for the transition. In atomic units the oscillator strength in the *dipole length approximation* is defined as

$$f_L = \frac{2}{3}|\langle \psi_i | \mathbf{r} | \psi_f \rangle|^2(E_f - E_i) \tag{A.6,4}$$

where $\mathbf{r}$ is the position operator. In the *dipole velocity approximation* the matrix element between states ψ_i and ψ_f is taken with respect to the momentum operator $\mathbf{p}$.

3. Show that the dipole velocity oscillator strength

$$f_V = \frac{\frac{2}{3}|\langle \psi_i | \mathbf{p} | \psi_f \rangle|^2}{E_f - E_i} \tag{A.6,5}$$

is identical to f_L for the exact wave function.

Consider in the following the hydrogen atom. Let us first see how well the virial theorem is fulfilled.

4. Determine the average kinetic and potential energy for the electron in the 1s, 2s, 2p, and 3p states by using the exact hydrogenic orbitals.

5. Calculate the average kinetic and potential energy for the electron in the 1s, 2s, 2p, and 3p states by using approximate orbitals given as Slater-type orbitals with screening parameters $\zeta_{1s} = 1.0$, $\zeta_{2s} = 0.5$, $\zeta_{2p} = 0.5$, and $\zeta_{3p} = \frac{1}{3}$ (Slater 1930). How well is the virial theorem fulfilled for the approximate orbitals?

Let us consider the equivalence of Eqs. (A.6,4) and (A.6,5) for the hydrogen atom.

6. Calculate the oscillator strength in both the dipole length and in the dipole velocity approximation for the transitions $1s \rightarrow 2s$, $1s \rightarrow 2p$, and $1s \rightarrow 3p$ of the hydrogen atom by using the exact hydrogenic orbitals. What is the sum of the oscillator strenghts for all the remaining transitions $1s \rightarrow np$ $(n = 4, 5, \ldots, \infty)$?

7. Calculate the same quantities as in question 6 by using the approximate Slater-type orbitals of question 5. How well is the equivalence between the oscillator strengths fulfilled for the approximate orbitals?

SOLUTIONS

A.1 1. $P(\frac{1}{4}) = \frac{2}{l} \int_0^{l/4} \sin^2 \frac{\pi n x}{l} dx = \frac{1}{4} - \frac{1}{2\pi n} \sin \frac{\pi n}{2}$

2.

$$P(\frac{1}{4}) = \frac{1}{4}, \qquad\qquad n \text{ even}$$

$$P(\frac{1}{4}) = \frac{1}{4} + \frac{(-1)^{(n+1)/2}}{2\pi n}, \qquad n \text{ odd}$$

3. The maximum probability is obtained for $n = 3$.

4. $P(\frac{1}{4}) \rightarrow \frac{1}{4}$

The classical limit gives $P(\frac{1}{4}) = \frac{1}{4}$. This limit is reached when $n \rightarrow \infty$.

A.2 1. Momentum eigenfunctions: $\psi_p(x) = \sqrt{1/l} \exp(ipx)$.

$$P(p) = |\langle \psi_p(x)|\psi_n(x)\rangle|^2 \qquad\qquad\qquad (A.2,I)$$

$$\langle \psi_p(x)|\psi_n(x)\rangle = \frac{\sqrt{2}}{2il} \int_0^l [\exp(-i(p - p_n)x) - \exp(-i(p \mid p_n)x)] dx \qquad (A.2,II)$$

$$= \frac{\sqrt{2} p_n}{l(p_n^2 - p^2)}[1 - (-1)^n \exp(-ipl)]$$

and $P(p)$ is obtained straightforwardly from (A.2,I-II).

2. $E_n = n^2\pi^2/2l^2$ and $E_n = p^2/2$. Thus

$$p = \frac{n\pi}{l} = p_n$$

3.

$$P(n\pi l^{-1} + \alpha) = \frac{4p_n^2}{l^2\alpha^2(2p_n + \alpha)^2}[1 - (-1)^n\cos(p_n l + \alpha l)]$$

$$= \frac{4}{l^2\alpha^2(2 + \alpha/p_n)^2}[1 - \cos(\alpha l)]$$

$$(A.2,III)$$

$$\cos(\alpha l) = 1 - \frac{\alpha^2 l^2}{2} + \mathcal{O}(\alpha^4)$$

which when inserted into (A.2,III) gives

$$\lim_{\alpha \to 0} P(n\pi l^{-1} + \alpha) = \frac{1}{2}$$

Notice that this probability is independent of n.
4. Using (A.2,III) we obtain

$$P(p) = P(p_n)\left(1 + \frac{\alpha}{2p_n}\right)^{-2}\frac{\sin^2(\alpha l/2)}{(\alpha l/2)^2}$$

This function has nodes for $\alpha = 2\pi m/l$, $m = \pm 1, \pm 2, \ldots$, and the maximum value is obtained for $\alpha = 0$, that is, at the classical momentum. The curves $P(p)$ for $n = 1$ and $n = \infty$ are plotted in Fig. A.2,I.

A.3 1. $\psi_1(x,0) = \sqrt{\frac{2}{l}} \sin\frac{\pi x}{l}$, $\psi_2(x,0) = \sqrt{\frac{2}{l}} \sin\frac{2\pi x}{l}$

$(l/2)|\psi_1(x,0)|^2$, $(l/2)|\psi_2(x,0)|^2$, and $(l/2)|\psi(x,0)|^2$ are plotted in Fig. A.3,I. The interference term between $\psi_1(x,0)$ and $\psi_2(x,0)$ in $\psi(x,0)$ enhances and reduces $(l/2)|\psi(x,0)|^2$ for small and large x values, respectively.
2. $\psi(x,t) = (1/\sqrt{2})[\psi_1(x,0)\exp(-iE_1 t) + \psi_2(x,0)\exp(-iE_2 t)]$
where
$$E_1 = \pi^2/2l^2 \text{ and } E_2 = 2\pi^2/l^2.$$

$$|\psi(x,t)|^2 = \frac{1}{2}\psi_1^2(x,0) + \frac{1}{2}\psi_2^2(x,0) + \psi_1(x,0)\psi_2(x,0)\cos\frac{3\pi^2 t}{2l^2}$$

$|\psi(x,t)|^2 = |\psi(x,0)|^2$ when $\cos(3\pi^2 t/2l^2) = 1$, that is, when $t = 4l^2 p/3\pi$, $p = 0, 1, 2, \ldots$.

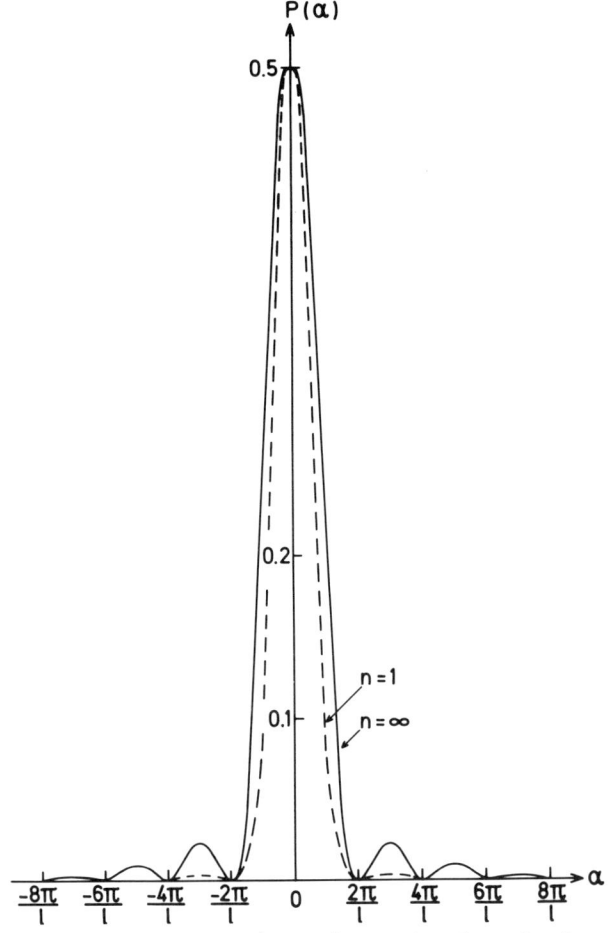

Fig. A.2,I. The probability function $P(p_n + \alpha)$ as a function of α for $n = 1$ and $n = \infty$.

3. $(l/2)|\psi(x,0)|^2$ was plotted in Fig. A.3, I.

$$|\psi(x,\frac{l^2}{3\pi})|^2 = \frac{1}{2}\psi_1^2(x,0) + \frac{1}{2}\psi_2^2(x,0)$$

$$|\psi(x,\frac{2l^2}{3\pi})|^2 = \frac{1}{2}(\psi_1(x,0) - \psi_2(x,0))^2$$

$(l/2)|\psi(x,t)|^2$ is plotted for $t = 0$, $l^2/3\pi$, and $2l^2/3\pi$ in fig. A.3,II. Note how the probability density switches back and forth within the box.

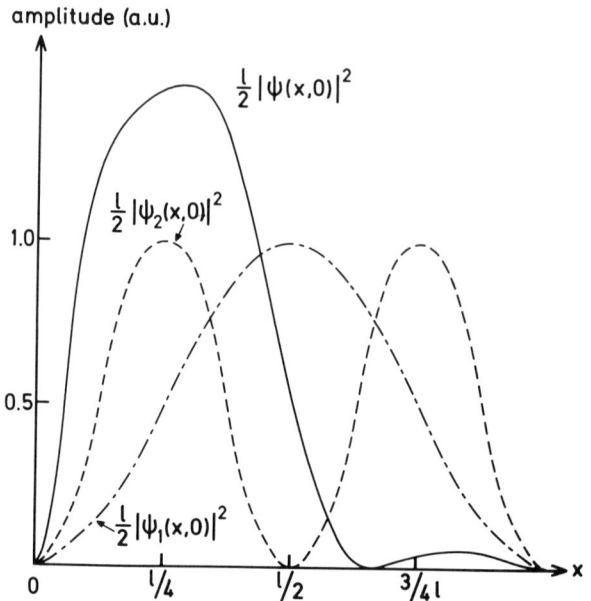

Fig. A.3,I. $(l/2)|\psi_1(x,0)|^2$, $(l/2)|\psi_2(x,0)|^2$, and $(l/2)|\psi(x,0)|^2$ as a function of x for $0 \le x < l$.

4.

$$\langle x \rangle = \frac{1}{2}\langle \psi_1(x,t) + \psi_2(x,t)|x|\psi_1(x,t) + \psi_2(x,t)\rangle$$

$$= \frac{l}{2} - \frac{16l}{9\pi^2}\cos\frac{3\pi^2 t}{2l^2}$$

Thus $\langle x \rangle$ varies with time.

5.

$$E = \frac{1}{2}\langle \psi_1(x,t) + \psi_2(x,t)|\hat{H}|\psi_1(x,t) + \psi_2(x,t)\rangle$$

$$= \frac{1}{2}(E_1 + E_2)$$

In contrast to $\langle x \rangle$, the energy E is independent of time and, as expected, equal to half the sum of the eigenvalues of the two stationary states.

A.4 1.

$$E_{nm} = \left(\frac{n^2}{a^2} + \frac{m^2}{b^2}\right)\frac{\pi^2}{2}, \qquad n, m = 1, 2, 3, \ldots$$

$$\psi_{nm} = \frac{2}{\sqrt{ab}}\sin\frac{\pi n x}{a}\sin\frac{\pi m y}{b}, \qquad n, m = 1, 2, 3, \ldots$$

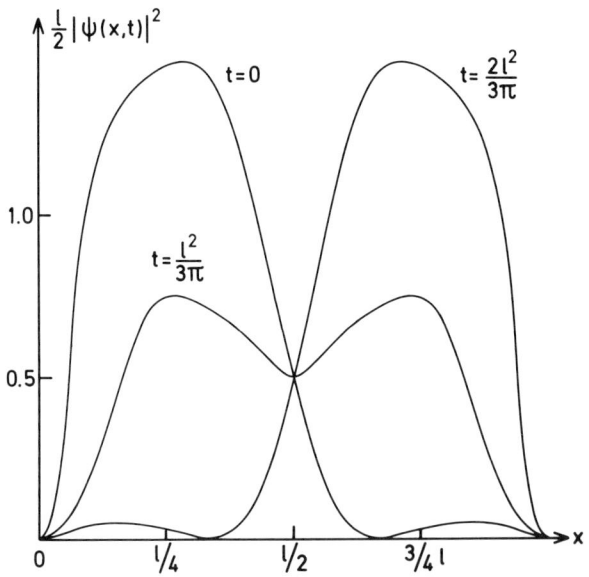

Fig. A.3,II. $(l/2)|\psi(x,t)|^2$ as a function of x for $0 \leqq x \leqq l$ when $t = 0$, $l^2/3\pi$, and $2l^2/3\pi$.

2. Possible ground-state total energies:

$$E_0^1 = 2E_{11} + 2E_{21} + E_{12} = \left(\frac{11}{a^2} + \frac{8}{b^2}\right)\frac{\pi^2}{2}$$

$$E_0^2 = 2E_{11} + 2E_{12} + E_{21} = \left(\frac{8}{a^2} + \frac{11}{b^2}\right)\frac{\pi^2}{2}$$

$$E_0^3 = 2E_{11} + 2E_{21} + E_{31} = \left(\frac{19}{a^2} + \frac{5}{b^2}\right)\frac{\pi^2}{2}$$

$$E_0^4 = 2E_{11} + 2E_{12} + E_{13} = \left(\frac{5}{a^2} + \frac{19}{b^2}\right)\frac{\pi^2}{2}$$

$$E_0^1 \leqq E_0^3 \quad \text{since} \quad \frac{a}{b} \leqq \sqrt{\frac{8}{3}}$$

$$E_0^2 \leqq E_0^4 \quad \text{since} \quad \frac{a}{b} \geqq \sqrt{\frac{3}{8}}$$

Either E_0^1 or E_0^2 is thus the energy of the ground state. When $a = b$, $E_0^1 = E_0^2$ and the ground state is degenerate. When $b < a$, E_0^1 is the energy of the ground state, and when $b > a$, E_0^2 is the energy of the ground state.

3.

$$E_{nm} = \left(\frac{n^2}{a^2} + \frac{m^2 a^2}{c^4} \right) \frac{\pi^2}{2}$$

Stationary points when $dE_{nm}/da = 0$ or $a^2 = c^2(n/m)$. The orbital energies at the stationary points are

$$E_{nm}^0 = \frac{\pi^2}{c^2} nm$$

Since

$$\frac{d^2 E_{nm}}{da^2} = \left(\frac{6n^2}{a^4} + \frac{2m^2}{c^4} \right) \frac{\pi^2}{2} > 0$$

all stationary points represent minima. The energy levels E_{12} and E_{21} cross when $a = b$ at the orbital energy $5\pi^2/2c^2$. The energy levels E_{11}, E_{21}, and E_{12} are plotted in Fig. A.4,I.

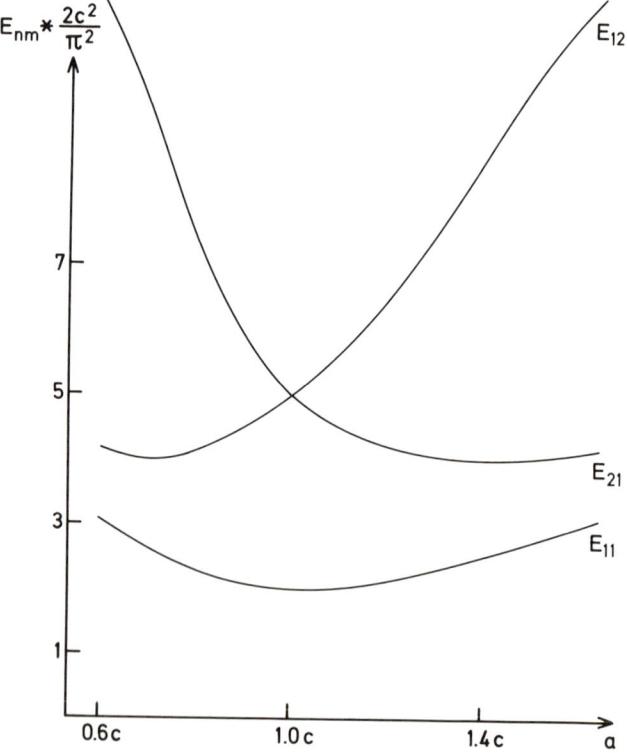

Fig. A.4,I. The energy levels E_{11}, E_{21}, and E_{12} as a function of a.

4.

$$a > c: \quad E_0^1 = \left(\frac{11}{a^2} + \frac{8a^2}{c^4}\right)\frac{\pi^2}{2}$$

$$a < c: \quad E_0^2 = \left(\frac{8}{a^2} + \frac{11a^2}{c^4}\right)\frac{\pi^2}{2}$$

Stationary properties:

$$a > c: \quad a = \left(\frac{11}{8}\right)^{1/4} c, \quad E_T = \frac{2\pi^2\sqrt{22}}{c^2} \quad \text{(minimum)}$$

$$a < c: \quad a = \left(\frac{8}{11}\right)^{1/4} c, \quad E_T = \frac{2\pi^2\sqrt{22}}{c^2} \quad \text{(minimum)}$$

When $a = b = c$, the total energy is

$$E_{abc} = \frac{19\pi^2}{2c^2}$$

and the energy lowering is thus

$$E_{abc} - E_T = \frac{1.176}{c^2} \text{ a.u.}$$

The ground-state total energy is plotted in Fig. A.4,II.

5. The minima for the orbital energy levels and the ground-state total energy occur at different locations. Determination of the minima for the ground-state total energy requires explicit consideration of the ground-state *wave function*.

A.5 1.

$$\hat{H} = -\frac{1}{2r^2}\frac{d}{dr}r^2\frac{d}{dr} - \frac{1}{r}$$

$$E_i(\alpha) = \frac{\langle \psi_i(r,\alpha)|\hat{H}|\psi_i(r,\alpha)\rangle}{\langle \psi_i(r,\alpha)|\psi_i(r,\alpha)\rangle}$$

2.

$$\alpha_1^{(0)} = 1, \qquad \alpha_2^{(0)} = \frac{8}{9\pi}, \qquad \alpha_3^{(0)} = \frac{5}{4}$$

3.

$$E_1(\alpha_1^{(0)}) = -0.5 \text{ a.u.}$$

$$E_2(\alpha_2^{(0)}) = -\frac{4}{3\pi} = -0.4244 \text{ a.u.}$$

$$E_3(\alpha_3^{(0)}) = -\frac{5}{16} = -0.3125 \text{ a.u.}$$

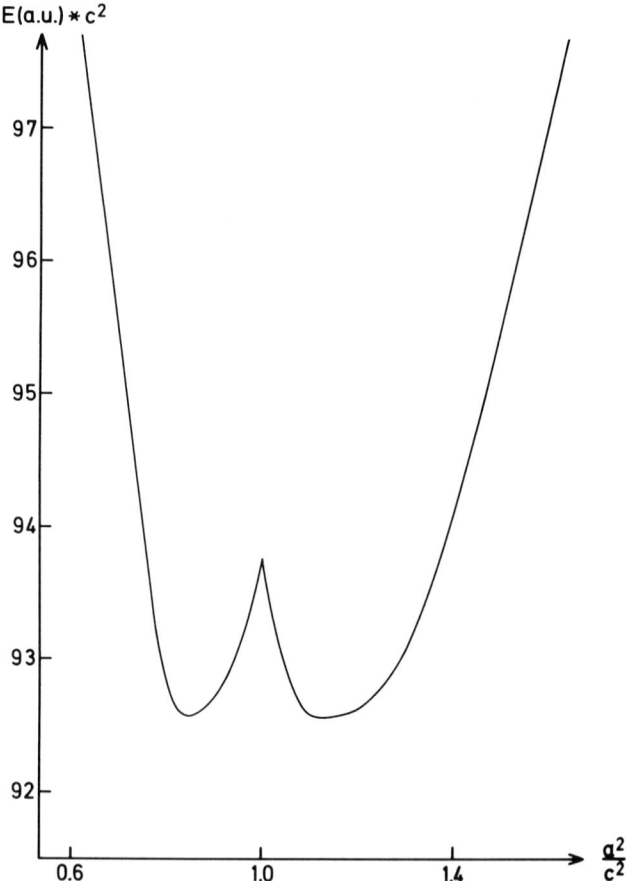

Fig. A.4,II. The ground-state total energy as a function of a^2/c^2.

4. Radially normalized functions:

$$\psi_1(r, \alpha_1^0) = 2 \exp(-r)$$

$$\psi_2(r, \alpha_2^0) = \frac{16}{3\pi\sqrt{3}} \exp\left(-\frac{8r^2}{9\pi}\right)$$

$$\psi_3(r, \alpha_3^0) = \frac{5\sqrt{15}}{8}\left(1 + \frac{5r}{4}\right)^{-2}$$

Because $\psi_1(r)$ is identical to the exact $1s$ function for hydrogen, it gives the exact energy.

For "medium" r values the Gaussian function looks more like the exact function than ψ_3 does. This is the region of space in which it is most important to simulate the exact wave function if we want a good ground-state total energy (E_2 is closer to E_1).

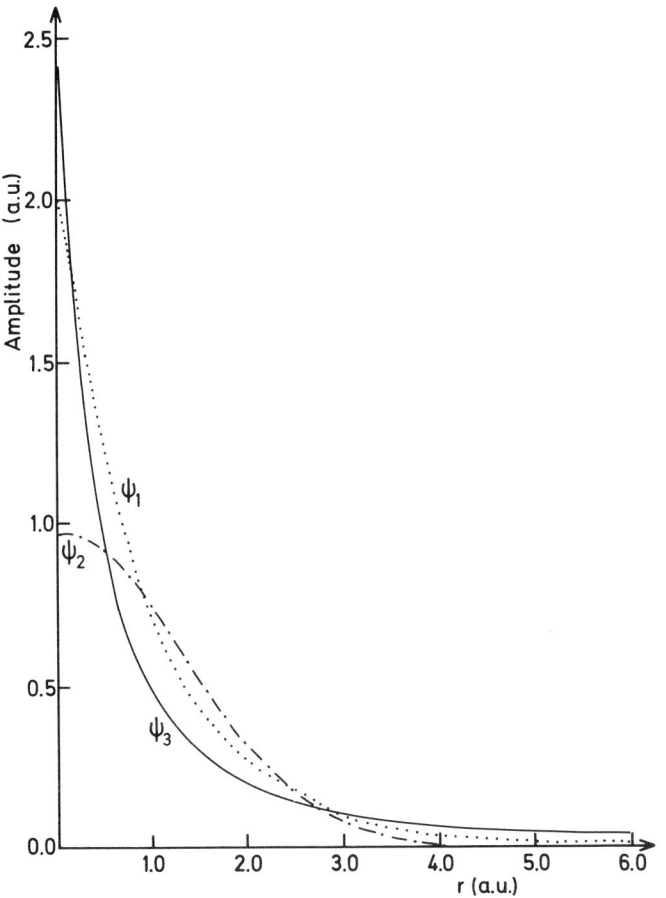

Fig. A.5,I. The normalized radial functions $\psi_i(r)$, $i = 1, 2, 3$, as a function of r.

5. Large values: never possible because the asymptotic behavior of each
   of the three functions is different.
   Small values:

$$\psi_1(r) \simeq 1 - r + \frac{r^2}{2} - \cdots$$

$$\psi_2(r) \simeq 1 - \alpha r^2 + \frac{\alpha^2 r^4}{2} - \cdots$$

$$\psi_3(r) \simeq 1 - 2\alpha r + 3\alpha^2 r^2 - \cdots$$

$\psi_2(r)$ can never have the same asymptotic behavior as
$\psi_1(r)$. ψ_1 and ψ_3 have the same behavior for $\alpha \simeq 0.5$.
However, $E_3(\alpha = 0.5) = -0.2$ a.u., which is a higher

energy than $E_3(\alpha_3^{(0)})$ and it is thus not energetically favorable to match the asymptotic behavior of the trial function and the exact function.

A.6 1.

$$[\hat{H}, x] = -\frac{\partial}{\partial x}[\frac{\partial}{\partial x}, x] = -\frac{\partial}{\partial x}$$

4. The virial theorem

$$\langle \hat{V}_n \rangle = 2E_n, \qquad \langle \hat{T}_n \rangle = -E_n$$

is fulfilled for the hydrogenic orbitals; that is,

$$\langle \hat{T}_{1s} \rangle_{exc} = 0.5 \text{ a.u.} \qquad\qquad \langle \hat{V}_{1s} \rangle_{exc} = -1.0 \text{ a.u.}$$

$$\langle \hat{T}_{2s} \rangle_{exc} = \langle \hat{T}_{2p} \rangle_{exc} = 0.125 \text{ a.u.} \quad \langle \hat{V}_{2s} \rangle_{exc} = \langle \hat{V}_{2p} \rangle_{exc} = -0.25 \text{ a.u.}$$

$$\langle \hat{T}_{3p} \rangle_{exc} = \frac{1}{18} \text{ a.u.} \qquad\qquad \langle \hat{V}_{3p} \rangle_{exc} = -\frac{1}{9} \text{ a.u.}$$

5. Normalized STO's:

$$\psi_{1s} = 2e^{-r} Y_{00}$$

$$\psi_{2s} = \frac{1}{2\sqrt{6}} re^{-r/2} Y_{00}$$

$$\psi_{2p} = \frac{1}{2\sqrt{6}} re^{-r/2} Y_{1m}, \qquad m = 0, \pm 1$$

$$\psi_{3p} = \frac{2\sqrt{2}}{81\sqrt{15}} r^2 e^{-r/3} Y_{1m}, \quad m = 0, \pm 1$$

ψ_{1s} and ψ_{2p} are identical to the exact hydrogenic orbitals and thus yield the same values for $\langle \hat{V} \rangle$ and $\langle \hat{T} \rangle$ as in question 4.

$$\langle \hat{V}_{2s} \rangle_{STO} = -\langle \psi_{2s} | \frac{1}{r} | \psi_{2s} \rangle = -0.25 \text{ a.u.} = \langle \hat{V}_{2s} \rangle_{exc}$$

$$\langle \hat{T}_{2s} \rangle_{STO} = -\frac{1}{2} \langle \psi_{2s} | \frac{1}{r^2} \frac{d}{dr} r^2 \frac{d}{dr} | \psi_{2s} \rangle = \frac{1}{24} \text{ a.u.} = \frac{1}{3} \langle \hat{T}_{2s} \rangle_{exc}$$

$$\langle \hat{V}_{3p} \rangle_{STO} = -\frac{1}{9} \text{ a.u.} = \langle \hat{V}_{3p} \rangle_{exc}$$

$$\langle \hat{T}_{3p} \rangle_{STO} = -\frac{1}{2} \langle \psi_{3p} | \frac{1}{r^2} \frac{d}{dr} r^2 \frac{d}{dr} + \frac{\mathbf{L}^2}{2r^2} | \psi_{3p} \rangle = \frac{7}{270} \text{ a.u.}$$

$$= \frac{7}{15} \langle \hat{T}_{3p} \rangle_{exc}$$

The virial theorem term is not fulfilled for the 2s and 3p Slater-type orbitals.

6. (1)

$$f_L(1s \rightarrow 2s) = f_V(1s \rightarrow 2s) = 0$$

(2)

$$\langle 1s|z|2p_z \rangle = \langle 1s|x|2p_x \rangle = \langle 1s|y|2p_y \rangle = \frac{256}{243\sqrt{2}} \text{ a.u.}$$

$$E_{2p} - E_{1s} = \frac{3}{8} \text{ a.u.}$$

$$f_L(1s \rightarrow 2p) = 3\frac{2}{3}|\langle 1s|z|2p_z \rangle|^2(E_{2p} - E_{1s}) = \frac{8192}{19683} = 0.4162$$

$$= f_V(1s \rightarrow 2p)$$

(3)

$$\langle 1s|z|3p_z \rangle = \langle 1s|x|3p_x \rangle = \langle 1s|y|3p_y \rangle = \frac{27}{64\sqrt{2}} \text{ a.u.}$$

$$E_{3p} - E_{1s} = \frac{4}{9} \text{ a.u.}$$

$$f_L(1s \rightarrow 3p) = f_V(1s \rightarrow 3p) = \frac{81}{1024} = 0.0791$$

(4)Application of the Thomas-Reice-Kuhn or f-sum rule gives

$$\sum_{n=4}^{\infty} f_L(1s \rightarrow np) = \sum_{n=4}^{\infty} f_V(1s \rightarrow np)$$

$$= 1 - f_L(1s \rightarrow 2p) - f_L(1s \rightarrow 3p) = 0.5047$$

7. (1)

$$f_L(1s \rightarrow 2s) = f_V(1s \rightarrow 2s) = 0$$

(2)Since $\psi_{2p}(STO) = \psi_{2p}(\text{exact})$, we have the same answer for the $1s \rightarrow 2p$ oscillator strength as in question 6.(2).

(3)

$$\langle \psi_{1s}|z|\psi_{3p_z} \rangle = \frac{9\sqrt{10}}{128} \text{ a.u.}$$

$$\langle \psi_{1s}|\frac{\partial}{\partial z}|\psi_{3p_z} \rangle = -\frac{3\sqrt{2}}{32\sqrt{5}} \text{ a.u.}$$

$$f_L^{STO}(1s \rightarrow 3p) = \frac{45}{1024} = 0.0439$$

$$f_V^{STO}(1s \rightarrow 3p) = \frac{81}{5120} = 0.0158$$

Notice the large difference between f_L^{STO} and f_V^{STO}. With the exact orbitals, these numbers were equal.

(4)

$$\sum_{n=4}^{\infty} f_L^{STO}(1s \rightarrow np) = 1 - f_L^{STO}(1s \rightarrow 2p) - f_L^{STO}(1s \rightarrow 3p)$$

$$= 0.5399$$

$$\sum_{n=4}^{\infty} f_V^{STO}(1s \rightarrow np) = 1 - f_V^{STO}(1s \rightarrow 2p) - f_V^{STO}(1s \rightarrow 3p)$$

$$= 0.5680$$

NON-EMPIRICAL METHODS

INTRODUCTION

In this section we demonstrate how *ab initio* calculations may be carried out on atoms and small diatomic molecules. We discuss, in problems B.1, B.6, and B.10, methods for evaluating the elementary integrals encountered in atomic and molecular *ab initio* calculations. These integrals are used to carry out standard variational (B.2, 3, 8), self-consistent field (SCF) (B.4, 9), configuration interaction (CI) (B.5, 9), and perturbational calculations (B.11, 12).

The problems concentrate on the calculation of ground-state total energies without (SCF) and with correlation effects included (CI and perturbational method). The calculation of ionization potentials (B.5), excitation energies (B.5, 7, 9), oscillator strengths (B.7, 9), dipole moments (B.10), and dispersion energies (B.13) has also been considered.

Most of the theoretical background needed to solve the problems is included in standard textbooks of quantum chemistry. A supplementary explanation is given in some of the problems. The self-consistent field Hartree–Fock method plays a central role in many of the problems. To avoid notational confusion we list below a few of the fundamental formulas in the SCF theory for a closed-shell system.

The SCF approach assumes that the wave function can be written as a *single* Slater determinant. The orbitals that are used to construct the Slater determinant are then determined by minimizing the total energy of the system. SCF calculations that do not impose symmetry restrictions on the wave function are referred to as unrestricted Hartree-Fock calculations. When symmetry restrictions are imposed on the wave function the resulting calculation is denoted a restricted Hartree-Fock calculation. When a restricted Hartree-Fock calculation is carried out on a closed-shell system the SCF orbitals $\{\psi_k\}$ are eigenfunctions of the Fock operator $\hat{F}$.

$$\hat{F}\psi_k = \varepsilon_k \psi_k \tag{B.0, 1}$$

$$\hat{F} = \hat{h} + \sum_{k}^{\text{occ}} (2\hat{J}_k - \hat{K}_k) \tag{B.0, 2}$$

where we sum over all doubly occupied (spatial) orbitals in the single determinant and where we have defined the Coulomb

$$\hat{J}_k(1)\psi_i(1) = \int d\tau_2 \psi_k^*(2) r_{12}^{-1} \psi_k(2)\psi_i(1) \tag{B.0, 3}$$

and exchange operator

$$\hat{K}_k(1)\psi_i(1) = \int d\tau_2 \psi_k^*(2) r_{12}^{-1} \psi_i(2)\psi_k(1) \tag{B.0, 4}$$

and $d\tau_2$ indicate the volume element $dx_2\, dy_2\, dz_2$.

The operator $\hat{h}$ in Eq. (B.0, 2) contains the one-electron terms in the Hamiltonian, that is, the kinetic energy and the nuclear-electronic attraction term. Using Eqs. (B.0, 1–2) we find that the SCF orbital energy is

$$\varepsilon_k = \langle \psi_k | \hat{F} | \psi_k \rangle = \langle k | \hat{h} | k \rangle + \sum_l^{occ} [2(kk\,|\,ll) - (kl\,|\,lk)] \tag{B.0, 5}$$

In the standard application of the SCF method (Roothaan 1951, Hall 1951) the SCF orbitals are expanded in a basis set of atomic orbitals, which in the subsequent problems we will assume are Slater-type orbitals, ϕ_μ. Thus

$$\psi_k = \sum_{\mu=1}^{M} \phi_\mu c_{\mu k} \tag{B.0, 6}$$

and the coefficients $c_{\mu k}$ are varied to obtain the minimum total energy. M is the number of basis functions. When the SCF solution is unchanged by the addition of one extra basis function, we have reached the *Hartree–Fock limit*.

Inserting the expanison (B.0, 6) in Eqs. (B.0, 1–4) gives the following matrix equation in the atomic orbital representation:

$$\mathbf{F}\mathbf{c}_k = \varepsilon_k \mathbf{S}\mathbf{c}_k \tag{B.0, 7}$$

where $\mathbf{c}_k$ is the column vector with elements $\{c_{\mu k}\}$ and $\mathbf{S}$ and $\mathbf{F}$ are matrices with elements

$$S_{\mu\nu} = \langle \phi_\mu | \phi_\nu \rangle = \langle \mu | \nu \rangle \tag{B.0, 8}$$

$$F_{\mu\nu} = \langle \mu | \hat{h} | \nu \rangle + \sum_{\rho\sigma} P_{\rho\sigma}[(\mu\nu|\rho\sigma) - \tfrac{1}{2}(\mu\sigma|\rho\nu)] \tag{B.0, 9}$$

The charge–bond order matrix is defined as

$$P_{\rho\sigma} = 2 \sum_k^{occ} c_{\rho k}^* c_{\sigma k} \tag{B.0, 10}$$

The SCF solution is determined iteratively. From an initial guess of the $c_{\rho k}$'s a zeroth-order value for $p_{\rho\sigma}$ and thus $F_{\mu\nu}$ can be obtained. Equation (B.0, 7) yields improved first-order values of $\mathbf{c}_k$, and the iterative procedure is established. The self-consistent values for ψ_k and ε_k are the SCF orbitals and orbital energies, respectively.

The total SCF energy, $E(\text{SCF})$, is the average value of the electronic Hamiltonian

$$\hat{H} = \sum_i \hat{h}(i) + \sum_{i>j} r_{ij}^{-1} + \sum_{\mu>\nu} \frac{Z_\mu Z_\nu}{R_{\mu\nu}} \qquad (\text{B.0, 11})$$

in the single determinantal SCF state. The last term in $\hat{H}$ represents the nuclear–nuclear repulsion that is present for molecules when $\hat{H}$ is given in the Born–Oppenheimer approximation. Expressed in terms of the SCF orbital energies and Coulomb and exchange integrals we have the following equivalent expressions for E (SCF) when the iterative cycle has *converged*:

$$E(\text{SCF}) = \sum_k^{occ} (\varepsilon_k + \langle k|\hat{h}|k \rangle) + \sum_{\mu>\nu} \frac{Z_\mu Z_\nu}{R_{\mu\nu}} \qquad (\text{B.0, 12})$$

$$E(\text{SCF}) = 2 \sum_k^{occ} \varepsilon_k - \sum_{kl}^{occ} [2(kk|ll) - (kl|lk)] + \sum_{\mu>\nu} \frac{Z_\mu Z_\nu}{R_{\mu\nu}} \qquad (\text{B.0, 13})$$

$$E(\text{SCF}) = 2 \sum_k^{occ} \langle k|\hat{h}|k \rangle + \sum_{kl}^{occ} [2(kk|ll) - (kl|lk)] + \sum_{\mu>\nu} \frac{Z_\mu Z_\nu}{R_{\mu\nu}} \qquad (\text{B.0, 14})$$

Only the last expression for E (SCF) represents an average value of $\hat{H}$ in a nonconverged SCF cycle.

It may be shown that the orbital energy ε_i is equal to

$$\varepsilon_i = E(\text{SCF}) - E_i^+ \qquad (\text{B.0, 15})$$

where E (SCF) is the total SCF energy for the state of interest and E_i^+ is the total single configurational energy of the ionized species for which an electron is removed from the SCF orbital ψ_i and all other orbitals are unchanged. Equation (B.0, 15) is known as *Koopmans' theorem* (Koopmans 1934). Koopmans' theorem thus tells that the orbital energies with opposite sign may be equated to an ionization potential provided that the ionization process can be adequately described by the removal of an electron from an SCF orbital without changing the wave functions of the residual electrons (the "frozen orbital" approximation).

In problem B.12, we perform a perturbational calculation where we use the SCF eigenvalue problem as zeroth order; that is, we write the Hamiltonian (B.0, 11) as

$$\hat{H} = \hat{H}_0 + \hat{V} \qquad (\text{B.0, 16})$$

where

$$\hat{H}_0 = \sum_i \hat{F}(i) + \sum_{\mu>\nu} \frac{Z_\mu Z_\nu}{R_{\mu\nu}} \qquad (\text{B.0, 17})$$

$$\hat{V} = \sum_{i>j} r_{ij}^{-1} - \sum_i \sum_k^{occ} \{2\hat{J}_k(i) - \hat{K}_k(i)\} \qquad (\text{B.0, } 18)$$

In deriving that form for the perturbation operator $\hat{V}$ we have used Eq. (B.0, 2). The formal expressions for the perturbation corrections to the SCF total energy and wave function in terms of $\hat{V}$ are given in problem B.11, and are used to determine the correlation energy for the H_2 molecule in problem B.12.

The most commonly used method for including electron correlation is the configuration interaction (CI) method, in which the wave function, Ψ, is written as a linear combination of Slater determinants, Ψ_k:

$$\Psi = \sum_k C_k \Psi_k \qquad (\text{B.0, } 19)$$

The expansion conefficients are determined variationally; that is, the coefficients C_k are determined from the eigenvalue problem

$$\mathbf{HC}_k = E_k \mathbf{C}_k \qquad (\text{B.0, } 20)$$

where

$$(\mathbf{H})_{kl} = \langle \Psi_k | \hat{H} | \Psi_l \rangle \qquad (\text{B.0, } 21)$$

is the configuration interaction matrix. To derive Eq. (B.0, 20) we have assumed that

$$\langle \Psi_k | \Psi_l \rangle = \delta_{kl}$$

If the expansion in Eq. (B.0, 19) contains all determinants that can be constructed within a given basis set, we talk about a *full* CI calculation.

In a CI calculation, the set of SCF orbitals is often used to construct the set of determinants Ψ_k in Eq. (B.0, 19) (see, e.g., problems B.5 and B.9). If Ψ_0 denotes the Hartree–Fock ground state and Ψ_1 a state that can be formed from Ψ_0 by a single replacement of an occupied SCF MO with a virtual SCF MO, it is easily shown that

$$\langle \Psi_0 | \hat{H} | \Psi_1 \rangle = 0$$

This equation is often referred to as the *Brillouin theorem*.

Textbooks that give sufficient background for solving the problems in this section.

Atkins (1970). Avery (1972). Ballhausen (1979). Levine (1974). Lowe (1978). Murrell and Harget (1972). Murrell, Kettle, and Tedder (1970). Pilar (1968). Slater (1968).

Textbooks that give a more elaborate treatment of the topics.

Cook (1974). McWeeny and Sutcliffe (1969). Pople and Beveridge (1970). Slater (1960).

Problem description

B.1 Atomic integrals for the He atom
B.2 Single-zeta orbital optimization for the He atom
B.3 Double-zeta variational calculation for the He$^+$ ion
B.4 Double-zeta SCF calculation for the He atom
B.5 Double-zeta CI calculation for the He atom
B.6 Calculation of two-center molecular integrals in a numerical stable fashion using Slater-type orbitals
B.7 H$_2^+$ ion: minimal basis ground-state potential energy curve and excitation energy and oscillator strength to the lowest excited state
B.8 H$_2^+$ ion: variational calculation using a "united atom" trial wave function
B.9 Minimal basis SCF and CI calculation on H$_2$
B.10 Dipole moments of LiH$^+$ and LiH
B.11 Lower bounds for perturbation expansions of the total energy
B.12 Perturbation calculation of the correlation energy for H$_2$
B.13 Van der Waals interaction between two hydrogen atoms

PROBLEMS

B.1 In this problem we will evaluate the *atomic* integrals that are needed to carry out an SCF double-zeta Slater-basis calculation on the ground state of the He atom. The normalized 1s Slater-type orbitals will be denoted ϕ_{1s_1} and ϕ_{1s_2}, with exponents ζ_1 and ζ_2, respectively.

1. Show that

$$S_{12} = \langle \phi_{1s_1} | \phi_{1s_2} \rangle = \frac{8\zeta_1^{3/2}\zeta_2^{3/2}}{(\zeta_1 + \zeta_2)^3}$$

$$h_{12} = \langle \phi_{1s_1} | -\frac{\nabla^2}{2} - \frac{2}{r} | \phi_{1s_2} \rangle = -\frac{4\zeta_1^{3/2}\zeta_2^{3/2}}{(\zeta_1 + \zeta_2)^2}\left(2 - \frac{\zeta_1\zeta_2}{\zeta_1 + \zeta_2}\right)$$

2. Derive the following expressions for the two-electron integrals.

$$(11|11) = \frac{5}{8}\zeta_1$$

$$(11|22) = \frac{\zeta_1^3\zeta_2(\zeta_1 + 4\zeta_2) + \zeta_1\zeta_2^3(\zeta_2 + 4\zeta_1)}{(\zeta_1 + \zeta_2)^4}$$

$$(12|12) = \frac{20\zeta_1^3\zeta_2^3}{(\zeta_1 + \zeta_2)^5}$$

$$(11|12) = \frac{16\zeta_1^{9/2}\zeta_2^{3/2}}{(3\zeta_1 + \zeta_2)^4}\left[\frac{12\zeta_1 + 8\zeta_2}{(\zeta_1 + \zeta_2)^2} + \frac{9\zeta_1 + \zeta_2}{2\zeta_1^2}\right]$$

3.   Evaluate the integrals in questions 1 and 2 together with h_{22}, h_{11}, $(22 \mid 21)$, and $(22 \mid 22)$, when $\zeta_1 = 1.45$ and $\zeta_2 = 2.91$ (Clementi and Roetti 1974).

B.2 The ground-state wave function for the He atom will in this problem be approximated by a single Slater determinant $|\phi_{1s}\bar{\phi}_{1s}|$ where ϕ_{1s} denotes a $1s$ Slater-type orbital with orbital exponent ζ.

1.   Use the atomic integrals in the previous problem to determine the ground-state total energy, $E(\zeta)$, of the He atom as a function of ζ.
2.   Determine the value of ζ that minimizes $E(\zeta)$.
3.   What is the minimum value of $E(\zeta)$?
The exact ground-state total energy of He is -2.9037 a.u.

B.3 In this problem we will perform a variational calculation of the total energy of He^+ by using a double-zeta Slater-basis set consisting of two $1s$ functions with exponents 1.45 and 2.91, respectively. The trial function ($=$orbital) is thus written as

$$\psi_{1s} = \sum_{i=1}^{2} c_i\phi_{1s_i}$$

where $\{\phi_{1s_1}, \phi_{1s_2}\}$ are the two Slater-type orbitals. The atomic integrals that are necessary to carry out the calculation are given in the solution to problem B.1.

1.   Determine the minimum ground-state total energy obtained by varying c_1 and c_2. Comment on the discrepancy between the calculated and exact total energy.
2.   What would the SCF ground-state total energy be in the same basis set?

B.4 We will carry out a double-zeta SCF calculation on the ground (^1S) state of helium, using a double-zeta Slater-type orbital basis set consisting of two $1s$ functions, ϕ_{1s_1} and ϕ_{1s_2}, with orbital exponents 1.45 and 2.91, respectively. The wave function for He is thus $\Psi(^1S) = |\psi_{1s}\bar{\psi}_{1s}|$ where the ψ_{1s} orbital is written as

$$\psi_{1s} = \sum_{i=1}^{2} c_i\phi_{1s_i} \qquad (B.4, 1)$$

and the variational parameters $\{c_i\}$ are determined in the SCF calculation. The atomic integrals that are necessary to carry out the calculation are given in the solution to problem B.1.

1.   Use ϕ_{1s_1} as the zeroth-order occupied molecular orbital; that is, put $c_1 = 1$ and $c_2 = 0$ in Eq. (B.4, 1). Form the zeroth-order Fock matrix, $\mathbf{F}^{(0)}$ [see Eq. (B.0, 9)].
2.   Solve the eigenvalue problem $\mathbf{F}^{(0)}\mathbf{c}_i^{(1)} = \varepsilon_i^{(1)}\mathbf{Sc}_i^{(1)}(i = 1, 2)$ to obtain the first approximation to the orbital energies $\varepsilon_1^{(1)}$, $\varepsilon_2^{(1)}$ and the eigenvector $\mathbf{c}_1^{(1)}$.
3.   Determine the ground-state SCF energy, $E^{(1)}$, at this step of the iterative procedure.

4. Use $c_1^{(1)}$ from question 2 to compute a new Fock matrix $\mathbf{F}^{(1)}$.

5. Determine the eigenvalues $\varepsilon_1^{(2)}$, $\varepsilon_2^{(2)}$ and eigenvector $c_1^{(2)}$ of $\mathbf{F}^{(1)}$.

6. Determine the ground-state SCF energy, $E^{(2)}$ at this step of the iterative procedure.

Continue the iterative process until convergence is obtained. Use as convergence criteria that the difference between the SCF total energy in two successive iterations must be less than 0.01 a.u.

7. What are the self-consistent values for ε_1, ε_2, c_1, and c_2, and what is the SCF ground-state energy?

8. Are the SCF ground-state energies evaluated in question 3, question 6, and so on, an upper bound to the exact ground-state energy?

Two-electron integrals in the SCF orbital basis are often needed in methods that go beyond SCF, that is, in the calculation of electronic correlation (see, e.g., the next problem).

9. Calculate the following two-electron integrals in the SCF basis:

$$(\psi_1 \psi_1 | \psi_1 \psi_1), \quad (\psi_1 \psi_2 | \psi_2 \psi_1)$$

where ψ_1 and ψ_2 denote the two SCF molecular orbitals.

B.5 In problem B.4 we performed an SCF calculation on the ground state of the He atom, using a double-zeta Slater-type orbital basis set. In this problem we will perform the subsequent full configuration interaction (CI) calculation, using the same basis set. At self-consistency (energy converged to 10^{-6} a.u.) we find that the SCF orbital energies are

$$\varepsilon_1 = -0.918 \text{ a.u.}, \qquad \varepsilon_2 = 2.821 \text{ a.u.}$$

The following integrals were obtained in the SCF basis in the previous problem:

$$(11|11) = 1.026 \text{ a.u.} \qquad (22|11) = 1.054 \text{ a.u.}$$

$$(22|22) = 1.141 \text{ a.u.} \qquad (11|12) = -0.327 \text{ a.u.}$$

$$(12|12) = 0.290 \text{ a.u.} \qquad (22|21) = -0.456 \text{ a.u.}$$

$$\langle 1|\hat{h}|1\rangle = -1.944 \text{ a.u.}, \quad \langle 1|\hat{h}|2\rangle = 0.327 \text{ a.u.}, \quad \langle 2|\hat{h}|2\rangle = 1.003 \text{ a.u.}$$

where $(ij|kl) = (\psi_i \psi_j | \psi_k \psi_l)$ and ψ_i denotes the SCF orbital i.

1. Determine all Slater determinants that can be constructed from the SCF molecular orbitals ψ_1, ψ_2 and that have the same spin and spatial symmetry (^1S) as the SCF ground-state determinant $|\psi_1 \bar{\psi}_1|$.

2. Use the Slater determinants from question 1 to set up the CI matrix for a *full* CI calculation for states of ^1S symmetry.

3. Determine the eigenvalues of the configuration interaction matrix.

4. What is the correlation energy for the He ground state within the considered basis set?

5. Calculate the ionization potential of He if we use single determinantal SCF states for both He and He^+. The single determinantal states for both He and He^+ should be constructed from the SCF molecular orbitals of He.

6. What is the ionization potential for He according to Koopmans' theorem? Why do we obtain the same result as in question 5?

In problem B.3, a full configuration interaction calculation on He^+ using the same basis gave a ground-state total energy of -1.980 a.u.

7. Determine the ionization potential for He if the total energies for both He and He^+ are determined from full CI calculations within the present basis set.

The excitation energy from the ground state of the helium atom to the *lowest* state of the same symmetry will be denoted by $\Delta E(^1S)$.

8. Determine $\Delta E(^1S)$ if the individual states are approximated by the single configurational SCF states in question 1.

9. Determine $\Delta E(^1S)$ if the individual total energies are determined from a full CI calculation.

The experimental total energy, ionization potential, and singlet $\Delta E(S^1)$ excitation energy are -2.9037 a.u., 24.580 eV, and 20.615 eV, respectively (Moore 1949).

10. Is the agreement between the calculated and the experimental results as you would expect? Comment on the trends in the calculated result. Discuss how you would improve the calculated total energy, ionization potential, and excitation energy.

B.6 Two center two-electron integrals in Slater-type orbital basis set calculation are often evaluated by using confocal elliptical coordinates (Harris and Michels 1976). The integrals are then expressed as sums of products of A and B integrals, defined as

$$A_m(x) = m!\, e^{-x} \sum_{j=0}^{m} \frac{x^{j-m-1}}{j!}, \qquad m = 0, 1, 2, \ldots \qquad (B.6, 1)$$

$$B_m(x) = (-)^{m+1} A_m(-x) - A_m(x), \qquad m = 0, 1, 2, \ldots \qquad (B.6, 2)$$

It is the purpose of the present problem to demonstrate how the $B_m(x)$ integrals may be evaluated in a numerically stable fashion.

1. Verify the relation

$$A_m(x) = A_0(x) + \frac{m A_{m-1}(x)}{x}, \qquad m = 1, 2, \ldots \qquad (B.6, 3)$$

2. Show that for $x > 0$

$$A_m(x) = \int_1^\infty \xi_m e^{-x\xi} d\xi \qquad (B.6, 4)$$

and that

$$B_m(x) = \int_{-1}^1 \eta^m e^{-x\eta} d\eta \qquad (B.6, 5)$$

The formula (B.6, 1) or (B.6, 3) can be used with confidence for the calculation of A integrals (Harris and Michels 1967). Formula (B.6, 2), however, is numerically unstable in some cases. We shall consider this problem in more detail in the following.

3. Expand the exponential in (B.6, 1) and determine the leading term of $A_m(x)$ for small $|x|$.

4. Use (B.6, 5) to determine the leading term of $B_m(x)$ for small $|x|$.

5. Use the leading-term formulas from questions 3 and 4 to calculate values of $A_m(x)$ and $B_m(x)$ for $m = 0, 1, 2$ and $x = 0.1$ and 0.5 .

6. On the basis of these results, discuss the cases in which formula (B.6, 2) may be numerically unstable.

We shall now consider another approach to the calculation of B integrals. To this end we introduce the *modified spherical Bessel functions of the first kind*, given by

$$i_n(x) = (\tfrac{1}{2}) \int_{-1}^{1} d\eta\, P_n(\eta) e^{-x\eta}, \qquad n = 0, 1, 2, \ldots \qquad (B.6, 6)$$

where we define the Legendre polynomials P_n via a generating function (Abramowitz and Stegun 1965)

$$(1 - \eta t + t^2)^{-\frac{1}{2}} = \sum_{j=0}^{\infty} P_j(\eta) t^j, \qquad |t| < 1 \qquad (B.6, 7)$$

Differentiation of (B.6, 7) with respect to t yields

$$P_0(\eta) = 1 \qquad (B.6, 8)$$

$$P_1(\eta) = \eta \qquad (B.6, 9)$$

$$(2n + 1)\eta P_n(\eta) = (n + 1)P_{n+1}(\eta) + n P_{n-1}(\eta), \qquad n = 1, 2, \ldots \qquad (B.6, 10)$$

7. Find expressions for B_0, B_1, B_2, and B_3 in terms of the modified spherical Bessel functions of the first kind.

It may be shown that for a given value of x, all $i_n(x)$ with n differing by a multiple of 2 have the same sign.

8. Show that the formulas found for B_0, B_1, B_2, and B_3 in question 7 are numerically stable.

This is a general result since it holds that in an expansion of η^n in Legendre polynomials all coefficients are positive [cf. Eqs. (B.6, 5) and (B.6, 6)].

The modified spherical Bessel functions of the first kind can be evaluated from various recurrence relations. We will examine the numerical stability of two such procedures. Differentiation of (B.6, 7) with respect to η yields (Abramowitz and Stegun 1965)

$$P'_{n+1}(\eta) + P'_{n-1}(\eta) = 2\eta P'_n(\eta) + P_n(\eta), \qquad n = 1, 2, \ldots \qquad (B.6, 11)$$

9. Use Eqs. (B.6, 10 and 11) to derive the relation

$$(2n + 1) \int_{-1}^{\eta} P_n(t)\, dt = P_{n+1}(\eta) - P_{n-1}(\eta), \qquad n = 1, 2, \ldots \qquad (B.6, 12)$$

10. Verify the following recurrence relation for the modified spherical Bessel functions:

$$(2n + 1)i_n(x) = x[i_{n+1}(x) - i_{n-1}(x)] \qquad (B.6, 13)$$

Formula (B.6, 13) gives rise to the same numerical difficulties as (B.6, 2) for the same range of n and x. By expressing the B integrals in terms of the Bessel functions, we have merely transferred the problem to the calculation of the i_n. However, these can be calculated by introducing the functions

$$r_n(x) = \frac{i_{n+1}(x)}{i_n(x)}, \qquad n = 0, 1, 2, \ldots \qquad (B.6, 14)$$

11.   Derive the recurrence relation

$$r_{n-1}(x) = \frac{-x}{2n + 1 - xr_n(x)}, \qquad n = 1, 2, \ldots \qquad \text{(B.6, 15)}$$

Because r_{n-1} tends to zero as $(2n + 1)^{-1}$ for large n, we may now approximate r_n by zero for some large value N of n and then use (B.6, 15) to find all r_n until we reach r_0. Because $xr_n(x)$ can be shown to be negative, the accuracy of r_n will increase when n gets smaller. To calculate the $i_n(x)$ we then start from an explicit value for $i_0(x)$, multiply by $r_0(x)$ to find $i_1(x)$, and so on until all desired $i_n(x)$ have been found. Experience shows that $N = 27 + |x|/1.95$ is a safe choice of N.

12.   Show that instead of using (B.6, 6) to find $i_0(x)$ we can use the safer formula

$$i_0(x) = \frac{e^{|x|}}{1 + |x| - xr_0(x)} \qquad \text{(B.6, 16)}$$

B.7 We will calculate the total electronic energy of the ground state of the H_2^+ molecular ion by using the bonding linear combination of hydrogenic 1s orbitals centered on atom A (ϕ_A) and atom B (ϕ_B) as the ground-state wave function. It is convenient first to evaluate some elementary integrals.

1.   Show that the overlap between ϕ_A and ϕ_B is

$$S(R) = \langle \phi_A | \phi_B \rangle = e^{-R}(1 + R + \frac{R^2}{3}) \qquad \text{(B.7, 1)}$$

where R is the internuclear separation.

2.   Show that

$$\varepsilon_{AA}(R) = \langle \phi_A | -\frac{1}{r_B} | \phi_A \rangle = -\frac{1}{R}[1 - e^{-2R}(1 + R)] \qquad \text{(B.7, 2)}$$

and that

$$\varepsilon_{AB}(R) = \langle \phi_A | -\frac{1}{r_B} | \phi_B \rangle = -e^{-R}(1 + R) \qquad \text{(B.7, 3)}$$

where r_B is the distance between nucleus B and the electron.

3.   Determine within the Born–Oppenheimer approximation the ground-state electronic energy (including the nuclear repulsion term), $E_0(R)$, for H_2^+ expressed as a function of R.

4.   Calculate $E_0(R)$ for $R = 2.4, 2.5, 2.6,$ and ∞ a.u.

5.   Determine with an accuracy of 0.1 a.u. the equilibrium distance, $R = R_{e'}$ for the H_2^+ ion.

6.   Determine the spectroscopic dissociation energy, D_e, for the H_2^+ ion.

The experimental equilibrium distance for H_2^+ is 2.00 a.u. and the spectroscopic dissociation energy D_e is 2.79 eV (Wind 1965).

7. Comment on the accuracy of the calculated equilibrium distance, R_e, and the calculated dissociation energy, D_e.

8. Show that

$$\Delta E(R) = \frac{2}{1 - S^2(R)}\left[(1 + R)\left(e^{-R} + \frac{S(R)}{R}e^{-2R}\right) - \frac{S(R)}{R}\right]$$

where $\Delta E(R)$ is the energy difference between the bonding ψ_g and antibonding ψ_u molecular orbital.

9. Determine the oscillator strength in the dipole length approximation for the excitation from ψ_g to ψ_u at the equilibrium internuclear separation obtained in question 5.

B.8 Consider the hydrogen molecular ion, H_2^+, and let R be the separation of the two hydrogen atoms. Assume that the bonding molecular orbital can be approximated by a $1s$ "united atom" *atomic* orbital

$$\psi(r) = ce^{-\zeta r}$$

with origin at the midpoint of the internuclear axis (c is a normalization constant).

We will now perform a variational calculation of the ground-state total energy of H_2^+, varying both ζ and R. It is convenient to use polar coordinates with the origin placed at the origin of $\psi(r)$.

1. Determine the total energy as a function of ζ and R for H_2^+ in the state $\psi(r)$. Assume initially that R is 2.00 a.u. (the experimental equilibrium separation, R_e).

2. Determine for $R = R_e$ the value of $\zeta(= \zeta_e)$ that minimizes the ground-state total energy in question 1. Calculate the total energy for $\zeta = \zeta_e$.

Let us now keep ζ fixed at $\zeta = \zeta_e$ and vary the internuclear separation, R.

3. Determine for $\zeta = \zeta_e$ the equilibrium distance and the corresponding ground-state energy using the energy expression from question 1.

We can continue the iterative process initiated in questions 2 and 3 until a self-consistent solution is obtained. Since the total energies obtained in the two iterations deviate by 10^{-3} a.u. (verify that!), we will consider the solutions obtained in question 3 as the converged result.

4. Use the results of question 3 to calculate the spectroscopic dissociation energy for H_2^+.

The most accurate total energy and dissociation energy for H_2^+ are -0.6026 a.u. and 2.793 eV, respectively (Wind 1965).

5. What is the reason for the large discrepancy between the computed and "experimental" dissociation energies?

B.9 In this problem we first carry out an SCF calculation on the H_2 molecule at the equilibrium internuclear separation $R = 1.40$ a.u., using a minimum basis. We then perform a full CI calculation, and finally we determine the oscillator strength for the transition that is dipole allowed. The minimum basis consists of a $1s$ Slater-type orbital, ϕ_A, with exponent 1 centered on hydrogen atom A and a $1s$ Slater-type orbital, ϕ_B, with exponent 1 centered on hydrogen atom B. The following integrals have been calculated by the methods outlined in problem B.6.

$$\langle \phi_A | \phi_B \rangle = S = 0.753$$

$$\langle \phi_A | \hat{h} | \phi_A \rangle = -1.110 \text{ a.u.}$$

$$\langle \phi_A | \hat{h} | \phi_B \rangle = -0.968 \text{ a.u.}$$

$$(\phi_A \phi_A | \phi_A \phi_A) = 0.625 \text{ a.u.}$$

$$(\phi_A \phi_B | \phi_A \phi_B) = 0.323 \text{ a.u.}$$

$$(\phi_A \phi_A | \phi_B \phi_B) = 0.504 \text{ a.u.}$$

$$(\phi_A \phi_A | \phi_A \phi_B) = 0.426 \text{ a.u.}$$

where $\hat{h}$ is the one-electron Hamiltonian and the notation for the two-electron integrals is defined in the Appendix.

1. Use symmetry arguments to show that the normalized SCF molecular orbitals are

$$\psi_g = (2 + 2S)^{-1/2}(\phi_A + \phi_B), \qquad \psi_u = (2 - 2S)^{-1/2}(\phi_A - \phi_B)$$

2. Calculate the numerical values of the one-electron integrals $\langle \psi_g | \hat{h} | \psi_g \rangle$, $\langle \psi_u | \hat{h} | \psi_u \rangle$, $\langle \psi_g | \hat{h} | \psi_u \rangle$ and the two-electron integrals $(gg|gg)$, $(gg|uu)$, $(gu|gu)$, $(gg|gu)$, $(uu|ug)$, and $(uu|uu)$.

3. Determine the SCF orbital energies ε_g and ε_u and the SCF ground-state total energy.

4. Determine all Slater determinants that can be constructed from the SCF molecular orbitals, ψ_g and ψ_u.

5. Form the linear combinations of the Slater determinants that transform according to irreducible representations of the point group for the H_2 molecule and that simultaneously are eigenfunctions of the total spin and its z-component.

6. Determine the eigenvalues and eigenvectors obtained from a full CI calculation on the H_2 molecule within the minimal basis set.

7. What is the ground-state correlation energy of H_2 within the considered basis set?

The oscillator strength for the dipole allowed transition from the ground state will now be evaluated.

8. Determine the matrix elements

$$\langle \phi_K | \mathbf{r} | \phi_L \rangle$$

where K and L can be either A or B and $\mathbf{r} = (x, y, z)$ is the position operator. Choose the midpoint of the internuclear axis as the origin of the coordinate system.

9. Determine the matrix elements

$$\langle \psi_i | \mathbf{r} | \psi_j \rangle = \mathbf{M}_{ij}$$

where $i, j = g, u$.

10. Do the values of M_{ij} depend on the choice of the origin of the coordinate system?
11. Determine the oscillator strength in the dipole length formulation for the dipole allowed transition when both the ground state and the excited state are approximated by single configurational states.
12. Determine the dipole length oscillator strength for the dipole allowed transition when both the ground state and the excited state are determined from the full CI calculation.

Kołos and Wolniewicz (1965, 1968, 1969) have performed the most accurate calculations on H_2 presently available in the literature. At $R = 1.40$ a.u. they calculated the lowest $^1\Sigma_u^+ \rightarrow {}^1\Sigma_g^+$ excitation energy to 12.75 eV. The oscillator strength was determined to 0.300 (Wolniewicz 1969).

13. Discuss the reasons for the discrepancy between these results and the present calculation.

B.10

In this problem we will calculate the dipole moment of the ground state of LiH and LiH^+. The ground-state wave functions are assumed to be single determinants constructed from the $\psi_{1\sigma}$ and $\psi_{2\sigma}$ molecular orbitals.

1. Determine the total dipole moment of the LiH molecule and of the LiH^+ ion using a coordinate system with an arbitrary origin. Express the results in terms of the position vectors of the nuclei and integrals over the electronic position operator, $\mathbf{r}$, over $\psi_{1\sigma}$ and/or $\psi_{2\sigma}$.

Let us make a constant displacement, $\mathbf{\Delta}$, of the origin of the coordinate system.

2. Show that the dipole moment for LiH is independent of the choice of origin of the coordinate system.
3. Show that the dipole moment of LiH^+ depends on the choice of the origin of the coordinate system.
4. Would the answer to questions 2 and 3 be the same if we used the exact wave function for LiH and LiH^+ ?

In the following we consider LiH, and for convenience choose the origin of the coordinate system to be the lithium atom, and let the z-axis point in the direction of the hydrogen atom. As the single determinantal state we will use the SCF ground state obtained in a minimum basis SCF calculation. The occupied orbitals of the minimum basis SCF calculation are

$$\psi_{1\sigma} = 0.9971\, \phi_{1sLi} + 0.0152\, \phi_{2sLi} + 0.0031\, \phi_{1sH}, \qquad \varepsilon_{1\sigma} = -2.450 \text{ a.u.}$$

$$\psi_{2\sigma} = 0.1461\, \phi_{1sLi} - 0.3632\, \phi_{2sLi} - 0.7831\, \phi_{1sH}, \qquad \varepsilon_{2\sigma} = -0.293 \text{ a.u.}$$

ϕ_{1sH} denotes a $1s$ Slater-type orbital with exponent 1 centered on the hydrogen atom, and ϕ_{1sLi} and ϕ_{2sLi} denote $1s$ and $2s$ Slater-type orbitals with exponents 2.69 and 0.64, respectively, centered on the Li atom.

5. Express the total dipole moment of LiH, μ_{LiH}, in terms of the internuclear separation R and the integrals

$$z_k = \langle \phi_{ksLi} | z | \phi_{1sH} \rangle, \qquad k = 1, 2$$

We will use the expression from question 5 to calculate the numerical value of the dipole moment for LiH. Consider first some elementary integrals that appear in the expression for μ_{LiH}.

6. Show that

$$z_2 = \frac{S_2}{0.64}\sqrt{\frac{5}{2}}$$

where

$$S_2 = \langle \phi_{3pzLi} | \phi_{1sH} \rangle$$

and ϕ_{3pzLi} is a normalized $3pz$ Slater-type orbital on Li with the same orbital exponent as ϕ_{2sLi}, that is, 0.64.

7. Show that

$$S_2 = \left(\frac{0.64^7}{30}\right)^{1/2}\frac{R^5}{8}(B_1(\beta)[A_2(\alpha) + A_4(\alpha)] - A_3(\alpha)B_0(\beta)$$
$$+ B_2(\beta)[A_1(\alpha) - A_3(\alpha)] - B_3(\beta)[A_0(\alpha) + A_2(\alpha)] + B_4(\beta)A_1(\alpha))$$

$$(\text{B.10, 1})$$

where $\alpha = (R/2)(1 + 0.64)$ and $\beta = (R/2)(1 - 0.64)$ and the A and B functions are defined in Eqs. (B.6, 4) and (B.6, 5).

8. Show that $S_2 = 0.4538$ for $R = 3.02$ a.u. Calculate z_2.

The overlap integral $S_1 = \langle \phi_{2pzLi} | \phi_{1sH} \rangle$ is evaluated in a similar fashion, and we find that $S_1 = 0.1242$ at $R = 3.02$ a.u. ϕ_{2pzLi} is a normalized $2pz$ Slater-type orbital on Li with the same exponent as ϕ_{1sLi}, that is, 2.69.

9. Show that z_1 is less than 0.05 a.u. for $R = 3.02$ a.u.
10. Show that contributions to the dipole moment of LiH from the ϕ_{1sLi} orbital can be neglected in a calculation that is accurate to 0.06 debye (D).
11. Show that contributions to the dipole moment of LiH from the molecular orbital $\psi_{1\sigma}$ can be neglected in a calculation that is accurate to 0.06 D.
12. Evaluate the dipole moment of LiH to an accuracy of 0.06 D for $R = 3.02$ a.u.

The experimental dipole moment for LiH is 5.83 D.

13. Comment on the accuracy of the calculated dipole moment of LiH.

B.11

In the present problem we examine whether or not an order by order calculation provides bounds to the ground state total energy. Initially we derive the formal expressions for the perturbation corrections to the wave function and the total energy for a nondegenerate state, $|k\rangle$. The Hamiltonian for the system can be written as

$$\hat{H} = \hat{H}_0 + \hat{V}$$

where $\hat{H}_0$ is the zeroth-order operator and $\hat{V}$ is the perturbation operator. We assume that the set of eigenfunctions $\{u_i^{(0)}\}$ and eigenvalues $\{E_i^0\}$ are known for $\hat{H}_0$, that is,

$$\hat{H}_0 u_i^{(0)} = E_i^{(0)} u_i^{(0)}$$

1. Show that a perturbation calculation gives the following corrections to the
    energy and wave function for the state $|k >$:

$$E_k^{(0)} = \langle u_k^{(0)}|\hat{H}_0|u_k^{(0)}\rangle \qquad \text{(B.11, 1)}$$

$$E_k^{(1)} = \langle u_k^{(0)}|\hat{V}|u_k^{(0)}\rangle \qquad \text{(B.11, 2)}$$

$$E_k^{(2)} = \langle u_k^{(0)}|\hat{V}|u_k^{(1)}\rangle \qquad \text{(B.11, 3)}$$

$$E_k^{(3)} = \langle u_k^{(1)}|\hat{V}|u_k^{(1)}\rangle - \langle u_k^{(0)}|\hat{V}|u_k^{(0)}\rangle\langle u_k^{(1)}|u_k^{(1)}\rangle \qquad \text{(B.11, 4)}$$

$$u_k^{(1)} = \sum_{m \neq k} u_m^{(0)} \frac{\langle u_m^{(0)}|\hat{V}|u_k^{(0)}\rangle}{E_k^{(0)} - E_m^{(0)}} \qquad \text{(B.11, 5)}$$

$u_k^{(i)}$ and $E_k^{(i)}$ refer to the ith-order corrections of the wave function and
energy of state $|k >$, respectively. In deriving the forgoing expressions
intermediate normalization of the wave function must be used; that is,

$$\langle u_k^{(0)}|u_k^{(i)}\rangle = \delta_{i0}, \qquad i = 0, 1, 2 \cdots$$

We will now consider the ground state, that is, $|k > = |0 >$.

2. Is $E_0^I = E_0^{(0)} + E_0^{(1)}$ an upper bound for E_0, the exact energy of the ground
    state?
3. Is $E_0^{II} = E_0^{(0)} + E_0^{(1)} + E_0^{(2)}$ an upper bound for E_0 ?
    Consider the energy function

$$E_k^{III} = \frac{\langle u_k^{(0)} + u_k^{(1)}|\hat{H}|u_k^{(0)} + u_k^{(1)}\rangle}{\langle u_k^{(0)} + u_k^{(1)}|u_k^{(0)} + u_k^{(1)}\rangle} \qquad \text{(B.11, 6)}$$

4. Is E_0^{III} an upper bound for E_0 ?
5. Show that

$$E_k^{III} = E_k^{(0)} + E_k^{(1)} + E_k^{(2)} + E_k^{(3)} + \mathcal{O}(4) \qquad \text{(B.11, 7)}$$

where $\mathcal{O}(4)$ denotes terms of at least fourth order in the perturbation.

6. Is the approximate form for E_0^{III}, where $\mathcal{O}(4)$ is neglected, an upper bound
    for E_0 ?

B.12

In the following we will perform a perturbation calculation of the ground-state
total energy of the H_2 molecule using the same minimal basis as in problem B.9.
As the zeroth-order solution we will use the SCF solution for the ground state of
H_2; that is, the zeroth-order Hamiltonian is [see Eqs. (B.0, 16–18)]

$$\hat{H}_0 = \sum_{i=1}^{2} \hat{F}(i)$$

where $\hat{F}(i)$ is the Fock operator for electron i. The zeroth-order states are thus
the set of Slater determinants that can be formed from the two SCF orbitals (see
problem B.9, question 4).

The SCF orbitals are determined by symmetry (see B.9, question 1) and will be denoted ψ_g and ψ_u for the bonding and antibonding molecular orbital, respectively. The orbital energies, ε_k, and the nonvanishing two-electron integrals are

$$\varepsilon_g = -0.619 \text{ a.u.}$$

$$\varepsilon_u = 0.401 \text{ a.u.}$$

$$(gg|gg) = 0.566 \text{ a.u.}$$

$$(gu|gu) = 0.140 \text{ a.u.}$$

$$(gg|uu) = 0.558 \text{ a.u.}$$

$$(uu|uu) = 0.582 \text{ a.u.}$$

Let $\Psi_0^{(i)}$ and $E_0^{(i)}$ denote the ith-order correction to the ground-state wave function and total energy, respectively.

1. Calculate $\Psi_0^{(1)}$, $E_0^{(0)}$, $E_0^{(1)}$, $E_0^{(2)}$, and $E_0^{(3)}$.
2. Determine the correlation energy from a perturbation calculation that is consistent through first, second, and third order in the perturbation, respectively.
3. Calculate

$$E_0^{III} = \frac{\langle \Psi_0^{(0)} + \Psi_0^{(1)} | \hat{H} | \Psi_0^{(0)} + \Psi_0^{(1)} \rangle}{\langle \Psi_0^{(0)} + \Psi_0^{(1)} | \Psi_0^{(0)} + \Psi_0^{(1)} \rangle}$$

and determine the corresponding correlation energy.
In the full CI calculation in problem B.9 we found a correlation energy of -0.015 a.u.
4. Comment on the accuracy obtained in the perturbation calculations and compare with E_0^{III}.

B.13

In this problem we will evaluate the long-range interaction (the van der Waals interaction) between two hydrogen atoms in their ground states. The hydrogen atoms are separated by a distance R where $R \gg 1.4$ a.u. In questions 3–4 and 5–9, we calculate an upper and a lower bound to the van der Waals interaction (dispersion) energy, respectively.

1. Show that the leading term in the long-range interaction operator $\hat{V}$ (i.e., the total Hamiltonian minus the Hamiltonians for the two isolated hydrogen atoms) can be written as

$$\hat{V} = \frac{1}{R^3}(x_1 x_2 + y_1 y_2 - 2z_1 z_2) \tag{B.13, 1}$$

The z-axis is chosen along the internuclear axis, and (x_i, y_i, z_i) are the electronic coordinates for electron i relative to nucleus i ($i = 1, 2$).
From a perturbation calculation we can determine a lower bound to the van der Waals interaction energy. The zeroth-order Hamiltonian H_0 is the sum of the Hamiltonians for hydrogen atoms centered on atom A and B. The perturbation

operator is given in Eq. (B.13, 1). The zeroth-order wave functions are thus the Slater determinants $|\phi_i \phi_j|$ where ϕ_i is a hydrogenic orbital centered on nucleus $i (i = A,B)$.

2.   Show that the first-order energy correction to the ground-state total energy, $E^{(1)}(R)$, vanishes.

Let $E^{(2)}(R)$ denote the second-order correction to the ground-state total energy.

3.   Show that

$$E^{(2)}(R) \geqq \frac{(\hat{V}^2)_{00}}{E_0 - E_{n*}} \qquad (\text{B.13, 2})$$

where $(\hat{V}^2)_{00}$ denotes the matrix element of the squared operator $\hat{V}^2$ in the zeroth-order ground state. E_{n*} is the energy for the lowest excited state Ψ_{n*} for which the matrix element $\hat{V}_{0n*}$ is different from zero.

4.   Show that the lower bound to the second-order energy correction, that is, the van der Waals interaction energy, is

$$E^{(2)}(R) \geqq -\frac{8}{R^6} \qquad (\text{B.13, 3})$$

From a combined perturbation–variation calculation we can determine an upper bound to the dispersion energy. First, however, we will prove some relations that turn out to be useful for that purpose.

5.   Show that a perturbation calculation through second order gives the following three coupled differential equations.

$$(\hat{H}_0 - E^{(0)})u^{(2)} = (E^{(1)} - \hat{V})u^{(1)} + E^{(2)}u^{(0)} \qquad (\text{B.13, 4})$$

$$(\hat{H}_0 - E^{(0)})u^{(1)} = (E^{(1)} - \hat{V})u^{(0)} \qquad (\text{B.13, 5})$$

$$(\hat{H}_0 - E^{(0)})u^{(0)} = 0 \qquad (\text{B.13, 6})$$

where $\hat{H} = \hat{H}_0 + \hat{V}$ and $u^{(i)}$ and $E^{(i)}$ are the i-order corrections to wave function and energy, respectively.

6.   Show that the second-order energy correction can be written as

$$E^{(2)} = 2\langle u^{(0)}|\hat{V} - E^{(1)}|u^{(1)}\rangle + \langle u^{(1)}|\hat{H}_0 - E^{(0)}|u^{(1)}\rangle$$

7.   Show that if $u^{(1)}$ is a real-valued function, then

$$J_2 \geqq E^{(2)}$$

where

$$J_2 = 2\langle u^{(0)}|\hat{V} - E^{(1)}|\chi\rangle + \langle\chi|\hat{H}_0 - E^{(0)}|\chi\rangle$$

and χ is an *arbitrary* real-valued function. (Hint: Set $\chi = u^{(1)} + v$ where v is an arbitrary real-valued function.)

Consider now again the determination of the dispersion energy of two hydrogen atoms.

8. Show that

$$\langle \phi_{1s} | x \hat{H} x | \phi_{1s} \rangle = 0$$

where $\hat{H}$ is the Hamiltonian for an isolated hydrogen atom and ϕ_{1s} is the $1s$ orbital for that atom. [Hint: Use the relation $[\hat{H}, x] = -(\partial/\partial x)$, Eq. (A.6, 2).] Assume that $\chi = c\hat{V}u^{(0)}$ where c is a real variational parameter, $\hat{V}$ is given in Eq. (B.13, 1), and $u^{(0)}$ is the zeroth-order ground-state wave function.

9. Show that this choice of χ leads to the following upper bound to the van der Waals interaction energy.

$$E^{(2)}(R) \leq -\frac{6}{R^6}$$

Kołos (1967) has calculated an accurate value for the van der Waals C_6 coefficient, that is, the coefficient to R^{-6}, of 6.499027 a.u. In an accurate calculation of the dispersion energy the perturbation treatment must be extended beyond second order and we obtain contributions to the dispersion energy that are proportional to R^{-8}, R^{-10}, and so on. Typical binding energies between atoms held together by van der Waals forces are of the order of $5kJ$ mole^{-1}

10. What is the size of a typical van der Waals radius if we assume that it suffices to consider the R^{-6} term? Compare the magnitude of the van der Waals radius with typical covalent bond lengths.

SOLUTIONS

B.1 2.

$$\phi_{1s_i} = \left(\frac{\zeta_i^3}{\pi}\right)^{1/2} e^{-\zeta_i r} = (4\zeta_i^3)^{1/2} e^{-\zeta_i r} Y_{00}, \qquad i = 1, 2 \quad \text{(B.1, I)}$$

$$r_{12}^{-1} = \sum_{l=0}^{\infty} \sum_{m=-l}^{l} \frac{4\pi}{2l+1} \frac{r_<^l}{r_>^{l+1}} Y_{lm}^*(\theta_1, \phi_1) Y_{lm}(\theta_2, \phi_2) \quad \text{(B.1, II)}$$

where $r_<$ means the smaller of r_1 and r_2 and $r_>$ means the larger. The angular integration in the two-electron integrals is nonvanishing only for the $l = 0$ term in the r_{12}^{-1} expansion. Thus

$$r_{12}^{-1} = \frac{4\pi}{r_>} Y_{00}^* Y_{00} = \frac{1}{r_>}$$

Consider, for example,

$$(11|12) = 2^4 \zeta_1^3 \zeta_2^3 \int_0^\infty r_1^2 \, dr_1 \, e^{-2\zeta_1 r_1} \int_0^\infty r_2^2 \, dr_2 \, r_>^{-1} e^{-2\zeta_2 r_2}$$

$$= 16\zeta_1^3 \zeta_2^3 \int_0^\infty r_1^2 e^{-2\zeta_1 r} \, dr_1 \left(\int_0^{r_1} r_2 e^{-2\zeta_2 r_2} \, dr_2 + \frac{1}{r_1} \int_{r_1}^\infty r_2^2 e^{-2\zeta_2 r_2} \, dr_2 \right)$$

A partial integration gives the results in question 2.

An alternative derivation of the two-electron integrals can be obtained by expressing the volume element $d\tau_1 \, d\tau_2$ in terms of interelectronic coordinates [see, e.g., Pilar (1968), Chap. 8-4].

3.

$$S_{11} = S_{22} = 1.00$$
$$S_{12} = S_{21} = 0.837$$
$$h_{11} = -1.850 \text{ a.u.}$$
$$h_{12} = h_{21} = -1.883 \text{ a.u.}$$
$$h_{22} = -1.586 \text{ a.u.}$$
$$(11|11) = 0.906 \text{ a.u.}$$
$$(21|11) = 0.903 \text{ a.u.}$$
$$(21|21) = 0.954 \text{ a.u.}$$
$$(22|11) = 1.183 \text{ a.u.}$$
$$(22|21) = 1.298 \text{ a.u.}$$
$$(22|22) = 1.819 \text{ a.u.}$$

B.2 1.

$$E(\zeta) = 2h_{11} + (11|11) = \zeta^2 - \frac{27}{8}\zeta$$

2.

$$\zeta = \frac{27}{16}$$

3. $E(27/16) = -(27/16)^2$ a.u. $= -2.8477$ a.u., that is, 98.07% of the exact total energy.

B.3 1.

$$\hat{H} = \hat{h} = -\frac{\nabla^2}{2} - \frac{2}{r}$$

The linear variational problem leads to the secular equation

$$\begin{vmatrix} \langle \phi_{1s_1} | \hat{h} | \phi_{1s_1} \rangle - E & \langle \phi_{1s_1} | \hat{h} | \phi_{1s_2} \rangle - ES_{21} \\ \langle \phi_{1s_2} | \hat{h} | \phi_{1s_1} \rangle - ES_{12} & \langle \phi_{1s_2} | \hat{h} | \phi_{1s_2} \rangle - E \end{vmatrix} \qquad \text{(B.3, I)}$$

$$= \begin{vmatrix} -1.850 - E & -1.883 - 0.837\,E \\ -1.883 - 0.837\,E & -1.586 - E \end{vmatrix} = 0$$

$E = -1.980$ a.u.   (lowest eigenvalue)

$E(\text{exact}) = -(Z^2/2) = -2.000$ a.u.

The difference is due to the use of a limited basis set. $E > E(\text{exact})$ according to the variational principle. The exact energy would have been obtained with the screening parameter $\zeta = 2.00$ and only one basis function. The relatively small discrepancy occurs because the exponents are optimized to give the lowest possible ground-state total energy for He (Clementi and Roetti 1974).

2. The total energy in question 1 was determined by solving the secular equation

$$\left| \langle \phi_{1s_i} | \hat{H} | \phi_{1s_j} \rangle - E S_{ij} \right| = 0 \qquad \text{(B.3, II)}$$

The SCF total energy is obtained from Eq. (B.0, 12) by solving

$$\left| \langle \phi_{1s_i} | \hat{F} | \phi_{1s_j} \rangle - E S_{ij} \right| = 0 \qquad \text{(B.3, III)}$$

Because He$^+$ is a one-electron system, all two-electron integrals are absent. Equations (B.3, II) and (B.3, III) are thus identical and E from Eq. (B.3, I) is also the SCF energy.

B.4 1.

$$\psi_{1s} = \psi_1^{(0)} = \phi_{1s_1}$$

$$p_{11}^{(0)} = 2$$

$$F_{11}^{(0)} = h_{11} + \sum_{\rho=1}^{2} \sum_{\sigma=1}^{2} p_{\rho\sigma}^{(0)} [(11|\rho\sigma) - \tfrac{1}{2}(1\sigma|\rho1)]$$

$$= h_{11} + (11|11) = -0.944 \text{ a.u.}$$

and similarly for the other elements.

$$\mathbf{F}^{(0)} = \begin{pmatrix} -0.944 & -0.980 \\ -0.980 & -0.174 \end{pmatrix}$$

2. The secular determinant

$$\begin{vmatrix} F_{11}^{(0)} - \varepsilon & F_{12}^{(0)} - \varepsilon S_{12} \\ F_{21}^{(0)} - \varepsilon S_{21} & F_{22}^{(0)} - \varepsilon \end{vmatrix} = 0$$

gives the eigenvalues

$$\varepsilon_1^{(1)} = -0.976 \text{ a.u.}, \qquad \varepsilon_2^{(1)} = 2.725 \text{ a.u.}$$

The eigenvector corresponding to $\varepsilon_1^{(0)}$ is

$$\psi_1^{(1)} = 0.851 \, \phi_{1s_1} + 0.173 \, \phi_{1s_2}$$

3. In each iteration, the total energy is most easily evaluated from Eq. (B.0, 12). Using that

$$\langle \psi_1^{(1)} | \hat{h} | \psi_1^{(1)} \rangle = -1.942 \text{ a.u.}$$

we find

$$E^{(1)} = -2.917 \text{ a.u.}$$

4.

$$\mathbf{F}^{(1)} = \begin{pmatrix} -0.885 & -0.943 \\ -0.943 & -0.127 \end{pmatrix}$$

5.

$$\varepsilon_1^{(2)} = -0.922 \text{ a.u.,} \qquad \varepsilon_2^{(2)} = 2.814 \text{ a.u.}$$

$$\mathbf{c}_1^{(2)} = \begin{pmatrix} 0.843 \\ 0.182 \end{pmatrix}$$

6. $E^{(2)} = -2.866$ a.u. using (B.0, 12).

7.

$$\varepsilon_1^{(3)} = -0.918 \text{ a.u.} \qquad \varepsilon_2^{(3)} = 2.823 \text{ a.u.}$$

$$\mathbf{c}_1^{(3)} = \begin{pmatrix} 0.842 \\ 0.183 \end{pmatrix}, \qquad \mathbf{c}_2^{(3)} = \begin{pmatrix} -0.621 \\ 1.819 \end{pmatrix}$$

$E^{(3)} = -2.862$ a.u. Thus $E^{(2)} - E^{(3)} < 0.01$ a.u. and the calculation has converged.

8. See the discussion following Eq. (B.0, 14).

9.

$$(\psi_1 \psi_1 | \psi_1 \psi_1) = 1.026 \text{ a.u.,} \qquad (\psi_1 \psi_2 | \psi_2 \psi_1) = 0.290 \text{ a.u.}$$

B.5 1.

$$\Psi_1 = |\psi_1 \bar{\psi}_1|$$

$$\Psi_2 = \frac{1}{\sqrt{2}} [|\psi_1 \bar{\psi}_2| - |\bar{\psi}_1 \psi_2|]$$

$$\Psi_3 = |\psi_2 \bar{\psi}_2|$$

2.

$$H_{11} = \langle \Psi_1 | \hat{H} | \Psi_1 \rangle = 2h_{11} + (11|11) = -2.862 \text{ a.u.}$$
$$H_{12} = \langle \Psi_1 | \hat{H} | \Psi_2 \rangle = 0 \qquad (\text{Brillouin's theorem})$$
$$H_{13} = \langle \Psi_1 | \hat{H} | \Psi_3 \rangle = (12|12) = 0.290 \text{ a.u.}$$
$$H_{22} = \langle \Psi_2 | \hat{H} | \Psi_2 \rangle = h_{11} + h_{22} + (22|11) + (12|21) = 0.403 \text{ a.u.}$$
$$H_{23} = \langle \Psi_2 | \hat{H} | \Psi_3 \rangle = \sqrt{2}[h_{12} + (12|22)] = -0.182 \text{ a.u.}$$
$$H_{33} = \langle \Psi_3 | \hat{H} | \Psi_3 \rangle = 2h_{22} + (22|22) = 3.247 \text{ a.u.}$$

3. Solution of the 3×3 secular determinant leads to the equation

$$E^3 - (H_{11} + H_{22} + H_{33})E^2$$
$$+ (H_{11}H_{22} + H_{11}H_{33} + H_{22}H_{33} - H_{13}^2 - H_{23}^2)E - H_{11}H_{22}H_{33}$$
$$+ H_{11}H_{23}^2 + H_{22}H_{13}^2 = E^3 + aE^2 + bE + c = 0$$

The roots of the third-order polynomial (see the Appendix) are determined from the parameters

$$a = -0.788, \qquad b = -9.254, \qquad c = 3.684$$
$$\alpha = 0.263, \qquad \beta = 1.776, \qquad \gamma = 0.560$$

to be

$$E_1 \text{ (CI)} = -2.876 \text{ a.u.}$$
$$E_2 \text{ (CI)} = 0.392 \text{ a.u.}$$
$$E_3 \text{ (CI)} = 3.272 \text{ a.u.}$$

4.

$$E \text{ (corr)} = E \text{ (CI)} - E \text{ (SCF)} = E_1 \text{ (CI)} - H_{11} = -0.014 \text{ a.u.}$$

5.

$$IP \text{ (He)} = E(\text{He}^+) - E \text{ (He)}$$
$$E(\text{He}^+) = h_{11} = -1.944 \text{ a.u.}$$
$$E \text{ (He)} = E \text{ (SCF)} = H_{11} = -2.862 \text{ a.u.}$$
$$IP \text{ (He)} = 0.918 \text{ a.u.}$$

6.

$$IP\ (\text{He}) = -\varepsilon_1 = 0.918\ \text{a.u.} = 24.99\ \text{eV}$$

Koopmans' theorem is derived under the assumption that we use the frozen orbital approximation.

7.

$$E(\text{He}^+) = -\ 1.980\ \text{a.u.}$$

$$E\ (\text{He}) = E_1\ (\text{CI}) = -\ 2.876\ \text{a.u.}$$

$$IP\ (\text{He}) = 0.896\ \text{a.u.} = 24.38\ \text{eV}$$

8.

$$\Delta E(^1S) = \langle \Psi_2 | \hat{H} | \Psi_2 \rangle - \langle \Psi_1 | \hat{H} | \Psi_1 \rangle = 88.84\ \text{eV}$$

9.

$$\Delta E(^1S) = E_2\ (\text{CI}) - E_1\ (\text{CI}) = 3.268\ \text{a.u.} = 88.92\ \text{eV}$$

10. The total energies and ionization potentials agree reasonably well with the experiments because the double-zeta basis set is optimized to describe the ground state. As expected, the CI treatment improves the agreement with experiment.

     The present basis set cannot describe excited states of He, and the excitation energies are thus unreasonably large at both levels of approximation. A reasonable description of the lowest excitation energy is not possible unless 2s functions are included in the basis set.

B.6 1.

$$A_m(x) = \frac{m}{x}(m-1)!\ e^{-x} \sum_{j=0}^{m-1} \frac{x^{j-(m-1)-1}}{j!} + m!\ e^{-x} \frac{1}{xm!}$$

$$= \frac{m}{x} A_m(x) + A_0(x) \qquad \text{Q.E.D.}$$

2.

(a)

$$\int_1^\infty e^{-x\xi}\ d\xi = \frac{e^{-x}}{x} = A_0(x)$$

Equation (B.6, 4) is thus valid for $m = 0$. Furthermore

$$\int_1^\infty \xi^m e^{-x}\ d\xi = A_0(x) + \frac{m}{x} \int_1^\infty \xi^{m-1} e^{-x\xi}\ d\xi$$

If we assume that (B.6, 4) is true for $m - 1$, the forgoing relation shows that it is also true for m [using Eq. (B.6, 3)]. Thus, Eq. (B.6, 4) follows by induction.

(b)

$$\int_{-1}^{1} e^{-x\eta} d\eta = -A_0(x) - A_0(-x) = B_0(x)$$

Equation (B.6, 5) is thus valid for $m = 0$. Furthermore

$$\int_{-1}^{1} \eta^{-m} e^{-x} d\eta = (-)^{m+1} A_0(-x) - A_0(x) + \frac{m}{x} \int_{-1}^{1} \eta^{m-1} e^{-x\eta} d\eta$$

$$\text{(B.6, I)}$$

Assume that (B.6, 5) is valid for $m - 1$, then (B.6, I) yields

$$\int_{-1}^{1} \eta^{m} e^{-x\eta} d\eta = (-)^{m+1} A_0(-x) - A_0(x)$$

$$+ \frac{m}{x} [(-)^{m+1} A_m(-x) - A_m(x)] = B_m(x)$$

where we have used the definition of $B_m(x)$ in Eq. (B.6, 2) and Eq. (B.6, 3). Equation (B.6, 5) then follows by induction.

3.

$$A_m(x) \sim m!\, x^{-m-1} \qquad \text{(B.6, II)}$$

4.

$$B_m(x) = \int_{-1}^{1} \eta^{m} (1 - x\eta + \tfrac{1}{2} x^2 \eta^2 - \cdots) d\eta$$

$$\sim \begin{cases} \dfrac{2}{m+1}, & m \text{ even} & \text{(B.6, III)} \\[2mm] \dfrac{2x}{m+2}, & m \text{ odd} & \text{(B.6, IV)} \end{cases}$$

5.

x	m	$A_m(x)$	$B_m(x)$
0.1	0	10	2
0.1	1	100	$-2/30$
0.1	2	2000	$2/3$
0.5	0	2	2
0.5	1	4	$-1/3$
0.5	2	16	$2/3$

6. $A_m(x)$ and $(-1)^{m+1}A_m(-x)$ are much larger than their difference, which determines $B_m(x)$. Equation (B.6, 2) is thus numerically dangerous and the problem is most serious when m gets larger and x gets smaller.

7.

$$B_0(x) = 2i_0(x)$$

$$B_1(x) = 2i_1(x)$$

$$B_2(x) = \frac{2}{3}[i_0(x) + 2i_2(x)]$$

$$B_3(x) = \frac{2}{5}[3i_1(x) + 2i_3(x)]$$

8. We only add terms with the same signs.
9. Differentiate (B.6, 10) with respect to η and multiply the equation by 2. Multiply (B.6, 11) by $2n + 1$ and subtract the two resulting equations, whereby we obtain

$$(2n + 1)P_n(\eta) = P'_{n+1}(\eta) - P'_{n-1}(\eta) \qquad \text{(B.6, V)}$$

which when integrated gives

$$(2n + 1)\int_{-1}^{\eta} P_n(t)\,dt = P_{n+1}(\eta) - P_{n-1}(\eta) - P_{n+1}(-1) + P_{n-1}(-1)$$

$P_n(-1) = (-)^n$, which can be shown by induction, using Eqs. (B.6, 8–10), and the relation in Eq. (B.6, 12) follows.

10.

$$(2n + 1)i_n(x) = \frac{2n + 1}{2}\int_{-1}^{1} P_n(\eta)e^{-x\eta}\,d\eta = x[i_{n+1}(x) - i_{n-1}(x)]$$

where we have used Eq. (B.6, V) and that $P_n(1) = 1$.
11. Equation (B.6, 13) gives $2n + 1 = xr_n(x) - (x/r_{n-1}(x))$ or

$$r_{n-1}(x) = \frac{-x}{2n + 1 - xr_n(x)} \qquad \text{Q.E.D.}$$

12.

$$i_0(x) = \frac{e^x - e^{-x}}{2x}$$

$$xr_0(x) = 1 - x\frac{e^x + e^{-x}}{e^x - e^{-x}}$$

$$x > 0: \qquad i_0(x)[1 + x - xr_0(x)] = e^x = e^{|x|}$$

$$x < 0: \qquad i_0(x)[1 - x - xr_0(x)] = e^{-x} = e^{|x|}, \qquad \text{Q.E.D.}$$

B.7 1. and 2.

Use confocal elliptical coordinates to derive Eqs. (B.7, 1–3).

3.

$$\hat{H} = -\frac{\nabla^2}{2} - \frac{1}{r_A} - \frac{1}{r_B} + \frac{1}{R}$$

$$\psi_g = (2 + 2S(R))^{-\frac{1}{2}}(\phi_A + \phi_B) \tag{B.7, I}$$

$$E_0(R) = \langle \psi_g | \hat{H} | \psi_g \rangle = E_{1s}(H) + \frac{1}{R} + \frac{\varepsilon_{AA}(R) + \varepsilon_{AB}(R)}{1 + S(R)} \tag{B.7, II}$$

where $E_{1s}(H) = -0.5$ a.u.

4. and 5.

R	$E_0(R)$
1.6	−0.5135
2.1	−0.5583
2.4	−0.5645
2.5 = R_e	−0.5648
2.6	−0.5645
2.9	−0.5609
6.0	−0.5091
∞	−0.5000

6.

$$D_e = E(\infty) - E(R_e) = 0.0648 \text{ a.u.} = 1.76 \text{ eV}$$

7. Both D_e and R_e are of the right order of magnitude, mainly because the ψ_g orbital predicts the correct dissociation product for H_2^+, namely, $H^+ + H$. The predicted chemical bond is, however, too "weak" (D_e is too small and R_e too large).

8.

$$\psi_u = (2 - 2S(R))^{-\frac{1}{2}}(\phi_A - \phi_B) \tag{B.7, III}$$

$$E_1(R) = \langle \psi_u | \hat{H} | \psi_u \rangle = -\frac{1}{2} + \frac{1}{R} + \frac{\varepsilon_{AA}(R) - \varepsilon_{AB}(R)}{1 - S(R)} \tag{B.7, IV}$$

and $\Delta E(R) = E_1^{(R)} - E_0^{(R)}$ follows from Eqs. (B.7, 1–3, II and IV).

9. $\Delta E(R_e) = 0.274$ a.u. $= 7.46$ eV.

Since the MO's are orthogonal, the choice of the origin of the coordinate system does not affect the electronic transition moment, M_{10}. Let us for convenience choose the origin at the center of the internuclear axis with the z-axis pointing toward nucleus A. The inversion symmetry gives

$$\langle \phi_A | z | \phi_B \rangle = \langle \phi_B | z | \phi_A \rangle = 0$$

$$\langle \phi_A | z | \phi_A \rangle = -\langle \phi_B | z | \phi_B \rangle = \frac{R}{2}$$

This implies [Eqs. (B.7, I and III)] that

$$M_{10} = \langle \psi_g | z | \psi_u \rangle = \frac{1}{2}(1 - S^2(R))^{-\frac{1}{2}} R = 1.406 \text{ a.u.}$$

$$f_{10} = \frac{2}{3} M_{10}^2 \Delta E(R_e) = 0.361$$

B.8   1.

$$\psi(r) = \left(\frac{\zeta^3}{\pi} \right)^{1/2} e^{-\zeta r}$$

$$\hat{H} = -\frac{\nabla^2}{2} - \frac{1}{r_A} - \frac{1}{r_B} + \frac{1}{R} \qquad \text{(B.8, I)}$$

$$\langle \psi(r) | -\frac{\nabla^2}{2} | \psi(r) \rangle = \frac{\zeta^2}{2} \qquad \text{(B.8, II)}$$

In the notation of Fig. B.8, I, r_B^{-1} can be written as

$$r_B^{-1} = \left(\frac{R^2}{4} + r^2 - Rr \cos \theta \right)^{-\frac{1}{2}}$$

Thus

$$\langle \psi(r) | -\frac{1}{r_B} | \psi(r) \rangle = \frac{2\zeta^3}{R} \int_0^\infty re^{-2\zeta r} \, dr \int_{r + R/2}^{|r - R/2|} u^{-\frac{1}{2}} \, du$$

$$\qquad \qquad \qquad \qquad \text{(B.8, III)}$$

$$= e^{-\zeta R} \left(\zeta + \frac{2}{R} \right) - \frac{2}{R}$$

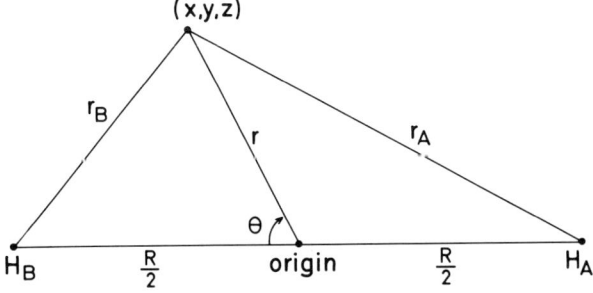

Fig. B.8,I. The coordinate system that is used for the H_2^+ molecule.

Equations (B.8, I–III) imply

$$E(\zeta) = \frac{\zeta^2}{2} - \frac{3}{R} + 2e^{-\zeta R}(\zeta + \frac{2}{R}) \qquad \text{(B.8, IV)}$$

2.

$$\partial E(\zeta)/\partial \zeta = 0 \quad \text{gives} \quad e^{-\zeta R} = \zeta/(2 + 2R\zeta). \qquad \text{(B.8, V)}$$

At $R = R_e = 2.00$ a.u. we obtain

$$e^{-2\zeta} = \frac{\zeta}{2 + 4\zeta} \qquad \text{(B.8, VI)}$$

which, solved numerically, within an accuracy of 10^{-4} a.u. yields $\zeta = \zeta_e = 0.912$ a.u.

$$E(\zeta_e) = -0.467 \ a.u. = -12.71 \text{ eV}.$$

3. At $\zeta = \zeta_e = 0.912$ a.u. $\partial E(R)/\partial R = 0$ gives

$$\exp(\zeta_e R) = \frac{2}{3}[2 + 2\zeta_e R + (\zeta_e R)^2]$$

which to an accuracy of 10^{-4} a.u. yields

$$R_e = 1.894 \text{ a.u.}, \qquad E(R_e) = -0.468 \text{ a.u.}$$

4.

$$D_e = E(\infty) - E(R_e) = \frac{\zeta_e^2}{2} - E(R_e) = 0.884 \text{ a.u.} = 24.05 \text{ eV}$$

5. The large difference is mainly caused by the fact that the trial function predicts an incorrect dissociation of H_2^+, namely, $H^+ + H^+ + e^-$.

B.9 1 From ϕ_A and ϕ_B we can construct two normalized symmetry orbitals in $D_{\infty h}$, the point group for H_2, namely

$$\psi_g = (2 + 2S)^{-1/2}(\phi_A + \phi_B) \qquad \text{(B.9, I)}$$

$$\psi_u = (2 - 2S)^{-1/2}(\phi_A - \phi_B) \qquad \text{(B.9, II)}$$

Since the symmetry orbitals have different symmetries, the Fock operator is diagonal in the representation spanned by ψ_g and ψ_u, and ψ_g and ψ_u are thus the SCF molecular orbitals.

2. Using Eqs. (B.9, I and II) and the numerical values of the atomic integrals, we obtain

$$\langle \psi_g | \hat{h} | \psi_g \rangle = -1.185 \text{ a.u.}$$
$$\langle \psi_g | \hat{h} | \psi_u \rangle = 0$$
$$\langle \psi_u | \hat{h} | \psi_u \rangle = -0.575 \text{ a.u.}$$

$$(gg|gg) = 0.566 \text{ a.u.}, \qquad (gg|uu) = 0.558 \text{ a.u.}$$
$$(gu|gu) = 0.140 \text{ a.u.}, \qquad (uu|uu) = 0.582 \text{ a.u.}$$
$$(gg|gu) = (uu|ug) = 0$$

3. Using Eq. (B.0, 5) we find

$$\varepsilon_g = \langle \psi_g | \hat{h} | \psi_g \rangle + (gg|gg) = -0.619 \text{ a.u.}$$

$$\varepsilon_u = \langle \psi_u | \hat{h} | \psi_u \rangle + 2(uu|gg) - (ug|ug) = 0.401 \text{ a.u.}$$

$$E_0 \text{ (SCF)} = \varepsilon_g + \langle \psi_g | \hat{h} | \psi_g \rangle + \frac{1}{R} = -1.090 \text{ a.u.}$$

4.

$$D_1 = |\psi_g \bar{\psi}_g|, \qquad M_S = 0$$
$$D_2 = |\psi_g \psi_u|, \qquad M_S = 1$$
$$D_3 = |\psi_g \bar{\psi}_u|, \qquad M_S = 0$$
$$D_4 = |\bar{\psi}_g \psi_u|, \qquad M_S = 0$$
$$D_5 = |\bar{\psi}_g \bar{\psi}_u|, \qquad M_S = -1$$
$$D_6 = |\psi_u \bar{\psi}_u|, \qquad M_S = 0$$

5.

$$\Psi_1(^1\Sigma_g^+) = D_1, \qquad \Psi_2(^1\Sigma_g^+) = D_6$$

$$\Psi_3(^1\Sigma_u^+) = \frac{1}{\sqrt{2}}(D_3 - D_4), \qquad \Psi_4(^3\Sigma_u^+, M_S = 1) = D_2$$

$$\Psi_5(^3\Sigma_u^+, M_S = 0) = \frac{1}{\sqrt{2}}(D_3 + D_4), \qquad \Psi_6(^3\Sigma_u^+, M_S = -1) = D_5$$

6. Because of the symmetry blocking of the CI matrix, only the $^1\Sigma_g^+$ block has nondiagonal elements, being a 2×2 matrix with elements

$$\langle \Psi_1 | \hat{H} | \Psi_1 \rangle = 2\langle \psi_g | \hat{h} | \psi_g \rangle + (gg|gg) = -1.804 \text{ a.u.}$$
$$\langle \Psi_2 | \hat{H} | \Psi_2 \rangle = 2\langle \psi_u | \hat{h} | \psi_u \rangle + (uu|uu) = -0.568 \text{ a.u.}$$
$$\langle \Psi_1 | \hat{H} | \Psi_2 \rangle = (gu|gu) = 0.140 \text{ a.u.}$$

The eigenvalues of the secular determinant are

$$E_1({}^1\Sigma_g^+) = -1.105 \text{ a.u.,} \qquad E_2({}^1\Sigma_g^+) = 0.162 \text{ a.u.}$$

where we have added the nuclear repulsion term, $1/R$. The corresponding CI wave functions are

$$\Psi_1 \text{ (CI)} = 0.994 \ \Psi_1 - 0.111 \ \Psi_2$$
$$\Psi_2 \text{ (CI)} = 0.111 \ \Psi_1 + 0.994 \ \Psi_2$$

The other symmetries give

$$E({}^1\Sigma_u^+) = \langle \Psi_3 | \hat{H} | \Psi_3 \rangle + \frac{1}{R}$$

$$= \langle \psi_g | \hat{h} | \psi_g \rangle + \langle \psi_u | \hat{h} | \psi_u \rangle + (gg|uu) + (gu|gu) + \frac{1}{R}$$

$$= -0.348 \text{ a.u.}$$

$$E({}^3\Sigma_u^+) - \frac{1}{R} = \langle \Psi_4 | \hat{H} | \Psi_4 \rangle = \langle \Psi_5 | \hat{H} | \Psi_5 \rangle = \langle \Psi_6 | \hat{H} | \Psi_6 \rangle$$

$$E({}^3\Sigma_u^+) = \langle \psi_g | \hat{h} | \psi_g \rangle + \langle \psi_u | h | \psi_u \rangle + (gg|uu) - (gu|gu) + \frac{1}{R}$$

$$= -0.628 \text{ a.u.}$$

7.

$$E(\text{corr}) = E(\text{CI}) - E\text{ (SCF)} = E_1({}^1\Sigma_g^+) - E_0(\text{SCF})$$

$$= -0.015 \text{ a.u.}$$

8. If the origin of the coordinate system is chosen to be at the center of the molecule with the z-axis pointing toward A, the only nonvanishing *atomic* integrals are

$$\langle \phi_A | z | \phi_A \rangle = \frac{R}{2}, \qquad \langle \phi_B | z | \phi_B \rangle = -\frac{R}{2}$$

9.

$$\langle \psi_g | z | \psi_u \rangle = \frac{R}{2}(1 - S^2)^{-1/2} = 1.064 \text{ a.u.}$$

10. When $\mathbf{r} \to \mathbf{r} + \boldsymbol{\Delta}$ where $\boldsymbol{\Delta}$ is a constant displacement we get

$$\langle \psi_u | \mathbf{r} + \boldsymbol{\Delta} | \psi_g \rangle = \langle \psi_u | \mathbf{r} | \psi_g \rangle$$

since $\langle \psi_u | \psi_g \rangle = 0$.

11. Initial state Ψ_1; final state Ψ_3.

$$E_{\text{final}} - E_{\text{initial}} = E(^1\Sigma_u^+) - E_0(\text{SCF}) = 0.742 \text{ a.u.} = 20.19 \text{ eV.}$$

$$M_{\text{if}} = \langle \Psi_1 | z(1) + z(2) | \Psi_3 \rangle = \sqrt{2} \langle \psi_g | z | \psi_u \rangle = 1.505 \text{ a.u.}$$

$$f_{\text{if}} = \frac{2}{3} M_{\text{if}}^2 (E_{\text{final}} - E_{\text{initial}}) = 1.120$$

12.

$$E_{\text{final}} - E_{\text{initial}} = E(^1\Sigma_u^+) - E_1(^1\Sigma_g^+) = 0.757 \text{ a.u.} = 20.60 \text{ eV.}$$

$$M_{\text{if}} = \langle \Psi_1(\text{CI}) | z(1) + z(2) | \Psi_3 \rangle = 1.329 \text{ a.u.}$$

$$f_{\text{if}} = 0.890$$

13. Basis functions with higher quantum numbers are needed for a better description of especially the excited state. Note the improvement in the CI transition moment $M_{\text{if}}(\text{CI}) = 1.329$ a.u. compared with $M_{\text{if}}(\text{SCF}) = 1.505$ a.u. The best value of Kołos and Wolniewicz is $M_{\text{if}} = 0.980$ a.u.

B.10 1.

$$\Psi(\text{LiH}, {}^1\Sigma^+) = |\psi_{1\sigma} \bar{\psi}_{1\sigma} \psi_{2\sigma} \bar{\psi}_{2\sigma}|$$

$$\Psi(\text{LiH}^+, {}^2\Sigma^+, M_S = \frac{1}{2}) = |\psi_{1\sigma} \bar{\psi}_{1\sigma} \psi_{2\sigma}|$$

The dipole moment (in atomic units) is

$$\mu = -\langle \Psi | \mathbf{r} | \Psi \rangle + \sum_A Z_A \mathbf{R}_B \qquad \text{(B.10, I)}$$

where Z_A and $\mathbf{R}_A$ are the charge and position, respectively, of nuclear A, and $\mathbf{r}$ is the position operator.

$$\mu_{\text{LiH}} = -2\langle \psi_{1\sigma} | \mathbf{r} | \psi_{1\sigma} \rangle - 2\langle \psi_{2\sigma} | \mathbf{r} | \psi_{2\sigma} \rangle + 3\mathbf{R}_{\text{Li}} + \mathbf{R}_{\text{H}} \qquad \text{(B.10, II)}$$

$$\mu_{\text{LiH}^+} = -2\langle \psi_{1\sigma} | \mathbf{r} | \psi_{1\sigma} \rangle - \langle \psi_{2\sigma} | \mathbf{r} | \psi_{2\sigma} \rangle + 3\mathbf{R}_{\text{Li}} + \mathbf{R}_{\text{H}} \qquad \text{(B.10, III)}$$

2. The dipole moment in the displaced coordinate system is

$$\mu'_{\text{LiH}} = -2\langle \psi_{1\sigma} | \mathbf{r} + \mathbf{\Delta} | \psi_{1\sigma} \rangle - 2\langle \psi_{s\sigma} | \mathbf{r} + \mathbf{\Delta} | \psi_{2\sigma} \rangle + 3(\mathbf{R}_{\text{Li}} + \mathbf{\Delta}) + \mathbf{R}_{\text{H}}$$

$$+ \mathbf{\Delta} = \mu_{\text{LiH}}$$

since the molecular orbitals are orthonormal.

3. From Eq. (B.10, III) we get

$$\mu'_{LiH^+} = -2\langle\psi_{1\sigma}|\mathbf{r} + \Delta|\psi_{1\sigma}\rangle - \langle\psi_{2\sigma}|\mathbf{r} + \Delta|\psi_{2\sigma}\rangle + 3(\mathbf{R}_{Li} + \Delta) + \mathbf{R}_H$$
$$+ \Delta = \mu_{LiH^+} + \Delta$$

4. The exact wave function may be written as a linear combination of Slater determinants

$$\Psi = \sum_i C_i \Psi_i, \qquad \langle\Psi_i|\Psi_j\rangle = \delta_{ij}, \qquad \langle\Psi|\Psi\rangle = 1$$

From (B.10, I) we get

$$\mu' = -\langle\Psi|\mathbf{r} + \Delta|\Psi\rangle + \sum_A Z_A(\mathbf{R}_A + \Delta)$$

$$= -\langle\Psi|\mathbf{r}|\Psi\rangle - N\Delta\langle\Psi|\Psi\rangle + \sum_A Z_A \mathbf{R}_A + \Delta \sum_A Z_A$$

$$= \mu + \Delta(\sum_A Z_A - N)$$

where N is the number of electrons. Thus the dipole moment is independent of the choice of origin when $\sum_A Z_A = N$ (noncharged systems).

5. From (B.10, II),

$$\mu_{LiH} = -2\langle\psi_{1\sigma}|z|\psi_{1\sigma}\rangle - 2\langle\psi_{2\sigma}|z|\psi_{2\sigma}\rangle + R \quad \text{(B.10, IV)}$$

Thus we have

$$\langle\psi_{1\sigma}|z|\psi_{1\sigma}\rangle = 0.00001\ R + 0.0062\ z_1 + 0.0001\ z_2 \quad \text{(B.10, V)}$$

$$\langle\psi_{2\sigma}|z|\psi_{2\sigma}\rangle = 0.6132\ R - 0.2288\ z_1 + 0.5688\ z_2 \quad \text{(B.10, VI)}$$

and obtain the dipole moment

$$\mu_{LiH} = -0.2266\ R + 0.4452\ z_1 - 1.1378\ z_2 \quad \text{(B.10, VII)}$$

6.

$$\phi_{2sLi} = 2\sqrt{\frac{0.64^5}{3}}\ re^{-0.64r}\ Y_{00}$$

$$\phi_{3pzLi} = \frac{2}{3}\sqrt{\frac{2\,(0.64)^7}{5}}\ r^2 e^{-0.64r}\ Y_{10}$$

We have

$$z\phi_{2sLi} = \frac{1}{0.64} \sqrt{\frac{5}{2}} \phi_{3pzLi}$$

and obtain

$$z_2 = \frac{S_2}{0.64} \sqrt{\frac{5}{2}}$$

7. Use confocal elliptical coordinates with Li as focus a and H as focus b. Then

$$z_a = r_a \cos\theta_a = -\frac{R}{2}(1 + \zeta\eta) \qquad (\text{B.10, VIII})$$

and

$$S_2 = \frac{(0.64)^{7/2}}{\pi} \left(\frac{2}{15}\right)^{1/2} \int r_a^2 \exp(-0.64\, r_a) \cos\theta_a \exp(-r_b)\, d\tau$$

Using the expressions in the Appendix for $d\tau$ and r_a and r_b in ellipsoidal coordinates together with the definitions of $A_n(x)$ and $B_n(x)$ in Eqs. (B.6, 4–5), we obtain Eq. (B.10, 1) straightforwardly. Note that a relabeling of the two focuses would change z_a into z_b and the expression for z_b would be different from that in Eq. (B.10, VIII). However, we would still obtain the expression for S_2 in Eq. (B.10, 1).

8.

$$\alpha = 2.4764, \qquad \beta = 0.5436, \qquad A_0(\alpha) = 0.0339$$

Equation (B.6, 3) gives

$$A_1(\alpha) = 0.0476, \qquad A_2(\alpha) = 0.0724,$$
$$A_3(\alpha) = 0.1217, \qquad A_4(\alpha) = 0.2305$$

Equations (B.6, 2 and 3) give the recurrence relation

$$B_n(\beta) = \frac{n}{\beta} B_{n-1}(\beta) - A_0(\beta) + (-)^{n+1} A_0(-\beta)$$

and

$$B_0(\beta) = 2.1000, \qquad B_1(\beta) = -0.3732, \qquad B_2(\beta) = 0.7268$$
$$B_3(\beta) = -0.2251, \qquad B_4(\beta) = 0.4430$$

and thus

$$S_2 = -0.4538$$

The minus sign on S_2 is caused by the arbitrariness in the sign of the volume element and must be changed; we find that

$$z_2 = 1.1207 \text{ a.u.}$$

9. $\phi_{1sLi} = 2.4892 \, e^{-2.69r}$

$\phi_{2pzLi} = 6.6959 \, r \cos \theta e^{-2.69r}$

$$z_1 = \langle \phi_{1sH} | z | \phi_{1sLi} \rangle = \frac{2.4892}{6.6958} \langle \phi_{1sH} | \phi_{2pzLi} \rangle = 0.0462 < 0.05 \text{ a.u.}$$

10. Using Eq. (B.10, VII) we see that the contribution to the dipole moment of LiH from ϕ_{1sLi} is $0.4452z_1$ a.u. $< 0.4452 \times 0.05$ a.u. $= 0.0226$ a.u. $= 0.0574$ D.
11. The contribution to the dipole moment from $\psi_{1\sigma}$ is (B.10, IV and B.10, V)

$$-2\langle \psi_{1\sigma} | z | \psi_{1\sigma} \rangle = -0.0009 \text{ a.u.} = -0.0023 \text{ D}$$

where we have used the values of z_2 and z_1 from questions 8 and 9, respectively.
12. The dipole moment of LiH is given in Eq. (B.10, VII) and to within an accuracy of 0.06 D we obtain

$$\mu_{LiH} \simeq -0.2266 \, R + 0.4452 \, z_2 - 1.1378 \, z_2 = 1.9595 \text{ a.u.}$$

$$= 4.9802 \text{ D}$$

13. An optimized minimum basis SCF calculation gives a fair description of ground-state properties.

B.11 1. The expressions for the perturbation corrections to the energy and the wave function are derived in all standard textbooks in quantum chemistry [see, e.g., Chapter 10-4 in Pilar (1968)].

2.

$$E_0^I = \langle u_0^{(0)} | \hat{H}_0 + \hat{V} | u_0^{(0)} \rangle$$

is the average value of $\hat{H}$ in the state $u_0^{(0)}$, that is, an upper bound to E_0.

3. E_0^{II} is not an upper bound to E_0, since it cannot be written as an average value of the Hamiltonian.

4. E_0^{III} is an upper bound to E_0.

$\cdot 5.$ $E_k^{III} = [E_k^{(0)} + E_k^{(1)} + 2E_k^{(2)}$

$\quad + \langle u_k^{(1)} | \hat{H}_0 | u_k^{(1)} \rangle + E_k^{(3)}$

$\quad + \langle u_k^{(0)} | \hat{V} | u_k^{(0)} \rangle \langle u_k^{(1)} | u_k^{(1)} \rangle][1 + \langle u_k^{(1)} | u_k^{(1)} \rangle]^{-1}$

$\quad = E_k^{(0)} + E_k^{(1)} + 2E_k^{(2)} - \langle u_k^{(1)} | u_k^{(1)} \rangle E_k^{(0)} - \langle u_k^{(1)} | u_k^{(1)} \rangle E_k^{(1)}$

$\quad + \langle u_k^{(1)} | \hat{H}_0 | u_k^{(1)} \rangle + E_k^{(3)} + \langle u_k^{(0)} | \hat{V} | u_k^{(0)} \rangle \langle u_k^{(1)} | u_k^{(1)} \rangle + \mathcal{O}(4)$

$\quad = E_k^{(0)} + E_k^{(1)} + E_k^{(2)} + E_k^{(3)} + \mathcal{O}(4)$

where we have used

$$(\hat{H}_0 - E_k^{(0)}) u_k^{(1)} + (\hat{V} - E_k^{(1)}) u_k^{(0)} = 0$$

and

$$\langle u_k^{(1)} | \hat{H}_0 - E_k^{(0)} | u_k^{(1)} \rangle = -\langle u_k^{(1)} | \hat{V} | u_k^{(0)} \rangle$$

6. Not an average value, thus not an upper bound.

B.12 1. We will use the notation of problem B.9. Using the expressions for $E_0^{(i)}$ and $u_0^{(i)}$ in Eqs. (B.11, 1–5) we find

$$E_0^{(0)} = \langle \Psi_1 | \hat{H}_0 | \Psi_1 \rangle + \frac{1}{R} = 2\varepsilon_g + \frac{1}{R} = -0.524 \text{ a.u.}$$

$$\Psi_0^{(1)} = \Psi_2 \frac{\langle \Psi_2 | \hat{V} | \Psi_1 \rangle}{\langle \Psi_1 | \hat{H}_0 | \Psi_1 \rangle - \langle \Psi_2 | \hat{H}_0 | \Psi_2 \rangle}$$

$$= \Psi_2 \frac{(gu|gu)}{2(\varepsilon_g - \varepsilon_u)} = -0.069 \, \Psi_2$$

where we have used the definition of $\hat{V}$ in Eq. (B.0, 18).

$E_0^{(1)} = \langle \Psi_1 | \hat{V} | \Psi_1 \rangle = -(gg|gg) = -0.566 \text{ a.u.}$

$E_0^{(2)} = \langle \Psi_1 | \hat{V} | \Psi_0^{(1)} \rangle = -0.069 \langle \Psi_1 | \hat{V} | \Psi_2 \rangle = -0.069 \, (gu|gu)$

$\quad = -0.010 \text{ a.u.}$

$E_0^{(3)} = (0.069)^2 [\langle \Psi_2 | \hat{V} | \Psi_2 \rangle - \langle \Psi_1 | \hat{V} | \Psi_1 \rangle \langle \Psi_2 | \Psi_2 \rangle]$

$\quad = (0.069)^2 [2(ug|ug) - 4(uu|gg) + (uu|uu) + (gg|gg)]$

$\quad = -0.004 \text{ a.u.}$

2. $E_{corr}(i)$ denotes the correlation energy obtained in a calculation that is consistent through order i in $\hat{V}$. Thus

$$E_{corr}(1) = E_0^{(0)} + E_0^{(1)} - E_{SCF} = 0$$

$$E_{corr}(2) = E_0^{(2)} = -0.010 \text{ a.u.}$$

$$E_{corr}(3) = E_0^{(2)} + E_0^{(3)} = -0.014 \text{ a.u.}$$

3.

$$E_0^{III} = \frac{\langle \Psi_1 - 0.069\, \Psi_2 | \hat{H} | \Psi_1 - 0.069\, \Psi_2 \rangle}{\langle \Psi_1 - 0.069\, \Psi_2 | \Psi_1 - 0.069\, \Psi_2 \rangle}$$

$$= \frac{2\varepsilon_g - (gg|gg) - 2 \times 0.069\,(gu|gu)}{1 + (0.069)^2}$$

$$+ \frac{(0.069)^2[2\varepsilon_u - 4(gg|uu) + 2(gu|gu) + (uu|uu)]}{1 + (0.069)^2} + \frac{1}{R}$$

$$= -1.103 \text{ a.u.}$$

$$E_{corr} = E_0^{III} - E_{SCF} = -0.013 \text{ a.u.}$$

This is approximately the same result as would be obtained in a third-order perturbation treatment, and the terms of $\mathcal{O}$ (4) which were disregarded in Eq. (B.11, 7) are thus of the order of 10^{-3} a.u.

4. The full CI calculation in problem B.9 gives the exact correlation energy within the considered minimal basis set. The second-order calculation gives 67% of the correlation energy and the third-order calculation 93%. 87% is obtained in the E_0^{III} calculation.

B.13 1. The zero-order Hamiltonian is

$$\hat{H}_0 = -\frac{1}{2}(\nabla_1^2 + \nabla_2^2) - \frac{1}{r_1} - \frac{1}{r_2} \qquad \text{(B.13, I)}$$

and the perturbation operator is

$$\hat{V} = \frac{1}{R} + \frac{1}{r_{12}} - \frac{1}{r_{1B}} - \frac{1}{r_{2A}} \qquad \text{(B.13, II)}$$

Using the nomenclature of Fig. B.13, I we find that

$$\frac{1}{r_{12}} = [(x_2 - x_1)^2 + (y_2 - y_1)^2 + (R + z_2 - z_1)^2]^{-1/2}$$

$$= \frac{1}{R}\left[1 + \frac{2(z_2 - z_1)}{R} + \frac{(x_2 - x_1)^2 + (y_2 - y_1)^2 + (z_2 - z_1)^2}{R^2}\right]^{-1/2}$$

$$\qquad \qquad \text{(B.13, III)}$$

$$\frac{1}{r_{1B}} = [x_1^2 + y_1^2 + (R - z_1)^2]^{-1/2} = \frac{1}{R}\left(1 - \frac{2z_1}{R} + \frac{r_1^2}{R^2}\right)^{-1/2} \quad \text{(B.13, IV)}$$

$$\frac{1}{r_{2A}} = [x_2^2 + y_2^2 + (R - z_2)^2]^{-1/2} = \frac{1}{R}\left(1 + \frac{2z_2}{R} + \frac{r_2^2}{R^2}\right)^{-1/2} \quad \text{(B.13, V)}$$

Using the Taylor expansion

$$(1 + \varepsilon)^{-1/2} = 1 - \frac{1}{2}\varepsilon + \frac{3}{8}\varepsilon^2 + \mathcal{O}(\varepsilon^3) \quad \text{(B.13, VI)}$$

where $\mathcal{O}(\varepsilon^3)$ denotes terms of order ε^3 or higher, and keeping terms through second order in r_i/R in Eq. (B.13, III–V) we obtain Eq. (B.13, 1).

2.

$$E^{(1)}(R) = \hat{V}_{00} = \langle |\phi_{1sA}\,\overline{\phi}_{1sB}|\,|\hat{V}|\,|\phi_{1sA}\,\overline{\phi}_{1sB}|\rangle = 0. \quad \text{(B.13, VII)}$$

where we have used the inversion symmetry of the H + H system.

3.

$$E^{(2)}(R) = \sum_{n \neq 0} \frac{|\hat{V}_{0n}|^2}{E_0 - E_n} \quad \text{(B.13, VIII)}$$

Since $E_0 - E_n$ is negative for all n, we obtain the inequality

$$E^{(2)}(R) \geqq \frac{1}{E_0 - E_{n*}} \sum_{n \neq 0} \hat{V}_{0n}\hat{V}_{n0}$$

$$= \frac{1}{E_0 - E_{n*}} \sum_n (\hat{V}_{0n}\hat{V}_{n0}) - (\hat{V}_{00})^2 = \frac{(\hat{V}^2)_{00}}{E_0 - E_{n*}} \quad \text{(B.13, IX)}$$

since $\hat{V}_{00} = 0$ and $\sum_n |n> <n| = 1$.

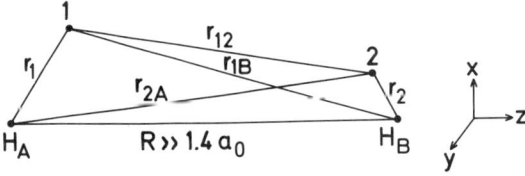

Fig. B.13,I. The coordinate system that is used to describe the interaction between two hydrogen atoms at an internuclear distance much larger than 1.4 a.u., the equilibrium distance of the H_2 molecule.

4. $E_0 = -1$ a.u. The state $|n^* >$ becomes

$$\Psi_{n^*} = |\phi_{2pA} \bar{\phi}_{2pB}| \qquad \text{and} \qquad E_{n^*} = -\frac{1}{4} \text{ a.u.}$$

$$\hat{V}^2 = \frac{1}{R^6}[x_1^2 x_2^2 + y_1^2 y_2^2 + 4z_1^2 z_2^2 + 2x_1 x_2 y_1 y_2 + \cdots]$$

$$(\hat{V}^2)_{00} = \langle |\phi_{1sA} \bar{\phi}_{1sB}| |\hat{V}^2| |\phi_{1sA} \bar{\phi}_{1sB}| \rangle$$

$$= \frac{1}{R^6}[|\langle \phi_{1s}|x^2|\phi_{1s}\rangle|^2 + |\langle \phi_{1s}|y^2|\phi_{1s}\rangle|^2 + 4|\langle \phi_{1s}|z^2|\phi_{1s}\rangle|^2]$$

$$= \frac{6}{R^6} \qquad\qquad\qquad\qquad (\text{B.13, X})$$

where we have used relations of the form

$$\langle \phi_{1s}|xy|\phi_{1s}\rangle = 0$$

$$\langle \phi_{1s}|x^2|\phi_{1s}\rangle = \langle \phi_{1s}|y^2|\phi_{1s}\rangle = \frac{1}{3}\langle \phi_{1s}|r^2|\phi_{1s}\rangle = 1 \text{ a.u.}$$

Thus

$$E^{(2)}(R) \geq -\frac{4}{3}\frac{6}{R^6} = -\frac{8}{R^6} \qquad \text{Q.E.D}$$

5. The perturbation equations are derived in all standard textbooks in quantum chemistry [see, e.g., Chapter 10-4 in Pilar (1968)].
6. From Eqs. (B.13, 4) and (B.13, 6) we get

$$E^{(2)} = -\langle u^{(0)}|E^{(1)} - \hat{V}|u^{(1)}\rangle \qquad (\text{B.13, XI})$$

Using Eq. (B.14, 5) together with Eq. (B.13, XI) we obtain

$$E^{(2)} = -\langle u^{(1)}|\hat{H}_0 - E^{(0)}|u^{(1)}\rangle$$

These two equations yield the identity

$$E^{(2)} = \langle u^{(1)}|\hat{H}_0 - E^{(0)}|u^{(1)}\rangle + 2\langle u^{(0)}|\hat{V} - E^{(1)}|u^{(1)}\rangle$$

7. We set $\chi = u^{(1)} + v$ and obtain

$$J_2 = E^{(2)} + \langle v|\hat{H}_0 - E^{(0)}|v\rangle - 2\langle u^{(0)}|E^{(1)} - \hat{V}|v\rangle$$
$$+ \langle u^{(1)}|\hat{H}_0 - E^{(0)}|v\rangle + \langle v|\hat{H}_0 - E^{(0)}|u^{(1)}\rangle$$
$$= E^{(2)} + \langle v|\hat{H}_0 - E^{(0)}|v\rangle \geqq E^{(2)}$$

8.

$$\hat{H} = -\frac{1}{2}\nabla^2 - \frac{1}{r}, \qquad \phi_{1s} = \pi^{-1/2}e^{-r}$$

$$\langle\phi_{1s}|x\hat{H}x|\phi_{1s}\rangle = -\frac{1}{2}\langle\phi_{1s}|x^2|\phi_{1s}\rangle - \langle\phi_{1s}|x\frac{\partial}{\partial x}|\phi_{1s}\rangle$$

$$= -\frac{1}{2}\langle\phi_{1s}|x^2|\phi_{1s}\rangle + \langle\phi_{1s}|\frac{x^2}{r}|\phi_{1s}\rangle = -\frac{1}{2} + \frac{1}{3}\langle\phi_{1s}|r|\phi_{1s}\rangle = 0$$

9.

$$J_2 = 2\langle u^{(0)}|\hat{V} - E^{(1)}|cVu^{(0)}\rangle + \langle u^{(0)}\hat{V}c|H_0 - E^{(0)}|c\hat{V}u^{(0)}\rangle$$

$$= (2c - E^{(0)}c^2)\langle u^{(0)}|\hat{V}^2|u^{(0)}\rangle$$

where we have used the relation in question 8. The extremum value of c is obtained from

$$\frac{\partial J_2}{\partial c} = (2 - 2E^{(0)}c)\langle u^{(0)}|\hat{V}^2|u^{(0)}\rangle = 0$$

which yields

$$J_2^{\text{min}} = -\frac{6}{R^6}$$

where we have used Eq. (B.13, X).

10. 5 kJ mole^{-1} = 1.904×10^{-2} a.u. An approximate van der Waals radius is thus obtained from the equation

$$\frac{7}{R^6} \approx 1.904 \times 10^{-3} \text{ a.u.}$$

$$R \approx 3.93 \text{ a.u.}$$

The equilibrium distance in H_2 is 1.4 a.u. and the van der Waals radius is thus considerably larger than the covalent radius.

C. SEMI-EMPIRICAL METHODS

INTRODUCTION

The problems in this section make use of extended Hückel methods in analyzing molecular electronic structure problems. The extended Hückel method, as well as most other semiempirical methods, may be interpreted as parametrized versions of the Hartree-Fock-Roothaan equation [Eq. (B.0,7)]. The various extended Hückel methods correspond to different parametrizations of the Fock operator, $\hat{F}$, and the overlap matrix S in the atomic orbital representation.

In problem C.1 we derive an extended Hückel method, the energy-weighted maximum-overlap (EWMO) method (Linderberg *et al.* 1976) in which $\hat{F}$ is parametrized as

$$F_{\mu\nu} \equiv \langle \phi_\mu | \hat{F} | \phi_\nu \rangle = \sum_\gamma S_{\mu\gamma} W_\gamma S_{\gamma\nu} \qquad \text{(C.0,1)}$$

where W_γ refers to the orbital energy of the atomic orbital ϕ_γ, and $S_{\gamma\nu}$ is the overlap $\langle \phi_\gamma | \phi_\nu \rangle$. The EWMO approximation is derived by optimizing a total energy expression of the form

$$E = \sum_\mu W_\mu q_\mu \qquad \text{(C.0,2)}$$

where q_μ is the occupation of atomic orbital ϕ_μ. The Wolfsberg-Helmholz approximation is shown to correspond to a simplified version of the EWMO approximation where the atomic overlap matrix elements are kept through first order. In problem C.2 we carry out an EWMO calculation on H_2O and perform an assignment of the photoelectron spectrum of H_2O based on the EWMO calculation.

Extended Hückel approximations may also be derived as a parametrized representation of a (unspecified) Hamiltonian matrix **H** with elements

$$H_{\mu\nu} = \langle \phi_\mu | \hat{H} | \phi_\nu \rangle \qquad \text{(C.0,3)}$$

The Hamiltonian matrix elements $H_{\mu\nu}$ are then formally parametrized according to a specific prescription, as done in, for example, problems C.3 and C.4.

The extended Hückel calculations have been used to carry out a conformation analysis of excited states of the methylene radical (C.3) and to perform an assignment of the photoelectron spectrum of H_2O_2 (C.4). A modification of the standard expression for the ground-state total energy in the extended Hückel method has been derived in problem C.6.

In problem C.5 we expand the total energy of a molecule in terms of the formal charge in the valence shell of the constituent atoms. The formal electronic charges are then optimized by using information about experimental ionization potentials. The method is applied to determine the ground-state of SF_6.

Textbooks that give sufficient background for solving the problems in this section

Avery (1972). Ballhausen and Gray (1965). Chandra (1974). Cook (1978). DeKock and Gray (1980). Gimarc (1979). Hanna (1969). La Paglia (1971). Levine (1974). Linderberg and Öhrn (1973). Lowe (1978). McWeeny (1979). Murrell, Kettle and Tedder (1970, 1978). Pilar (1968).

Textbooks that give a more elaborate treatment of the topics

Dewar (1969). Murrell and Harget (1972). Pople and Beveridge (1970).

Problem Description

   C.1 The energy-weighted maximum-overlap and the Wolfsberg-Helmholz approximations
   C.2 The photoelectron spectrum of H_2O from an energy-weighted maximum-overlap calculation
   C.3 Conformation analysis of the ground and excited states of CH_2 in an extended Hückel calculation
   C.4 The photoelectron spectrum of H_2O_2 in an extended Hückel calculation
   C.5 Population analysis of AB_Z complexes
   C.6 Ground-state total energies in extended Hückel approximations

PROBLEMS

C.1 In this problem we derive two extended Hückel models that explicitly consider the overlap between the atomic orbitals, namely, the energy-weighted maximum-overlap (EWMO) approximation (Linderberg et al. 1976), and the Wolfsberg-Helmholz approximation. The present derivation of the EWMO approximation illustrates most of the principles encountered when deriving the Hartree-Fock-Roothaan equations.

The set of molecular orbitals $\{\psi_r\}$ is given as linear combinations of the set of atomic orbitals $\{\phi_\mu\}$, that is,

$$\psi_r = \sum_\mu \phi_\mu c_{\mu r} \quad \text{or equivalently} \quad \phi_\mu = \sum a^*_{\mu r} \psi_r \quad \text{(C.1,1)}$$

The molecular orbitals are orthonormal, whereas the atomic orbitals are normalized but nonorthogonal:

$$\langle \phi_\mu | \phi_\nu \rangle \equiv \delta_{\mu\nu} + S_{\mu\nu} = (1 + S)_{\mu\nu} \quad \text{(C.1,2)}$$

1. Show that

$$\mathbf{a} = (1 + S)\mathbf{c} \quad \text{(C.1,3)}$$

where the columns in the square matrix $\mathbf{c}$ are the molecular orbital coefficients defined in Eq. (C.1,1).

We will now assume that the total electronic energy can be calculated from the equation

$$E_0 = \sum_\mu W_\mu q_\mu \quad \text{(C.1,4)}$$

where W_μ is the atomic orbital energies and q_μ is the probability of observing an electron in atomic orbital ϕ_μ, that is

$$q_\mu = \sum_r n_r |\langle \psi_r | \phi_\mu \rangle|^2 \quad \text{(C.1,5)}$$

where n_r is the occupation number of the molecular spin orbital ψ_r, that is, $n_r = 1$ if ψ_r is occupied and $n_r = 0$ otherwise. The formation of a molecule will cause a change in the total energy

$$\Delta E = \sum_\mu W_\mu (q_\mu - q_\mu^0) \quad \text{(C.1,6)}$$

where q_μ and q_μ^0 are the probabilities of observing an electron in atomic orbital ϕ_μ before and after the molecular formation has taken place, respectively. We will determine the molecular orbital coefficients $c_{\mu r}$ that maximize ΔE. To assure that the molecular orbitals can be kept orthonormal under the variation of the molecular orbital coefficients $c_{\mu r}$ we maximize the function

$$E(\mathbf{c}) = \Delta E - \sum_{rs} \langle \psi_r | \psi_s \rangle \lambda_{sr} \quad \text{(C.1,7)}$$

where $\{\lambda_{rs}\}$ is a set of Lagrangian multipliers. Thus, the coefficient matrix $\mathbf{c}$ is determined by requiring that

$$\delta E(\mathbf{c}) = 0 \quad \text{(C.1,8)}$$

2. Show that

$$\delta(\Delta E) = \sum_{\substack{\mu r \\ \nu}} n_r W_\nu [a^*_{\nu r}(\delta_{\nu\mu} + S_{\nu\mu})\delta c_{\mu r} + \delta c^*_{\mu r}(\delta_{\mu\nu} + S_{\mu\nu})a_{\nu r}] \quad \text{(C.1,9)}$$

3. Show that the matrix equation

$$(1 + S)Wa = a\lambda \qquad (C.1,10)$$

satisfies Eq. (C.1,8).

The atomic orbital energy parameters W_μ are negative, and we define a diagonal matrix α such that

$$W = -\alpha\alpha \qquad (C.1,11)$$

4. Show that the off-diagonal Lagrangian multiplier λ_{rs} ($r \neq s$) in Eq. (C.1,10) can be eliminated and that the resulting matrix eigenvalue equation becomes

$$-\alpha(1 + S)\alpha a' = a'\varepsilon \qquad (C.1,12)$$

where ε is a diagonal matrix the diagonal elements of which can be interpreted as molecular orbital energies.

For a set of nonorthogonal atomic orbitals the matrix form of the Hartree-Fock-Roothaan eigenvalue equations [Eq. (B.0,7)] is given as

$$Fc = (1 + S)c\varepsilon \qquad (C.1,13)$$

where F is the Fock matrix and c is the set of LCAO coefficients.

5. Show that Eq. (C.1,12) corresponds to a model in which the Fock matrix F can be identified as

$$F = (1 + S)W(1 + S) \qquad (C.1,14)$$

Equation (C.1,14) is known as the energy-weighted maximum-overlap approximation to the Fock matrix.

6. Show that through first order in S, F in Eq. (C.1,14) can be written as

$$F = W + SW + WS \qquad (C.1,15)$$

Equation (C.1,15) is known as the Wolfsberg-Helmholz approximation to the Fock matrix.

C.2 In this problem we will perform an assignment of the H_2O photoelectron spectrum based on an energy-weighted maximum-overlap (EWMO) calculation. The EWMO model was derived in problem C.1. We will include the $1s$ orbitals on H and the $2p$ orbitals on O in the calculation. The $2p$ orbital energy of oxygen is -16.16 eV. The overlap integral between the $1s$ orbital on hydrogen A and the $1s$ orbital on hydrogen B is $\langle \phi_{1sA} | \phi_{1sB} \rangle = 0.372$. The overlap integral between a $2p\sigma$ orbital directed along an OH bond and the hydrogen $1s$ orbital is $\langle \phi_{2p\sigma} | \phi_{1s} \rangle = 0.351$. The HOH bond angle is $104.45°$ (Dressler and Ramsay 1959).

1. Determine the overlap matrix $1 + S$ in the atomic orbital representation [see Eq. (C.1,2)].

2. Determine the EWMO approximation to the Fock matrix F in the atomic orbital representation [see Eq. (C.1,14)]

$$F = (1 + S)W(1 + S)$$

W is a diagonal matrix the diagonal elements of which are the atomic orbital energies [see Eq. (C.1,4)].

3. Determine the normalized symmetry orbitals.
4. Determine the molecular orbital energies for H_2O by using the EWMO model [use Eq. (C.1,13) with **F** as defined in Eq. (C.1,14) to determine the EWMO orbital energies].

The photoelectron spectrum of water is given in Fig. C.2,1 (Turner *et al.* 1970). The spectrum has three peaks, all of which show considerable vibrational structure.

5. Perform an assignment of the photoelectron spectrum of H_2O based on the EWMO calculation.

The three peaks in the photoelectron spectrum of H_2O are experimentally (increasing binding energy) assigned to ionizations out of molecular orbitals of b_1, a_1, and b_2 symmetry.

6. Is the experimental assignment in agreement with the one obtained from the EWMO calculation? If not, indicate the main reason for the discrepancy.
7. Discuss, based on the eigenfunctions of the EWMO calculation, the character of the orbitals from which the electron is ionized. Do you expect the three electronic states of H_2O^+ that result form ionizing an electron out of each of the three highest occupied molecular orbitals of the EWMO calculation to have a larger, smaller, or about the same HOH bond angle as the HOH bond angle of the H_2O molecule?

C.3 In this problem we carry out a conformation analysis on the three lowest states of the methylene radical CH_2 by using an extended Hückel model. We include in the extended Hückel calculation the $2p$ orbital on C and the $1s$ orbital on H, and we use the coordinate system given in Fig. C.3,1.

  The parametrization of the extended Hückel calculation will be as follows:

$$\alpha = \langle \phi_{2piC} | \hat{H} | \phi_{2piC} \rangle, \qquad i = x, y, z$$

$$\alpha = \langle \phi_{1si} | \hat{H} | \phi_{1si} \rangle, \qquad i = A, B$$

$$\langle \phi_{1sA} | \hat{H} | \phi_{1sB} \rangle = 0$$

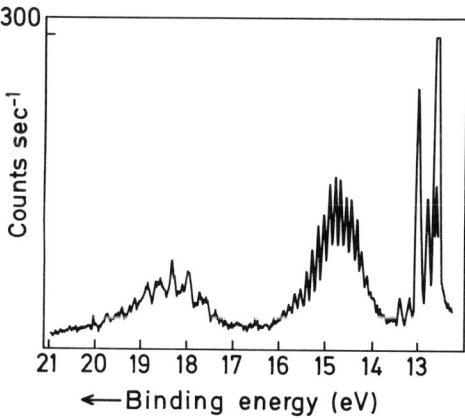

Fig. C.2,1. The photoelectron spectrum of H_2O. (Reprinted by permission of John Wiley and Sons, Ltd. from Turner *et al.* (1970))

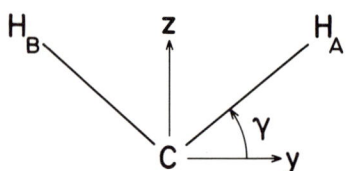

Fig. C.3,1. Coordinate system chosen for the CH$_2$ radical.

and

$$\beta = \langle \phi_{2p\sigma} | \hat{H} | \phi_{1sA} \rangle$$

where $\hat{H}$ is the Hamiltonian for the CH$_2$ radical. ϕ_{1sA} and ϕ_{1sB} denote the 1s-orbitals on the two hydrogen atoms and $\phi_{2p\sigma}$ is a 2p orbital on carbon directed toward a 1s orbital. All overlap integrals are set equal to zero in the extended Hückel calculation.

1. Determine the normalized symmetry orbitals.
2. Determine the normalized molecular orbitals and the corresponding orbital energies as a function of α, β, and γ where γ is defined in Fig. C.3,1.

Assume now that $\alpha = 0.0$ and $\beta = -2.0$ eV .

3. Sketch the orbital energies as a function of γ for $0° \leq \gamma \leq 90°$.

The three lowest states of CH$_2$ have, in an *ab initio* calculation (Harding and Goddard 1977), been determined to be $^3B_1$ (0.0 eV), $^1A_1$ (0.48 eV), and $^1B_1$ (1.56 eV) where the numbers in the parentheses are the relative total energies of the states.

4. Use the result of question 3 to estimate the range of γ angles that may give the relative energies determined by Harding and Goddard for the three states. Determine the electronic configurations and a Slater determinantal wave function for each of the states.

We will use the extended Hückel calculation to get a numerical estimate of the geometry of the three states. We will consider only a change of the HCH angle and thus assume that the CH bond lengths are the same in the three states.

5. Determine the terms in the total electronic energy of the $^3B_1$, $^1A_1$, and $^1B_1$ states that depend on α, β, and γ. Include the nuclear repulsion term that depends on γ in the total energy expression. Assume that the CH bond length is 2.08 a.u.

Assume now that α is independent of γ and that $\beta = -2.0$ eV .

6. Determine the value of γ that will minimize the total energies of the $^3B_1$, $^1B_1$, and $^1A_1$ states, respectively.

The experimental values for the HCH angle are 135° for the $^3B_1$ and the $^1B_1$ states and 105° for the $^1A_1$ state (Bernheim *et al.* 1970, Wasserman *et al.* 1970, Herzberg and Johns 1971).

7. Comment on the accuracy of the results obtained in question 6.

C.4 In this problem we perform an assignment of the photoelectron spectrum of the hydrogen peroxide molecule H$_2$O$_2$ based on an extended Hückel-type approximation. We assume that the OH bonds are perpendicular to the OO bond and that the angle between the two OH bonds is 90°. We include the $2p$ orbitals on O and the $1s$ orbital on hydrogen in the calculation and we parametrize the integrals as follows:

a. Overlap integrals between orbitals centered on nonneighboring atoms are zero. The overlap integrals between orbitals centered on neighboring atoms are

$$S_{p\sigma} = \langle \phi_{2p\sigma} | \phi_{2p\sigma} \rangle$$

$$S_{p\pi} = \langle \phi_{2p\pi} | \phi_{2p\pi} \rangle$$

$$S_{sp\sigma} = \langle \phi_{1sH} | \phi_{2p\sigma} \rangle$$

b. Nondiagonal matrix elements of the Hamiltonian are equal to a constant K multiplied by the corresponding overlap integrals.

c. Diagonal matrix elements of the Hamiltonian matrix are equal to a constant α for all the valence orbitals.

1.   Determine the normalized symmetry orbitals.
2.   Determine the molecular orbital energies as a function of K, α, $S_{p\sigma}$, $S_{p\pi}$, and $S_{sp\sigma}$.

The overlap integrals $S_{p\sigma}$, $S_{p\pi}$, and $S_{sp\sigma}$ may be evaluated by using Slater-type orbitals, and we find that $S_{p\sigma} = 0.38$, $S_{p\pi} = 0.18$, and $S_{sp\sigma} = 0.54$.

The constant K appearing in the off-diagonal matrix elements of the Hamiltonian matrix, H_{AB} may be determined from the Wolfsberg-Helmholz approximation where

$$H_{AB} = k(H_{AA} + H_{BB}) \frac{S_{AB}}{2}$$

k is a constant that usually is chosen between 1 and 3. We assume $k = 2$.

3.   Determine the molecular orbital energies as a function of α.

The photoelectron spectrum of hydrogen peroxide is given in Fig C.4,1 (Osafune and Kimura 1974). Traces of H_2O_2 and O_2 are present in the spectrum and the peaks that originate from these impurities are marked in the figure.

The peaks in the photoelectron spectrum have been assigned to five electronic states of $H_2O_2^+$, which are labeled 1,2,3,4, and 5. The states labeled 1 and 2 are degenerate in the extended Hückel calculation.

4.   Perform an assignment of the photoelectron spectrum of H_2O_2 based on the extended Hückel calculation. Use the assignment of the state labeled 3 to get an estimate of α and use this estimate to evaluate the positions of the other peaks.

5.   Indicate for all the ionization processes the part of the H_2O_2 molecule in which the ejected electron was predominantly localized (e g , in the $p\sigma(OO)$ bond).

C.5 In this problem we perform a population analysis of a compound or complex of the form AB_Z . We consider initially a free atom A that has q electrons in the valence shell. The ground-state total energy of atom A is a function of q and will be denoted by $E(q)$.

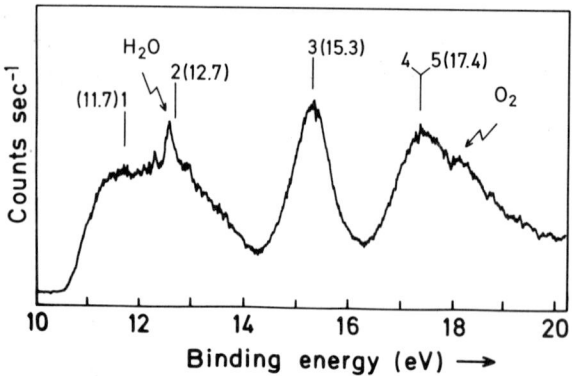

Fig. C.4,1. The photoelectron spectrum of hydrogen peroxide. (Reprinted from Chem. Phys. Lett. by permission of North-Holland Publishing Company and the authors).

1.   Show that a Taylor expansion of $E(q)$ through second order in q can be written in the form

$$E(q) = E_0 + qE_1 + \frac{1}{2}q(q-1)E_2 + \mathcal{O}(q^3) \qquad \text{(C.5,1)}$$

where $\mathcal{O}(q^3)$ denotes terms of order q^3 or higher.

2.   Give a physical interpretation of the coefficients E_0, E_1, and E_2.

Let I_A and I_{A+} denote the ionization potentials of A and A^+, respectively.

3.   Express E_1 and E_2 in terms of I_A, I_{A+}, and q.

Consider now the complex AB_Z, where the central atom, A, is surrounded by Z ligands B, and where the total number of valence electrons for the complex is N. Assume that the total energy for the complex AB_Z can be determined by adding the total energies of the individual atoms.

4.   Determine the valence charges on the central atom, q_A, and on the ligands, q_B, which minimize the total energy for the complex AB_Z. Express the results in terms of the constants E_i^x ($i = 1, 2$ and $x = A, B$), N and Z.

(Note that E_i^A and E_i^B are atomic quantities and are thus only a function of the ionization potentials and q of the isolated atom.)

We will now consider the compound SF_6. The experimental ionization potentials for S and F are (Moore 1949)

$$I_S = 10.36 \text{ eV}, \qquad I_{S^.} = 23.4 \text{ eV}$$

$$I_F = 17.42 \text{ eV}, \qquad I_{F^.} = 34.98 \text{ eV}$$

5.   What is the formal charge on F and on S?

6.   Is the sign on the formal charge as you would expect?

A Hartree-Fock calculation on SF_6 gave a formal charge on F of $-0.318e$ (Hay 1977).

C.6 In semiempirical molecular orbital methods, such as Hückel and extended Hückel methods, the ground-state total energy is usually estimated as

$$E_0 = \sum_r n_r \varepsilon_r \tag{C.6,1}$$

where n_r is the occupation number of molecular spin orbital ψ_r; that is, $n_r = 1$ if ψ_r is an occupied molecular orbital and $n_r = 0$ if ψ_r is an unoccupied molecular spin orbital. ε_r is the molecular orbital energy. In this problem we will show that this expression can be modified in a simple fashion which brings it more in agreement with Hartree-Fock total energies.

Let V_{nn} and V_{ee} denote the average value of the nuclear and electron repulsion operator with respect to the exact Hartree-Fock ground state.

1.   If we assume that the Hartree-Fock total energy can be approximated as in Eq. (C.6,1), where ε denotes the Hartree-Fock orbital energies, show then that this assumption implies that

$$V_{nn} \approx V_{ee} \tag{C.6,2}$$

It has been shown that the Hartree-Fock total energy for the ground state is well approximated by the relation (Ruedenberg 1977)

$$E_{HF} = \frac{3}{7}(V_{ne} + 2V_{nn}) \tag{C.6,3}$$

where V_{ne} denotes the Hartree-Fock nuclear-electronic attraction energy.

2.   Use Eq. (C.6,3) and the virial theorem to show that the potential energy, V, then can be expressed as

$$V = 6(V_{nn} - V_{ee}) \tag{C.6,4}$$

3.   If the assumptions in question 1 were fulfilled, which implications would this have on the model considered in question 2?

4.   Show that the Hartree-Fock total energy in Eq. (C.6,3) can be rewritten as

$$E_{HF} = \frac{3}{2} \sum_r n_r \varepsilon_r \tag{C.6,5}$$

5.   Evaluate, using Table C.6,1, the Hartree-Fock ground-state total energies for He, CH_4, and FH using both the approximate total energy expressions of Eqs. (C.6,1) and (C.6,5). Compare the result with the Hartree-Fock total energies given in Table C.6,1.

SOLUTIONS

C.1 1.

$$\langle \psi_r | \phi_{p'} \rangle = a_{p'r}^*$$

Thus

$$a_{p'r} = \langle \phi_{p'} | \psi_r \rangle = \sum_\mu \langle \phi_{p'} | \phi_\mu \rangle c_{\mu r} = \sum_\mu (\delta_{p'\mu} + S_{p'\mu}) c_{\mu r}$$

Table C.6,1
Orbital energies and ground-state total energies for He, CH$_4$ and HF

Molecule/atom	Orbital energy (a. u.)	Hartree–Fock total energy (a. u.)
He	$\varepsilon_{1s} = -0.92$	-2.86
CH$_4$	$\varepsilon_{1a_1} = -11.21$	-40.17
	$\varepsilon_{2a_1} = -0.94$	
	$\varepsilon_{1t_2} = -0.54$	
HF	$\varepsilon_{1\sigma} = -26.29$	-100.07
	$\varepsilon_{2\sigma} = -1.60$	
	$\varepsilon_{3\sigma} = -0.77$	
	$\varepsilon_{1\pi} = -0.65$	

2.

$$\delta(\Delta E) = \delta[\sum_{\mu} W_{\mu}(q_{\mu} - q_{\mu}^0)] = \delta \sum_{\mu} W_{\mu} q_{\mu}$$

$$= \sum_{r\mu} W_{\mu} n_r [a_{\mu r}^*(\delta a_{\mu r}) + a_{\mu r}(\delta a_{\mu r}^*)] \qquad (C.1,I)$$

$$= \sum_{\substack{r\mu \\ \nu}} n_r W_{\nu} [a_{\nu r}^*(\delta_{\nu\mu} + S_{\nu\mu})\delta c_{\mu r} + \delta c_{\mu r}^*(\delta_{\mu r} + S_{\mu r})a_{\nu r}]$$

3.

$$\delta \sum_{rs} \langle \psi_r | \psi_s \rangle \lambda_{sr} = \sum_{rs} [\delta c_{\mu r}^* a_{\mu s} \lambda_{sr} + a_{\mu r}^* \lambda_{sr} \delta c_{\mu s}] \qquad (C.1,II)$$

Equation (C.1,8) is fulfilled when the coefficients to $\delta c_{\mu r}^*$ and $\delta c_{\mu r}$ are equal to zero. From Eqs. (C.1,I-II) we see that the coefficient to $\delta c_{\mu r}^*$ is

$$\sum_{\nu} n_r(\delta_{\mu r} + S_{\mu r})W_{\nu}a_{\nu r} - \sum_{s} a_{\mu s}\lambda_{sr} = 0 \qquad (C.1,III)$$

The complex conjugate of Eq. (C.1,III) is equal to the coefficient to $\delta c_{\mu r}$.

When $n_r = 1$, Eq. (C.1,III) becomes identical to Eq. (C.1,10), whereas when $n_r = 0$, the coefficients $a_{\mu r}$ (and thus $c_{\mu r}$) are undetermined. We may thus *define* **c** from Eq. (C.1,10) also in the case when $n_r = 0$.

4. Using Eqs. (C.1,10-11) we have

$$-\alpha(1 + S)\alpha\alpha = \alpha a\lambda \qquad (C.1,IV)$$

Defining

$$\mathbf{a}'' = \alpha\mathbf{a} \qquad (C.1,V)$$

and multiplying Eq. (C.1,IV) from the right by a unitary matrix **U**
we obtain

$$-\alpha(1 + S)\alpha a'' U = a'' U U^\dagger \alpha U \qquad\qquad (C.1,VI)$$

Since λ is Hermitian, **U** can be chosen as the unitary matrix that
diagonalizes λ, that is, $U^\dagger \lambda U = \varepsilon$.
Denoting

$$a' = a'' U$$

implies that Eqs. (C.1,VI) and (C.1,12) are identical.

5. Upon definition of a new matrix $c = \alpha a'$, Eq. (C.1,12) becomes

$$-\alpha(1 + S)c = \alpha^{-1} c \varepsilon$$

or

$$-(1 + S)\alpha\alpha(1 + S)c = (1 + S)c\varepsilon \qquad\qquad (C.1,VII)$$

Equation (C.1,14) then follows from Eqs. (C.1,11) and (C.1,13).

6.

$$F = (1 + S)W(1 + S) = W + SW + WS + \mathcal{O}(S^2).$$

C.2 1. We use the coordinate system given in Fig. C.2,I. In the atomic
basis $\{\phi_{2p_zO}, \phi_{2p_yO}, \phi_{2p_zO}, \phi_{1sA}, \phi_{1sB}\}$ the symmetric overlap matrix
becomes

$$1 + S = \begin{pmatrix} 1.0 & 0.0 & 0.0 & 0.215 & 0.215 \\ & 1.0 & 0.0 & 0.278 & -0.278 \\ & & 1.0 & 0.0 & 0.0 \\ & & & 1.0 & 0.372 \\ & & & & 1.0 \end{pmatrix}$$

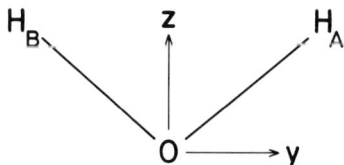

Fig. C.2,I. Coordinate system chosen for H_2O.

2.

$$\mathbf{F} = \begin{pmatrix} -17.418 & 0.0 & 0.0 & -7.489 & -7.489 \\ & -18.264 & 0.0 & -6.869 & 6.869 \\ & & -16.16 & 0.0 & 0.0 \\ & & & -17.489 & -9.624 \\ & & & & -17.489 \end{pmatrix}$$

3.

$$\chi_1(a_1) = \phi_{2p_zO}, \qquad \chi_2(a_1) = 0.604\,(\phi_{1sA} + \phi_{1sB})$$

$$\chi_3(b_2) = \phi_{2p_yO}, \qquad \chi_4(b_2) = 0.892\,(\phi_{1sA} - \phi_{1sB})$$

$$\chi_5(b_1) = \phi_{2p_xO}$$

4. a_1 symmetry:
    The molecular orbital energies are determined from the secular determinant

$$\begin{vmatrix} -17.418 - E & -9.047 - 0.260E \\ -9.047 - 0.260E & -19.783 - E \end{vmatrix} = 0$$

$$\varepsilon_{1a_1} = -22.11 \text{ eV}; \qquad \varepsilon_{2a_1} = -12.75 \text{ eV}$$

b_2 symmetry:

$$\begin{vmatrix} -18.264 - E & -12.254 - 0.496E \\ -12.254 - 0.496E & -12.516 - E \end{vmatrix} = 0$$

$$\varepsilon_{1b_2} = -19.32 \text{ eV}, \qquad \varepsilon_{2b_2} = -5.39 \text{ eV}$$

b_1 symmetry:

$$\varepsilon_{1b_1} = -16.16 \text{ eV}$$

5. The photoelectron spectrum contains three peaks that must be assigned to the three lowest electronic states. The EWMO calculation gives the following assignment: 12.7 eV (b_1); 15.0 eV (b_2); 18.5 eV (a_1). The EWMO orbital energies are shifted 1-4 eV relative to the experimentally observed peaks.

6. The peaks at 15.0 eV and 18.5 eV have been assigned incorrectly in the EWMO calculation. This is mainly because the 2s orbital on oxygen (which transforms as a_1) has not been included in the calculation. If the 2s orbital on oxygen were included in the calculation, the ε_{1a_1} orbital energy would change and be numerically smaller. The correct assignment would then be obtained from the EWMO calculation.

7.

$$\psi_{1b_1} = \phi_{2pxO}$$

$$\psi_{1a_1} = 0.44\,(\phi_{1sA} + \phi_{1sB}) + 0.52\,\phi_{2pzO}$$

$$\psi_{1b_2} = 0.28\,(\phi_{1sA} - \phi_{1sB}) + 0.81\,\phi_{2pyO}$$

The ψ_{1b_1} orbital represents a nonbonding oxygen atomic orbital. $\psi_{1b_1}^2$ is thus a lone pair located on the oxygen atom. ψ_{1a_1} has OH and HH bonding character, while ψ_{1b_2} has OH bonding and HH antibonding character. Since the ψ_{1b_1} orbital is nonbonding, the electronic state that results from ionizing an electron out of the ψ_{1b_1} orbital will have about the same HOH bond angle as the HOH bond angle in H_2O. The ψ_{1a_1} orbital has HH bonding character and the ionized state therefore loses HH bonding character and gives an HOH bond angle that is larger than that of H_2O. The ψ_{1b_2} orbital has HH antibonding character and the ionized state will therefore get a smaller HOH bond angle than the one in H_2O. Accurate *ab initio* calculations give bond angles of $112.5°$ for the $^2B_1$, $180°$ for the $^2A_1$, and $57.7°$ for the $^2B_2$ states (Meyer 1971) in agreement with the predictions of the EWMO calculation.

C.3 1.

$$x_1(a_1) = \phi_{2pzC}, \qquad x_2(a_1) = \frac{1}{\sqrt{2}}(\phi_{1sA} + \phi_{1sB}), \qquad x_3(b_2) = \phi_{2pyC},$$

$$x_4(b_2) = \frac{1}{\sqrt{2}}(\phi_{1sA} - \phi_{1sB}), \qquad x_5(b_1) = \phi_{2pxC}$$

2.

$$\psi_{1b_1} = \phi_{2pxC}, \qquad\qquad \varepsilon_{1b_1} = \alpha$$

$$\psi_{1b_2} = \frac{1}{2}(\phi_{1sA} - \phi_{1sB} + \sqrt{2}\,\phi_{2pyC}) \qquad \varepsilon_{1b_2} = \alpha + \sqrt{2}\,\beta\cos\gamma$$

$$\psi_{2b_2} = \frac{1}{2}(\phi_{1sA} - \phi_{1sB} - \sqrt{2}\,\phi_{2pyC}) \qquad \varepsilon_{2b_2} = \alpha - \sqrt{2}\,\beta\cos\gamma$$

$$\psi_{1a_1} = \frac{1}{2}(\phi_{1sA} + \phi_{1sB} + \sqrt{2}\,\phi_{2pzC}) \qquad \varepsilon_{1a_1} = \alpha + \sqrt{2}\,\beta\sin\gamma$$

$$\psi_{2a_1} = \frac{1}{2}(\phi_{1sA} + \phi_{1sB} - \sqrt{2}\,\phi_{2pzC}) \qquad \varepsilon_{2a_1} = \alpha - \sqrt{2}\,\beta\sin\gamma$$

3.

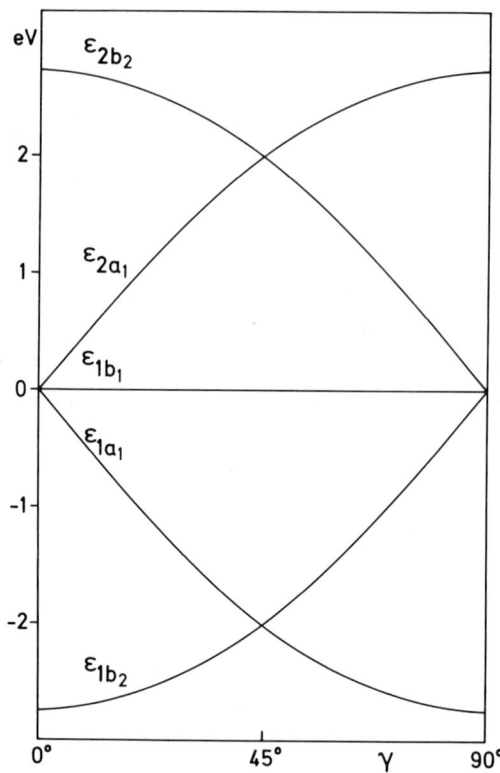

Fig. C.3,I. The orbital energies of CH_2 as a function of the HCH angle γ.

4. Figure C.3,I shows that γ must be close to 0 to give an electronic ground state of $^3B_1$ symmetry. For small γ values, relatively small total energy differences are expected between the $^1A_1$ and $^1B_1$ excited states and the $^3B_1$ ground state.

$$^3B_1: \quad \psi^2_{1b_2} \psi^1_{1a_1} \psi^1_{1b_1}$$

$$\Psi(^3B_1, M_S = 0) = \frac{1}{\sqrt{2}} [|\psi_{1b_2} \overline{\psi_{1b_2}} \psi_{1a_1} \overline{\psi_{1b_1}}| + |\psi_{1b_2} \overline{\psi_{1b_2}} \overline{\psi_{1a_1}} \psi_{1b_1}|]$$

$$^1B_1: \quad \psi^2_{1b_2} \psi^1_{1a_1} \psi^1_{1b_1}$$

$$\Psi(^1B_1) = \frac{1}{\sqrt{2}} [|\psi_{1b_2} \overline{\psi_{1b_2}} \psi_{1a_1} \overline{\psi_{1b_1}}| - |\psi_{1b_2} \overline{\psi_{1b_2}} \overline{\psi_{1a_1}} \psi_{1b_1}|]$$

$$^1A_1: \quad \psi^2_{1b_2} \psi^2_{1a_1} \qquad \Psi(^1A_1) = |\psi_{1b_2} \overline{\psi_{1b_2}} \psi_{1a_1} \overline{\psi_{1a_1}}|$$

5.

$$E({}^1B_1) = E({}^3B_1) = 4\alpha + \sqrt{2}\beta(2\cos\gamma + \sin\gamma) + \frac{1}{4.16\cos\gamma}$$

$$E({}^1A_1) = 4\alpha + 2\sqrt{2}\beta(\cos\gamma + \sin\gamma) + \frac{1}{4.16\cos\gamma}$$

6.

$$\frac{\partial E({}^1B_1)}{\partial\gamma} = \beta\sqrt{2}(-2\sin\gamma + \cos\gamma) + \frac{\sin\gamma}{4.16\cos^2\gamma} = 0$$

or

$$\cos\gamma = 2\sin\gamma + \frac{2.3126\sin\gamma}{\cos^2\gamma} \qquad (C.3,I)$$

where we have inserted $\beta = -2/27.2107$ a.u. Equation (C.3,I) can be written as

$$\tan\gamma = \frac{1}{2 + \dfrac{2.3126}{\cos^2\gamma}}$$

This equation may be used to determine an iterative solution of γ.

γ(guess)	γ(calc.)
0	13.06°
13.06°	12.70°
12.70°	12.72°
12.72°	12.72°

Thus

$$\gamma = 12.72°, \qquad \sphericalangle\,HCH = 154.56°$$

Similarly

$$\frac{\partial E({}^1A_1)}{\partial\gamma} = \beta 2\sqrt{2}(-\sin\gamma + \cos\gamma) + \frac{\sin\gamma}{4.16\cos^2\gamma} = 0$$

gives

$$\gamma = 22.93°, \qquad \sphericalangle\,HCH = 134.14°$$

7. The extended Hückel calculation gives a qualitatively correct prediction of the HCH angles in the three states: The angles are of the right order of magnitude, γ is the same for ${}^1B_1$ and ${}^3B_1$ and the Hückel calculation also predicts that $\gamma({}^1B_1)$ decreases about 20°

relative to $\gamma(^1\!A_1)$. However, quantitatively the HCH angles are far too large. A change of the β parameter for example, to -3 eV (2-3 eV is the commonly used range for $|\beta|$) would not improve the agreement. To describe the HCH angle reliably we would need to include the $2s$ orbital on C in the calculation and to consider the correlation effects explicitly.

C.4 1. H_2O_2 belongs to the point group C_2. We use the coordinate systems given in Fig. C.4,I.

$$\chi_1(a) = \frac{1}{\sqrt{2}}(\phi_{sA} + \phi_{sB}), \qquad \chi_2(a) = \frac{1}{\sqrt{2}}(\phi_{p_xA} + \phi_{p_xB})$$

$$\chi_3(a) = \frac{1}{\sqrt{2}}(\phi_{pzA} - \phi_{pzB}), \qquad \chi_4(a) = \frac{1}{\sqrt{2 + 2S_{p\sigma}}}(\phi_{p_yA} - \phi_{p_yB})$$

$$\chi_5(b) = \frac{1}{\sqrt{2}}(\phi_{sA} - \phi_{sB}), \qquad \chi_6(b) = \frac{1}{\sqrt{2}}(\phi_{p_xA} - \phi_{p_xB})$$

$$\chi_7(b) = \frac{1}{\sqrt{2}}(\phi_{pzA} + \phi_{pzB}), \qquad \chi_8(b) = \frac{1}{\sqrt{2 - 2S_{p\sigma}}}(\phi_{p_yA} + \phi_{p_yB})$$

2. The secular determinant of a symmetry (x_1 to x_4) is

$$\begin{vmatrix} \varepsilon - \alpha & (\varepsilon - K)S_{sp\sigma} & 0 & 0 \\ (\varepsilon - K)S_{sp\sigma} & \varepsilon - \alpha & (\varepsilon - K)S_{p\pi} & 0 \\ 0 & (\varepsilon - K)S_{p\pi} & \varepsilon - \alpha & 0 \\ 0 & 0 & 0 & \varepsilon - \dfrac{\alpha + KS_{p\sigma}}{1 + S_{p\sigma}} \end{vmatrix} = 0$$

$$\varepsilon_1 = \alpha, \qquad \varepsilon_{2,3} = \frac{\alpha \mp K\sqrt{S_{p\pi}^2 + S_{sp\sigma}^2}}{1 \mp \sqrt{S_{p\pi}^2 + S_{sp\sigma}^2}}, \qquad \varepsilon_4 = \frac{\alpha + KS_{p\sigma}}{1 + S_{p\sigma}}$$

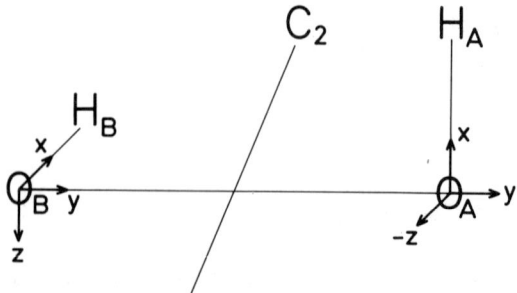

Fig. C.4,I. The geometry and choice of coordinate system for H_2O_2.

The secular determinant of b symmetry (χ_5 to χ_8) becomes identical to the one of a symmetry, except for the single element (χ_4, χ_4), which must be replaced by the (χ_8, χ_8), that is, $\epsilon - (\alpha - KS_{p\sigma})/(1 - S_{p\sigma})$ corresponding to the root

$$\varepsilon_5 = \frac{\alpha - KS_{p\sigma}}{1 - S_{p\sigma}}$$

ε_1, ε_2, and ε_3 are thus doubly degenerate. Note that the degeneracy is a consequence of the parametrization and not caused by the symmetry.

3.

$$H_{AB} = k(H_{AA} + H_{BB})\frac{S_{AB}}{2} = KS_{AB}$$

$$K = 2\alpha$$

which gives the orbital energy diagram for H_2O_2 in Fig. C.4,II.

4. The states (1,2) correspond to ionization out of orbital ε_1, state 3 to ionization out of orbital ε_4, and the states (4,5) to ionization out of ε_3.

$$\varepsilon_4 = 1.28\,\alpha = 15.3 \text{ eV}, \qquad \varepsilon_3 = 16.26 \text{ eV}, \qquad \varepsilon_1 = 11.95 \text{ eV} .$$

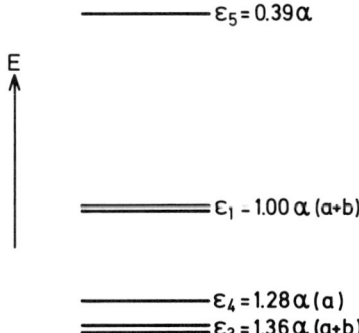

Fig. C.4,II. Orbital energy diagram for H_2O_2.

5.

$$\epsilon_1, \quad \psi_1(a) = 0.316\,\chi_1 - 0.949\,\chi_3$$
$$\psi_1(b) = 0.316\,\chi_5 - 0.949\,\chi_7$$
$$\epsilon_4, \quad \psi_4(a) = \chi_4$$
$$\epsilon_3, \quad \psi_3(a) = 0.543\,\chi_1 + 0.567\,\chi_2 + 0.149\,\chi_3$$
$$\psi_3(b) = 0.543\,\chi_5 + 0.567\,\chi_6 + 0.149\,\chi_7$$

States 1 and 2 of $H_2O_2^+$ originate from an ionization out of mainly the two nonbonding $p\pi$ symmetry orbitals on oxygen (symmetry-adapted lone pairs on oxygen). State 3 of $H_2O_2^+$ originates from an ionization out of the $p\sigma(OO)$ bond and states 4 and 5 from an ionization out of mainly the symmetry-adapted $\sigma(OH)$ bonds.

C.5   1.

$$E(q) = E(q = 0) + \left(\frac{\partial E}{\partial q}\right)_{q=0} q + \frac{1}{2}\left(\frac{\partial^2 E}{\partial q^2}\right)_{q=0} q^2 + \cdots \quad \text{(C.5,I)}$$

We can thus identify

$$E_0 = E(q = 0), \qquad E_1 = \left(\frac{\partial E}{\partial q}\right)_{q=0} - \frac{1}{2}\left(\frac{\partial^2 E}{\partial q^2}\right)_{q=0},$$
$$E_2 = \left(\frac{\partial^2 E}{\partial q^2}\right)_{q=0} \qquad\qquad\qquad\qquad \text{(C.5,II)}$$

2. E_0 is the total energy of the core electrons. E_1 is an orbital energy parameter. E_2 is an average electronic pair interaction energy of the valence electrons [$q(q-1)/2$ is the number of electron pairs].

3.

$$I_A = E(q_A - 1) - E(q_A) \qquad \text{(C.5,III)}$$
$$I_{A'} = E(q_A - 2) - E(q_A - 1) \qquad \text{(C.5,IV)}$$

From Eq. (C.5,1) we obtain

$$E_2 = I_{A'} - I_A \qquad \text{(C.5,V)}$$
$$E_1 = (1 - q)I_{A'} + (q - 2)I_A \qquad \text{(C.5,VI)}$$

4.

$$N = q_A + Zq_B \qquad \text{(C.5,VII)}$$

From Eqs. (C.5,1) and (C.5,VII) we find that the total energy of AB_Z is

$$E_{AB_Z} = E_A + ZE_B$$

$$= E_0^A + (N - Zq_B)E_1^A + (N - Zq_B)(N - Zq_B - 1)\frac{E_2^A}{2}$$

$$+ZE_0^B + ZE_1^B q_B + Zq_B(q_B - 1)\frac{E_2^B}{2}$$

(C.5,VIII)

The total energy is minimal when

$$\frac{\partial E_{AB_Z}}{\partial q_B} = 0$$

which gives

$$q_B = \frac{E_1^A - E_1^B + NE_2^A + \frac{1}{2}(E_2^B - E_2^A)}{E_2^B + ZE_2^A}$$

(C.5,IX)

q_A is obtained from Eqs. (C.5,IX) and (C.5,VII).

5. We have seven valence electrons on F and six on S. Thus $N = 48$. The number of valence electrons on F and S in SF_6 are determined from Eqs. (C.5,V-VII)

$$q_F = 7.050 \quad \text{and} \quad q_S = 5.700$$

and the formal charge is

$$\rho_F = -0.050 \, e \quad \text{and} \quad \rho_S = 0.30 \, e$$

6. Flourine is more electronegative than S and the charge distribution obtained is thus as expected and furthermore is in qualitative agreement with the Hartree-Fock results.

C.6 1. The exact Hartree-Fock total energy can be written as [see Eq. (B.0,13)]

$$E_{HF} = \sum_r n_r \varepsilon_r + V_{nn} - V_{ee}$$

(C.6,I)

and Eq. (C.6,2) follows straightforwardly when Eq. (C.6,1) is E_{HF}.

2.

$$E_{HF} = T + V = \frac{1}{2}V$$

(C.6,II)

Thus

$$V = 2E_{HF} = \frac{6}{7}(V_{ne} + 2V_{nn}) \qquad (C.6,III)$$

The total potential energy may alternatively be written as

$$V = V_{nn} + V_{ee} + V_{ne} \qquad (C.6,IV)$$

Inserting V_{ne} from Eq. (C.6,IV) in Eq. (C.6,III) gives Eq. (C.6,4).

3. Inserting Eq. (C.6,2) in Eq. (C.6,4) gives $V = 0$. The virial theorem then implies that $T = 0$, which is meaningless. The assumptions made in questions 1 and 2 are thus not compatible.

4. Using Eqs. (C.6,I-II) and Eq. (C.6,4) we obtain

$$E_{HF} = \sum_r n_r \varepsilon_r + \frac{1}{6}V = \sum_r n_r \varepsilon_r + \frac{1}{3}E_{HF}$$

which yields Eq. (C.6,5).

5.

He: $\sum_r n_r \varepsilon_r = -1.84$ a.u., $\qquad \frac{3}{2}\sum_r n_r \varepsilon_r = -2.76$ a.u.

CH$_4$: $\sum_r n_r \varepsilon_r = -27.54$ a.u., $\qquad \frac{3}{2}\sum_r n_r \varepsilon_r = -41.31$ a.u.

FH: $\sum_r n_r \varepsilon_r = -59.92$ a.u., $\qquad \frac{3}{2}\sum_r n_r \varepsilon_r = -89.88$ A.u.

The assumptions leading to Eq. (C.6,5) give total energies that are much more in agreement with calculated Hartree-Fock total energies than the simpler sum of orbital energies in Eq. (C.6,1).

D.
QUALITATIVE ATOMIC ORBITAL THEORY

INTRODUCTION

In this section we use *qualitative* atomic orbital theory to describe the electronic structure of some atomic systems. In qualitative atomic orbital theory, extensive use is made of symmetry arguments (constants of motion) and model considerations, but calculations of the *ab initio* type are not carried out. Some examples of *ab initio* atomic structure calculations are included in section B.

One of the main objectives of qualitative atomic orbital theory is to describe the splitting of electronic configurations into terms, levels, and so on. A similar splitting occurs in molecular structure calculations (examples are in section E) or when crystal field theory is applied to inorganic complexes (see section G).

When an operator commutes with the atomic Hamiltonian, that is , is a *constant of motion*, common eigenfunctions exist between the constant of motion and the Hamiltonian, and we can then characterize energy eigenvalues with quantum numbers describing the eigenstates of the constant of motion. If the square and the z-component of the total orbital angular momentum operator

$$\mathbf{L} = \sum_{i=1}^{N} \mathbf{l}_i \qquad \text{(D.0, 1)}$$

and of the spin angular momentum operator

$$\mathbf{S} = \sum_{i=1}^{N} \mathbf{s}_i \qquad \text{(D.0, 2)}$$

commute with the Hamiltonian, we can characterize the eigenstates of the atoms by the L, M_L and S, M_S quantum numbers. This is the so called LS or Russel–Saunders coupling scheme. The eigenstates are denoted *terms* and are written as

$$^{2S+1}L \qquad [\text{term}] \qquad \text{(D.0, 3)}$$

The term with the lowest energy is obtained from Hund's rules: (1) maximum spin multiplicity $(2S + 1)$ and (2) maximum L if two terms have the same value of $2S + 1$.

$\mathbf{L}^2$ and $\mathbf{S}^2$ commute with the Hamiltonian only if spin–orbit coupling is neglected. The total angular momentum operators $\mathbf{J}^2$ and $\mathbf{J}$, where

$$\mathbf{J} = \mathbf{L} + \mathbf{S} \qquad\qquad (\text{D.0, 4})$$

commute with the Hamiltonian when spin–orbit coupling is included. In the LS coupling scheme where spin–orbit coupling is assumed to be small, L and S may still be considered to be "approximately good" quantum numbers and an eigenstate for the Hamiltonian (a *level*) is denoted

$$^{2S+1}L_J \qquad [LSJ \text{ level}] \qquad\qquad (\text{D.0, 5})$$

The LS coupling scheme is used in problems D.1, D.2, and D.3.

For atoms with a large value of Z (the atomic number) the spin–orbit interaction becomes so large that it is often more appropriate to use another coupling scheme, the jj coupling scheme, in which we first couple the orbital and the spin angular momentum of the individual electron to obtain

$$\mathbf{j}_i = \mathbf{l}_i + \mathbf{s}_i$$

The individual electron angular momenta, $\mathbf{j}_i$, are then coupled to obtain the total J quantum number, that is, the *levels*. The notation for the jj coupled levels is

$$(n_1 l_{1j_1} n_2 l_{2j_2})_J \qquad [j_1 j_2 \text{ level}] \qquad\qquad (\text{D.0, 6})$$

If we perform a jj coupling of, for example, a $\cdots 7p^2$ electronic configuration, one of the jj levels will be $(7p_{1/2} 7p_{3/2})_1$. An example of how to determine the relative energy ordering of jj levels is given in problem D.3.

The optical spectrum of Bi^+ is assigned to problem D.3. To perform this assignment it is necessary to know the appropriate selection rules for the dipole transitions. These rules are not usually included in textbooks in quantum chemistry, and we shall therefore summarize them here. In the LS coupling scheme the selection rules are (Sobelman 1979)

(1) $\Delta S = 0$ $\qquad\qquad\qquad\qquad\qquad\qquad\qquad\qquad$ (D.0, 7)

(2) $\Delta L = 0, \pm 1, \qquad L_1 + L_2 \geq 1$ $\qquad\qquad\qquad\qquad$ (D.0, 8)

(3) $\sum_i |l_i| = 1 \qquad$ one electron transition $\qquad\qquad$ (D.0,9)

(4) $\Delta J = 0, \pm 1, \qquad J_1 + J_2 \geq 1$ $\qquad\qquad\qquad\qquad$ (D.0, 10)

(5) even $\leftrightarrow$ odd (o), $\qquad$ parity $\qquad\qquad\qquad\qquad\qquad$ (D.0, 11)

where $\Delta S = S_2 - S_1$, and so on. Rules 3, 4 and 5 are strictly valid and rules

1 and 2 are valid to the extent that the spin–orbit coupling is small, an assumption that is normally true when interpreting spectra in the LS coupling scheme. It should be recalled that the parity of an N-electron atomic state is given as

$$p = (-1)^{l_1 + l_2 + \cdots + l_N} \tag{D.0, 12}$$

where l_i is the individual orbital angular momentum quantum number.

In the jj coupling scheme rules 4 and 5 in Eqs. (D.0, 10) and (D.0, 11) are still valid and rules 1, 2 and 3 are replaced by

(6) $\Delta j = 0, \pm 1, \qquad j_1 + j_2 \geq 1$ for the "jumping" electron $\qquad$ (D.0, 13)

(7) $\Delta j = 0$ $\qquad$ for all other electrons $\qquad\qquad\qquad\qquad$ (D.0, 14)

These last two rules mean that we can promote only one electron at a time and that the quantum number for that electron must fulfill Eq. (D.0, 13).

In qualitative atomic orbital theory, the term energies are often expressed in terms of the so called Slater F_k and G_k integrals. An example of the use of the F_k and G_k integrals is given in problem D.2. For alkali atoms, which look like pseudo-hydrogenic systems, the term energies can be obtained from the simple Coulomb approximation as described in problem D.4.

In the last problem we discuss the accuracy of the ground-state total energies obtained from independent particle models (D.5).

Textbooks that give sufficient background for solving the problems in this section

Atkins (1970). Avery (1972). Chandra (1974). DeKock and Gray (1980). Eyring, Walter, and Kimball (1944). Karplus and Porter (1970). La Paglia (1971). Levine (1974). Lowe (1978). Murrell, Kettle, and Tedder (1970). Pilar (1968). Tinkham (1964).

Textbooks that give a more elaborate treatment of the topics

Condon and Shortley (1935). Mizushima (1970). Slater (1960, 1968). Sobelman (1979).

Problem Description

D.1 Assignment of the optical spectrum of Ti^+ using the LS coupling scheme

D.2 Term splitting of an nd^2 valence electronic configuration expressed in terms of F_k integrals

D.3 Assignment of the optical spectrum of Bi^+ using both the LS and the jj coupling schemes

D.4 The Coulomb approximation for the alkali atoms: ionization potentials and excitation energies

D.5 Ground-state total energies for closed-shell atoms in an independent-
particle model

PROBLEMS

D.1 The experimental total energies of the lowest electronic states of Ti^+ (Ti II) are
(in cm^{-1}): 0, 94, 225, 393, 908, 984, 1087, 1216, 4629, and 4898 (Moore 1949),
where the ground-state total energy is set equal to zero. Using the LS coupling
scheme, we can assign these electronic states to some of the lowest levels that
originate from the ground-state electronic configuration [Kr] $3d^2 4s$ and from the
excited electronic configuration [Kr] $3d^3$, where [Kr] denotes the ground-state
electronic configuration of krypton.

1.   Show by constructing a microstate diagram that an electronic configuration
     d^2 gives the following terms: ^3F, ^3P, ^1G, ^1D, and ^1S.
2.   Determine the terms that originate from the ground-state electronic
     configuration [Kr] $3d^2 4s$.
3.   Use Hund's rules to determine the lowest two terms originating from the
     [Kr] $3d^2 4s$ electronic configuration when it is known that the term splitting
     caused by the coupling between the $3d^2$ and the $4s$ electron is very small
     compared to the term splitting of the $3d^2$ electronic configuration.
4.   Determine the lowest term originating from the [Kr] $3d^3$ electronic
     configuration.
5.   Determine the multiplet (spin–orbit) splitting of the three terms that were
     determined in questions 3 and 4.
6.   Perform an assignment of the measured spectrum to the $|LSJ\rangle$ multiplet
     levels of question 5. Discuss the quality of this assignment based on the LS
     coupling scheme, for example, by investigating how well Landé's interval
     rule is fulfilled.

D.2 In this problem we will calculate the term splitting of an atomic nd^2 valence
electronic configuration. A d^2 electronic configuration gives rise to five terms,
^3F, ^3P, ^1G, ^1D, and ^1S (see problem D.1, question 1). The energy splitting of these
terms is caused by the electronic repulsion operator

$$\hat{V} = \sum_{i>j} r_{ij}^{-1} \qquad (D.2, 1)$$

The splitting can be determined from first-order perturbation theory to be

$$\langle \Psi_0 | \hat{V} | \Psi_0 \rangle$$

where Ψ_0 refers to a wave function for one of the terms ^3F, ^3P, ^1G, ^1D, or ^1S.

When Ψ_0 refers to a single determinantal wave function, it can be shown
(Condon and Shortly 1935, Chap. VI.9) that the term splitting is only caused by
a fraction of $\langle \Psi_0 | \hat{V} | \Psi_0 \rangle$ namely

$$\langle \Psi_0 | \hat{V} | \Psi_0 \rangle' = \sum_{i>j=1}^{N'} [\langle ij | r_{12}^{-1} | ij \rangle - \langle ij | r_{12}^{-1} | ji \rangle] \qquad (D.2, 2)$$

where N' is the number of electrons in the *open* shell (d^2 in the present case).

Thus, i and j denote the open shell orbitals. Note that Eq.(D.2, 2) differs from the normal Slater-Condon result since we only sum over the open shell. It is shown in most standard textbooks of atomic physics (see, e.g., Tinkham 1964, Chap. 6-7) that

$$\langle ij|r_{12}^{-1}|ij\rangle = \sum_{k=0}^{\infty} a^k(l^i m_l^i, l^j m_l^j) F^k(n^i l^i, n^j l^j) \tag{D.2, 3}$$

$$\langle ij|r_{12}^{-1}|ji\rangle = \delta_{m_s^i m_s^j} \sum_{k=0}^{\infty} b^k(l^i m_l^i, l^j m_l^j) G^k(n^i l^i, n^j l^j) \tag{D.2, 4}$$

where we have introduced the notation

$$a^k(l^i m_l^i, l^j m_l^j) = c^k(l^i m_l^i, l^i m_l^i) c^k(l^j m_l^j, l^j m_l^j) \tag{D.2, 5}$$

$$b^k(l^i m_l^i, l^j m_l^j) = [c^k(l^i m_l^i, l^j m_l^j)]^2 \tag{D.2, 6}$$

All integrals $F^k \geq 0$ and for equivalent electrons, as for example in the d^2 configuration, $F^k = G^k$. The quantum numbers n, l, m_l, and m_s have their usual meaning. The c^k coefficients are tabulated for various electronic configurations by Condon and Shortley (1935), and are given in Table D.2,1 for the cases in which $l^i = 2$ and $l^j = 2$, that is, the nonvanishing c^k coefficients needed to evaluate $\langle \Psi_0|\hat{V}|\Psi_0\rangle'$ when Ψ_0 is constructed from d orbitals alone. If a single determinantal wave function is known, Eqs. (D.2, 2–6) can be used to determine the term energy in terms of the F^k and (sometimes) G^k functions.

1. Use the microstate diagram in Table D.1, I to show that the term energies for ^3F and ^1G are

$$E(^3F) = F_0 - 8F_2 - 9F_4 \tag{D.2, 7}$$

$$E(^1G) = F_0 + 4F_2 + F_4 \tag{D.2, 8}$$

where $F_0 = F^0$, $F_2 = F^2/49$, and $F_4 = F^4/441$.

According to Table D.1, I the two nonvanishing determinants with $M_L = 1$ and $M_S = 1$ are $|d_2 d_{-1}|$ and $|d_1 d_0|$.

Table D.2,1
$c^k(l^i m_l^i, l^j k_l^j)$ **for the d^2 Electronic Configuration**

l^i, l^j	m_l^i	m_l^j	$k = 0$	2	4
d, d	± 2	± 2	1	$-\sqrt{4/49}$	$+\sqrt{1/441}$
	± 2	± 1	0	$+\sqrt{6/49}$	$-\sqrt{5/441}$
	± 2	0	0	$-\sqrt{4/49}$	$+\sqrt{15/441}$
	$+1$	$+1$	1	$+\sqrt{1/49}$	$-\sqrt{16/441}$
	± 1	0	0	$+\sqrt{1/49}$	$+\sqrt{30/441}$
	0	0	1	$+\sqrt{4/49}$	$+\sqrt{36/441}$
	± 2	∓ 2	0	0	$+\sqrt{70/441}$
	± 2	∓ 1	0	0	$-\sqrt{35/441}$
	± 1	∓ 1	0	$-\sqrt{6/49}$	$-\sqrt{40/441}$

2.   Show that

$$E(^3P) + E(^3F) = \langle |d_2 d_{-1}||\hat{V}||d_2 d_{-1}|\rangle + +\langle |d_1 d_0||\hat{V}||d_1 d_0|\rangle \quad \text{(D.2, 9)}$$

Equation (D.2, 9) is an example of the so-called *Slater sum rule* (Slater 1968).

3.   Show that

$$E(^3P) = F_0 + 7F_2 - 84F_4 \qquad \text{(D.2, 10)}$$

$$E(^1D) = F_0 - 3F_2 + 36F_4 \qquad \text{(D.2, 11)}$$

4.   Determine the condition that F_2 and F_4 must fulfill in order to satisfy Hund's rules for the ground-state term.

Hund's rules can in many cases be used only to determine the ground-state term and *not* the relative energies of the excited states. Atomic ions with $3d^2$ valence electronic configurations show such deviations from Hund's rules (Moore 1949):

Table D.2,2
Experimental Term Energies[a] for Ions with $3d$ Valence Electronic Configurations

Ion	$E(^3F)$	$E(^1D)$	$E(^3P)$	$E(^1G)$
Ti^{2+}	0.0	8,473	10,536	14,399
V^{3+}	0.0	10,960	13,121	18,389
Cr^{4+}	0.0	13,200	15,500	22,060
Mn^{5+}	0.0	15,336	17,782	25,511

[a] All energies are in cm^{-1} and are measured relative to the $E(^3F)$.

5.   Use the term energies for ^3F, ^1D, and ^3P to compute F_0, F_2, and F_4 and then calculate $E(^1G)$ from Eq. (D.2, 8) Do this for all ions Ti^{2+} to Mn^{5+} Comment on the relative accuracy of $E(^1G)$ of the atoms in the isoelectronic sequence.

D.3 The experimental total electronic energies of the nine lowest states of Bi^+ (Bi II) are (in cm^{-1}): 0, 13,324, 17,030, 33,963, 44,173, 69,133, 69,598, 88,769, and 89,883 (Moore 1949) where the ground-state total energy is set equal to zero. These electronic states have been assigned to the ground-state electronic configuration [Ba] $6s^2 6p^2$ (the first five states) and to the excited electronic configuration [Ba] $6s^2 6p^1 7s^1$ (the last four states). [Ba] denotes the ground-state electronic configuration of barium. We will investigate whether the assignment of the spectrum of Bi II is best described within the *LS* coupling scheme or within the *jj* coupling scheme.

Consider initially the *LS* coupling scheme.

1.   Determine the terms that originate from the ground and excited electronic configuration given above.

2.   Determine the multiplet (spin–orbit) splitting of the terms in question 1.

3. Perform an assignment of the experimental electronic spectrum to the lowest nine $|LSJ\rangle$ levels. Assume normal multiplet splitting for both configurations. Is the Landé interval rule fulfilled?

Consider the electronic excitations from the $|LSJ\rangle$ levels originating from the [Ba] $6s^2 6p^2$ configuration to $|LSJ\rangle$ levels originating from the [Ba] $6s^2 6p^1 7s^1$ configuration.

4. Which of these electronic excitations are dipole allowed?

We will now consider the jj coupling scheme.

5. Determine the total angular momentum quantum numbers (j) for an electron in an s orbital and an electron in a p orbital.

6. Determine within the jj coupling scheme the states of Bi II that originate from the [Ba] $6s^2 6p^2$ and [Ba] $6s^2 6p^1 7s^1$ electronic configurations.

Consider excitations between states that originate from the p^2 and from the excited sp configuration.

7. Which of these electronic transitions are dipole allowed? Which coupling scheme (LS or jj) predicts the most absorption lines?

In order to perform an assignment of the jj coupled states to the measured spectrum it is useful first to correlate the $|LSJ\rangle$ levels and the $|j_1 j_2 J\rangle$ levels. The energy *ordering* of $|j_1 j_2 J\rangle$ *levels* must be the same as that obtained for the $|LSJ\rangle$ *levels* where Hund's rules can be used (see question 3). The reason for this correlation is that J is a "good" quantum number for both coupling schemes. Use also that $|j_1 j_2 J\rangle$ levels with the same values of j_1 and j_2 are well separated from levels with different values of j_1 and/or j_2.

8. Correlate the levels in question 6 (jj coupling) with those obtained in question 2 (LS coupling).

9. Use the result of the previous question to perform an assignment of the nine lowest states in the jj coupling scheme.

10. Which of the assignments, the one from question 3 or the one from question 9, gives the most consistent description of the measured spectrum?

D.4 In this problem we will investigate the validity of the so-called *Coulomb approximation* (Sobelman 1979) for the alkali atoms Li, Na, and so on. In this approximation, the term energies of the alkali atoms are determined from the hydrogenlike formula

$$E_{nl} = E_{core} - \frac{Z_{net}^2}{2(n^*)^2}$$

where Z_{net} and E_{core} are the net charge and energy, respectively, of the ion core, that is, the noble-gas-like core obtained when the single valence electron with quantum numbers n, l is removed. E_{core} is assumed to be constant for all atoms in an isoelectronic series. n^* is an effective principal quantum number, which with good accuracy can be represented in the form

$$n^* = n - \delta_l$$

where n is the principal quantum number and δ_l is the *quantum defect* or *Rydberg correction*, which depends only on the quantum number l.

The ionization potentials of the isoelectronic Li series are 5.390 eV (Li),

18.206 eV (Be^+), 37.920 eV (B^{2+}), and 64.476 eV (C^{3+}) (Moore 1949).
1.   Use these ionization potentials to determine the quantum defect δ_s for Li, Be^+, B^{2+}, C^{3+}.

2.   Why does δ_s decrease from Li to C^{3+}?

Excitation of Li from its ground state into the excited electronic configuration $1s^1 2p^1$ requires an energy of 1.848 eV (Moore 1949).
3.   Determine the quantum defect δ_p.
4.   Give a physical argument for the difference between δ_s and δ_p.

The experimental excitation energies for the $2s \rightarrow 3p$ and $2s \rightarrow 4p$ excitations of Li are 3.834 eV and 4.522 eV, respectively (Moore 1949).
5.   Use the Coulomb approximation to compute these excitation energies.

D.5 We will consider the ground state of a neutral atom with nuclear charge Z. We will assume that all subshells with the same principal quantum number n are filled; that is, we are considering closed-shell atoms like He and Ne. Furthermore, we assume that the electrons are noninteracting. The orbital energies of the atom then become hydrogenic; that is, they are identical to the orbital energies of a hydrogen atom with nuclear charge Z.
1.   Show that the forgoing assumptions lead to the relation

$$Z = n_o(n_o + 1)\frac{(2n_o + 1)}{3}$$

where n_o is the principal quantum number of the highest occupied atomic orbital.
2.   Determine the ground-state energy in terms of n_o.

The Hartree–Fock ground-state energies for He and Ne are −2.86 and −128.5 a.u., respectively (Clementi and Roetti 1974).
3.   Calculate the ground-state energy for He and Ne, using the energy expression from question 2. Comment on the discrepancy between those energies and the Hartree–Fock energies.

The effect of electronic screening is often approximately taken into account by using screened hydrogenic orbitals rather than exact hydrogenic orbitals. In a screened hydrogenic orbital the orbital exponent is $Z_{nl} = Z - S_{nl}$ where S_{nl} is the screening constant (Slater 1930).
4.   Use Slater's screening constants to calculate the ground-state energies of He and Ne. Comment on the accuracy of this approach relative to Hartree–Fock and the nonscreened approach.

SOLUTIONS

D.1 1. The $M_L = \sum_{i=1}^{2} m_s \geq 0$ and $M_S = \sum_{i=1}^{2} m_{l_i} \geq 0$ part of the microstate diagram is given in Table D.1, I.
2. $^4F, \, ^2F, \, ^4P, \, ^2P, \, ^2G, \, ^2D, \, ^2S$.
3. $^4F, \, ^2F$, since ^3F is the ground-state term for a d^2 configuration.
4. ^4F

Table D.1.I
Microstate Diagram for a d^2 Configuration

$M_L \setminus M_S$	1	0
4		$\|d_2 \bar{d}_2\|$ (^1G)
3	$\|d_2 d_1\|$ (^3F)	$\|d_2 \bar{d}_1\|\|\bar{d}_2 d_1\|$ $(^1G, ^3F)$
2	$\|d_2 d_0\|$ (^3F)	$\|d_2 \bar{d}_0\|\|\bar{d}_2 d_0\|\|d_1 \bar{d}_1\|$ $(^1G, ^3F, ^1D)$
1	$\|d_2 d_{-1}\|\|d_1 d_0\|$ $(^3F, ^3P)$	$\|\bar{d}_2 d_{-1}\|\|d_2 \bar{d}_{-1}\|$
		$\|\bar{d}_1 d_0\|\|d_1 \bar{d}_0\|$ $(^1G, ^3F, ^1D, ^3P)$
0	$\|d_2 d_{-2}\|\|d_1 d_{-1}\|$ $(^3F, ^3P)$	$\|\bar{d}_2 d_{-2}\|\|d_2 \bar{d}_{-2}\|\|d_0 \bar{d}_0\|$
		$\|d_1 \bar{d}_{-1}\|\|\bar{d}_1 d_{-1}\|$ $(^1G, ^3F, ^1D, ^3P, ^1S)$

5. $^4F_{9/2}, \ ^4F_{7/2}, \ ^4F_{5/2}, \ ^4F_{3/2}$

   $^2F_{7/2}, \ ^2F_{5/2}$

6. The lowest 10 states fall in groups of 4, 4, and 2 states with about the same energies. That must correspond to the three terms ^4F, ^4F, and ^2F, respectively, and we have the assignment in Fig. D.1, I. Consider the multiplet splitting of the ground-state term (^4F):

$$\frac{^4F_{5/2} - ^4F_{3/2}}{5/2} = 37.6 \text{ cm}^{-1}$$

$$\frac{^4F_{7/2} - ^4F_{5/2}}{7/2} = 37.4 \text{ cm}^{-1}$$

$$\frac{^4F_{9/2} - ^4F_{7/2}}{9/2} = 37.3 \text{ cm}^{-1}$$

The Landé rule is thus fulfilled quite accurately for the ground-state term. The same holds for the lowest excited ^4F term, and the LS coupling scheme gives a good description of this part of the Ti II spectrum.

D.2 1. The determinant $\|d_2 d_1\|$ is the wave function for ^3F with $M_S = 1$ and $M_L = 3$; that is,

$$E(^3F) = \langle \|d_2 d_1\| \| \hat{V} \| d_2 d_1\| \rangle$$

$$= \sum_{k=0}^{4} [a^k(22, 21) F^k(nd, nd) - b^k(22, 21) F^k(nd, nd)]$$

$$= F_0 - 8F_2 - 9F_4 \qquad \text{Q.E.D}$$

$$E(^1G) = \langle \|d_2 \bar{d}_2\| \| \hat{V} \| d_2 \bar{d}_2\| \rangle = F_0 + 4F_2 + F_4 \qquad \text{Q.E.D.}$$

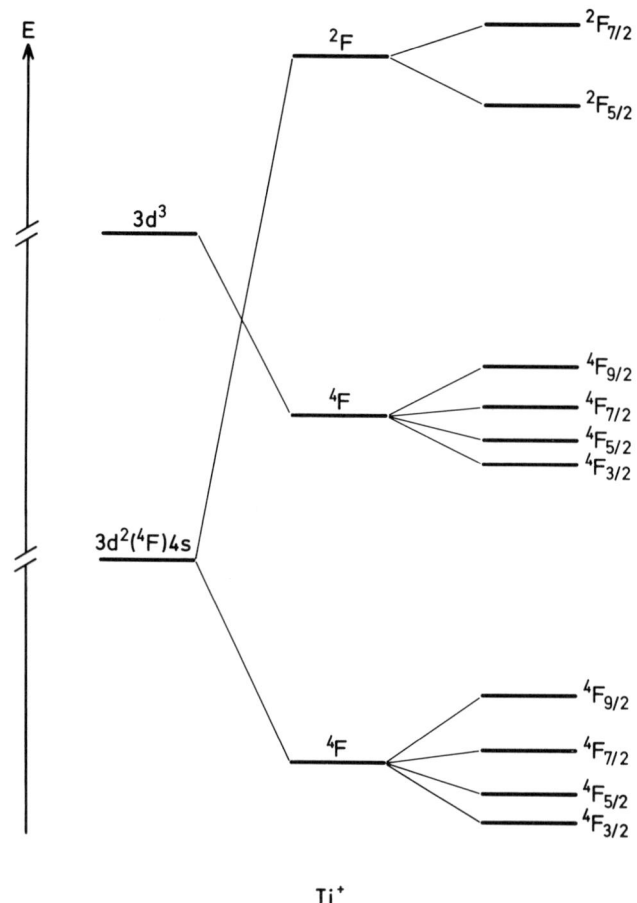

Fig. D.1, I. Term and level diagram for a $3d^2(^4F)4s^1$ and a $3d^3$ electronic configuration in the Russell–Saunders coupling scheme.

2. The determinants with $M_S = M_L = 1$ do not mix energywise with determinants with other M_S and M_L values. We may form one linear combination of $|d_2 d_{-1}|$ and $|d_1 d_0|$ that represents a wave function for the ^3P state and another combination that is an eigenfunction for ^3F. In this representation, $\hat{V}$ is diagonal and the sum of the diagonal elements is $E(^3P) + E(^3F)$. The invariance of the trace under unitary transformations then gives that this sum must be equal to the sum of diagonal elements in the original basis, $\{|d_2 d_{-1}|, |d_1 d_0|\}$ and Eq. (D.4, 9) follows.

3.

$$\langle |d_2 d_{-1}||\hat{V}||d_2 d_{-1}|\rangle = F_0 - 2F_2 - 39F_4$$
$$\langle |d_1 d_0||\hat{V}||d_1 d_0|\rangle = F_0 + F_2 - 54F_4$$

Using Eqs. (D.2, 7 and 9) we find that

$$E(^3P) = F_0 + 7F_2 - 84F_4 \qquad \text{Q.E.D.}$$

Using the $M_S = 0$ and $M_L = 2$ block of Table D.1, I and the Slater sum rule, we find

$$E(^1G) + E(^1D) + E(^3F) = \langle |d_2 \bar{d}_0||\hat{V}||d_2 \bar{d}_0|\rangle + \langle |\bar{d}_2 d_0||\hat{V}||\bar{d}_2 d_0|\rangle$$
$$+ \langle d_1 \bar{d}_1 ||\hat{V}||d_1 \bar{d}_1|\rangle \qquad \text{(D.2, I)}$$

However,

$$\langle d_2 \bar{d}_0||\hat{V}||d_2 \bar{d}_0|\rangle = \langle |\bar{d}_2 d_0||\hat{V}||\bar{d}_2 d_0|\rangle = F_0 - 4F_2 + 6F_4$$
$$\langle |d_1 \bar{d}_1 ||\hat{V}||d_1 \bar{d}_1|\rangle = F_0 + F_2 + 16F_4$$

and Eq. (D.2,11) follows from Eqs. (D.2, I) and (D.2, 7–8).

4. According to Hund's rules ^3F must be the ground-state term. Since $F_k \geq 0$, Eqs. (D.2, 7–10) show that ^3F certainly lies lower than both ^1G and ^1D.

$E(^3F) \leq E(^3P)$ requires that

$$F_2 \geq 5F_4$$

5.

Table D.2,I
F_k Integrals for Ions with $3d^2$ Valence Electronic Configurations

ion	F_0 (cm^{-1})	F_2 (cm^{-1})	F_4 (cm^{-1})	$E(^1G)$
Ti^{2+}	9,092	1056.76	70.8714	13,390 (−7.0%)
V^{3+}	11,608	1345.19	94.0905	17,083 (−7.1%)
Cr^{4+}	13,890	1607 14	114.7619	20,433 (−7.4%)
Mn^{5+}	16,070	1857.51	134.4095	23,634 (−7.4%)

The single determinantal description is almost equally good for all $3d^2$ ions. It is best by a small margin for the least charged ions that have a less contracted charge cloud and thus a less correlated motion of the electrons.

D.3 1. [Ba] $6s^2 6p^2$; ^3P, ^1D, ^1S
    [Ba] $6s^2 6p^1 7s^1$; $^3P^o$, $^1P^o$ (o denotes odd parity).
  2. $^3P_0$, $^3P_1$, $^3P_2$, $^1D_2$, $^1S_0$;
    $^3P_0^o$, $^3P_1^o$, $^3P_2^o$, $^1P_1^o$
  3. Using Hund's rule we obtain the assignment in Fig. D.3, I by
assigning the experimentally measured states to $|LSJ\rangle$ levels in
order of increasing energy (normal multiplet splitting is assumed
also for the $6p^1 7s^1$ configuration).
    Since $\Delta E(^3P_1 \rightarrow {}^3P_0)/1 = 13{,}324 \text{ cm}^{-1}$ and $\Delta E(^3P_2 \rightarrow {}^3P_1)/2$
$= 1853 \text{ cm}^{-1}$, the Landé interval rule is far from fulfilled for the ^3P
multiplet splitting. The same is true for the $^3P^o$ multiplet splitting.
Thus, the Bi II spectrum is not well described within the LS coupling
scheme.

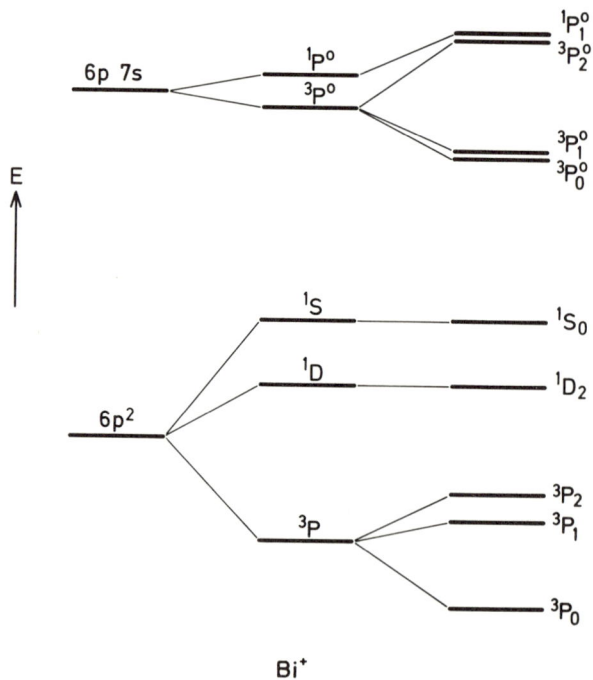

Fig. D.3, I. Term and level diagram for a $6p^2$ and a $6p^1 7s^1$ electronic configuration in the Russell–Saunders coupling scheme.

4. Using Eqs. (D.0, 7–10) we have the following allowed transitions:

$$^3P_0 \rightarrow {}^3P_1^o; \qquad {}^3P_1 \rightarrow {}^3P_0^o, {}^3P_1^o, {}^3P_2^o$$

$$^3P_2 \rightarrow {}^3P_1^o, {}^3P_2^o; \qquad {}^1D_2 \rightarrow {}^1P_1^o; \qquad {}^1S_0 \rightarrow {}^1P_1^o$$

5. s electron: $j = \frac{1}{2}$

   p electron: $j = \frac{1}{2}, \frac{3}{2}$

6. The jj coupling of the two $6p$ electrons gives $(6p_{1/2}6p_{1/2})_0$ [$(6p_{1/2}6p_{1/2})_1$ vanishes as a result of the Pauli exclusion principle, since $m_{j1} = m_{j2} = \frac{1}{2}$ in order to have $J = 1$], $(6p_{1/2}6p_{3/2})_1$, $(6p_{1/2}6p_{3/2})_2$, $(6p_{3/2}6p_{3/2})_2$, $(6p_{3/2}6p_{3/2})_0$ [$(6p_{3/2}6p_{3/2})_3$, and $(6p_{3/2}6p_{3/2})_1$ vanish as a result of the Pauli exclusion principle].

   The jj coupling of a ps configuration gives the following states: $(6p_{1/2}7s_{1/2})_0$, $(6p_{1/2}7s_{1/2})_1$, $(6p_{3/2}7s_{1/2})_2$, $(6p_{3/2}7s_{1/2})_1$.

7. Using Eqs. (D.0,10–11) and (D.0, 13–14) we have the following allowed transition:

$$(6p_{1/2}6p_{1/2})_0 \rightarrow (6p_{1/2}7s_{1/2})_1$$

$$(6p_{1/2}6p_{3/2})_1 \rightarrow (6p_{1/2}7s_{1/2})_0, (6p_{1/2}7s_{1/2})_1, (6p_{3/2}7s_{1/2})_2, (6p_{3/2}7s_{1/2})_1$$

$$(6p_{1/2}6p_{3/2})_2 \rightarrow (6p_{1/2}7s_{1/2})_1, (6p_{3/2}7s_{1/2})_2, (6p_{3/2}7s_{1/2})_1$$

$$(6p_{3/2}6p_{3/2})_2 \rightarrow (6p_{3/2}7s_{1/2})_2, (6p_{3/2}7s_{1/2})_1$$

$$(6p_{3/2}6p_{3/2})_0 \rightarrow (6p_{3/2}7s_{1/2})_1$$

Thus, the jj coupling scheme predicts 11 allowed transitions, whereas the LS coupling scheme gave eight allowed transitions (see question 4).

8. The correlation for the $6p^2$ configuration is displayed in Fig. D.3, II, while Fig. D.3, III gives the correlation for the $6p^1 7s^1$ configuration.

9. See Fig. D.3, IV.

10. Landé's interval rule was not fulfilled in the LS coupling scheme (question 3), and the splitting between the three ^3P_J levels is as large as the splitting between the ^1D and ^1S terms. For the $6p^1 7s^1$ configuration it is even worse, so L and S are not even approximately good quantum numbers.

    However, the splitting between, for example, the $(6p_{1/2}6p_{3/2})_1$ and $(6p_{1/2}6p_{3/2})_2$ is small compared with the energy difference between, for

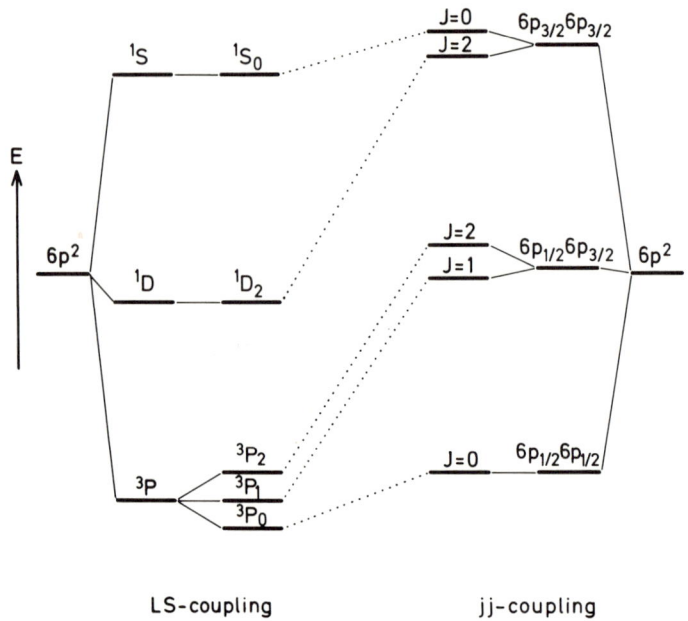

D.3, II. The relation between levels in a Russell–Saunders coupling scheme and a *jj* coupling scheme for a $6p^2$ electronic configuration.

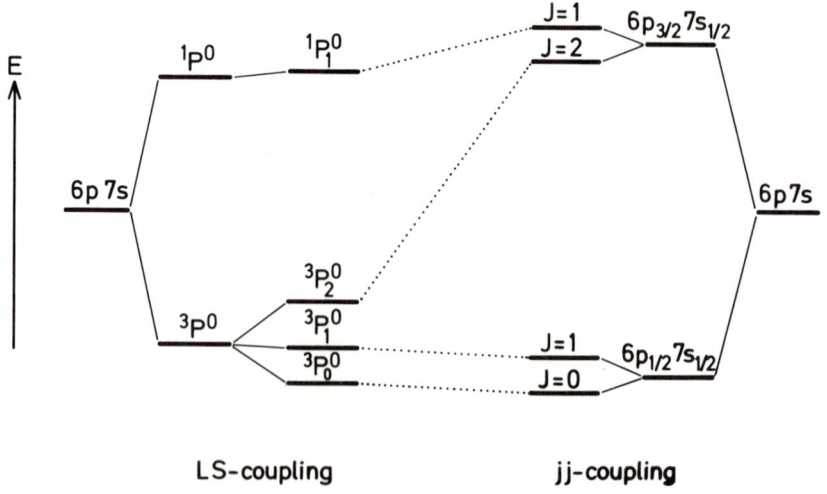

Fig. D.3, III. The relation between levels in a Russell–Saunders coupling scheme and a *jj* coupling scheme for a $6p^1 7s^1$ electronic configuration.

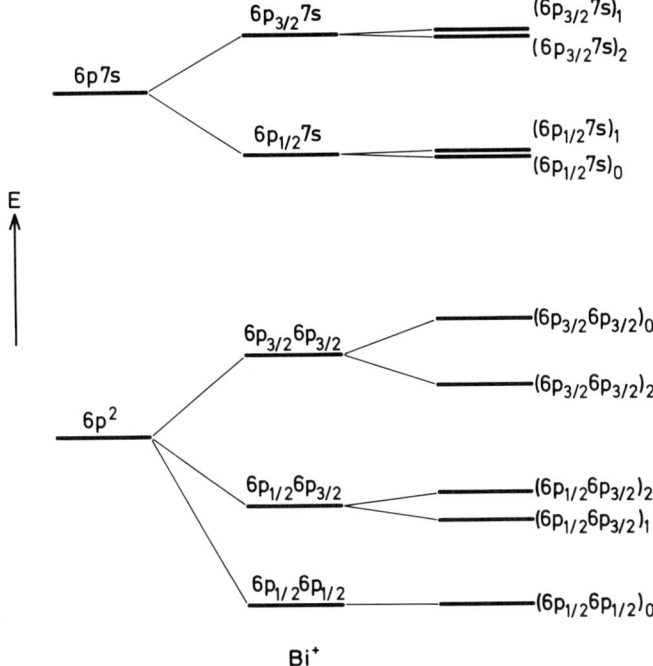

Fig. D.3, IV. Energy level diagram for a $6p^2$ and a $6p^1 7s^1$ electronic configuration in the *jj* coupling scheme.

example, $(6p_{1/2} 6p_{3/2})$ and $(6p_{3/2} 6p_{3/2})$ states, indicating that j_1 and j_2 are approximately good quantum numbers. The small splitting in J is even better for the $6p^1 7s^1$ electronic configuration.

Thus, *jj* coupling is definitely best for Bi II, especially for the excited electronic configuration, indicating that the energy splitting caused by the electronic repulsion is smaller than that caused by the spin–orbit interaction. The electrostatic interaction is smallest for nonequivalent electrons, and the *jj* coupling scheme is therefore best for the $6p^1 7s^1$ electronic configurations.

D.4 1. The electron is removed from the $2s$ orbital and the ionization potentials, $IP(2s)$, are calculated from the expression

$$IP(2s) = \frac{Z_{net}^2}{2(n^*)^2}$$

where

$$n^* = 2 - \delta_s$$

Table D.4,I
Quantum Defects, δ_s, for the Li Isoelectronic Series

Atom/ion	Li I	Be II	B III	C IV
Z_{net}	1	2	3	4
n^*	1.5888	1.7289	1.7970	1.8375
δ_s	0.4112	0.2711	0.2030	0.1625

2. Large ion charge implies that the $1s$ orbitals become contracted; that is, the $1s$ electrons will almost screen with their whole charge. The potential seen by the optical $2s$ electron will thus be more and more Coulomblike when Z_{net} gets larger. Thus, δ_s becomes smaller when Z_{net} increases.

3. The excitation energy $\Delta E(2s \rightarrow 2p)$ is

$$\Delta E(2s \rightarrow 2p) = -\tfrac{1}{2}[(2 - \delta_p)^{-2} - (2 - \delta_s)^{-2}]$$

$$\delta_p = 0.0401$$

4. The reason for the decrease $\delta_s \rightarrow \delta_p$ is analogous to that discussed in question 2: The larger l is, the larger is the probability for observing the "optical electron" at large distances from the nucleus and the more Coulomblike is the potential seen by the optical electron.

5.

$$\Delta E(2s \rightarrow 3p) = -\tfrac{1}{2}[(3 - \delta_p)^{-2} - (2 - \delta_s)^{-2}] = 3.837 \text{ eV}$$

$$\Delta E(2s \rightarrow 4p) = 4.522 \text{ eV}$$

Thus, the Coulomb approximation describes the experimental spectrum very well.

D.5 1. The number of electrons is

$$Z = 2 \sum_{n=1}^{n_0} \sum_{l=0}^{n-1} \sum_{m_l=-l}^{l} 1 = 2 \sum_{n=1}^{n_0} n^2 = n_0(n_0 + 1)\frac{2n_0 + 1}{3} \quad \text{(D.5, I)}$$

Use, for example, induction to prove the last equality.

2. The number of electrons with principal quantum number n is

$$N_n = 2n^2$$

The orbital energy of such an electron is

$$\varepsilon_n = \frac{-Z^2}{2n^2}$$

The ground-state energy becomes

$$E = \sum_{n=1}^{n_0} \varepsilon_n N_n = -n_0 Z^2 = -n_0^3 (n_0 + 1)^2 \frac{(2n_0 + 1)^2}{9}$$

where we have used Eq. (D.5, I).

3. $E_{He} = -4$ a.u., $E_{Ne} = -200$ a.u.

Both the Hartree–Fock method and the present hydrogenic appproximation are independent-particle models. In the former, an electron moves in the average potential from all the other electrons. This average electron–electron interaction, which is absent in the hydrogenic approximation, is thus very important for obtaining a realistic independent-particle model description.

4.

$$Z_{1s}^{He} = 1.70; \qquad E_{He} = -2.89 \text{ a.u.}$$

$$Z_{1s}^{Ne} = 9.7 \text{ and } Z_{2s}^{Ne} = Z_{2p}^{Ne} = 5.85; \qquad E_{Ne} = -128.31 \text{ a.u.}$$

The screened results are in very close agreement with the HF energies.

QUALITATIVE MOLECULAR ORBITAL THEORY

INTRODUCTION

A qualitative description of the electronic structure of molecular systems can often be obtained from simple molecular orbital (MO) arguments without resorting to explicit MO calculations. In *qualitative molecular orbital theory* all physical information is obtained from an MO diagram, that is, a diagrammatic representation of the relative ordering of MO energy levels.

Some simple rules that may be useful when constructing an MO diagram are listed later. In problems E.1 and E.2 calculations based on simple model systems are performed in order to justify some of the rules [(D-F) and H]. Rules A-K are based on the results of a large number of MO calculations. They should be considered as *guidelines* when constructing the MO diagrams. Exceptions may occur and will be discussed whenever they appear.

The symmetry of the molecular states can be determined from the MO diagram by occupying the MO's according to the "aufbau principle" (E.4-16). According to Koopmans' theorem (see the introduction to section B) the ionization potential for ionizing an electron out of an occupied MO is approximately equal to the orbital energy of the occupied MO with opposite sign. In problems E.6-10 we have used this fact to carry out an assignment of some photoelectron spectra. Correlation diagrams (i.e., diagrams displaying the correlation between the symmetry of the MO's and/or states and the symmetry of the atomic orbitals and/or states in the separated atom and united atom limits) are constructed in problems E.3-4 and E.12-14.

Qualitative information about the electronic structure of molecules may in some cases be obtained by using perturbation arguments. The electronic structure of an "unperturbed" system that is similar in structure to the system of interest is assumed to be known. We may, for example, know an MO diagram (E.5,9,11-12) or know some states (E.15-16) of the unperturbed system. The perturbation can then be a moderate-sized change in the geometry of the molecule (a distortion), and the changes in the MO diagram can be determined according to rule K. The perturbations may also be, for example, external charges, approaching the unperturbed atom or molecule (E.15-16).

We will now turn to the rules used in the construction of the MO diagrams. These rules are similar to those discussed in Chapter 14 of Lowe

(1978) and in Chapter 1 of Gimarc (1979). Further justification and applications of the rules may be found in these two textbooks. The rules are as follows.

(A) Consider the valence orbitals of the constituent atoms and assign to the valence atomic orbitals an orbital energy approximately the size of the ionization potential of the atomic valence shell. The atomic orbital energies may thus be, for example, a set of Hartree-Fock atomic orbital energies.

(B) Divide the molecule for which the MO diagram is to be constructed into two fragments in such a way that molecular symmetry orbitals can be formed for each fragment. A molecular symmetry orbital is an orbital that transforms according to an irreducible representation of the molecular point group. A symmetry orbital may thus be a single atomic orbital or a linear combination of atomic orbitals of a given fragment. For example, NH_3 (C_{3v} molecular point group symmetry) is divided into $N + 3 \times H$. The atomic orbitals of N are symmetry orbitals themselves, while the symmetry orbitals for the $3 \times H$ system consist of appropriate linear combinations of the three $1s$ atomic orbitals on H.

(C) The orbital energy of a symmetry orbital is equal to the orbital energy of the atomic orbitals used to construct the symmetry orbital, provided that the distance between the atoms constituting the fragment is much larger than the normal covalent bond length. If the bond lengths of the fragment are of the size of the covalent bond lengths, symmetry orbital energies are determined by constructing an MO diagram for the fragment. Thus C_2H_2 is divided into a $2 \times C$ and a $2 \times H$ fragment. The orbital energies for the symmetry orbitals of the $2 \times H$ fragment become the $1s$ atomic orbital energy, whereas an MO diagram must be constructed in order to obtain the approximate location of the orbital energies for the symmetry orbitals of the $2 \times C$ fragment.

(D) The MO diagram can now be constructed. The MO diagram displays how the orbital energies of the symmetry orbitals change position when MO's are formed as a result of the molecular binding. Symmetry orbitals of the two fragments with the same symmetry interact (have nonvanishing matrix elements over the molecular Hamiltonian) and thus form *bonding* and *antibonding* MO's. The relative energies of these MO's are determined by rules E-H.

(E) If only one symmetry orbital exists of a given symmetry, this orbital forms a *nonbonding* MO with the same energy as the symmetry orbital [see the discussion after rule (K)].

(F) If two symmetry orbitals exist of a given symmetry, the splitting in MO orbital energy between the bonding and antibonding orbital is (a) proportional to the overlap between the symmetry orbitals, and (b) inversely proportional to the energy difference between the orbital energies of the two symmetry orbitals.

(G) Three symmetry orbitals of the same symmetry combine to give one strongly bonding and one strongly antibonding MO. The orbital energy of the third MO is located between the orbital energies of the three symmetry orbitals. The exact location of this orbital energy depends on the nature of the considered problem.

(H) When bonding and antibonding orbitals are formed, the energy gain obtained in forming the bonding MO is always smaller than the energy loss that is a result of forming the antibonding MO.

(I) Molecular states are obtained by filling the MO's with valence electrons according to the aufbau principle. The symmetry of the molecular states is determined by taking the direct product of the irreducible representations according to which the occupied MO's transform. A completely filled orbital always gives a totally symmetric state (Unsöld's theorem).

(J) Molecules with similar structure (BH_2, CH_2, etc.) have qualitatively similar MO diagrams and differ mainly in the occupancy of the MO's.

(K) When the molecular point group symmetry is distorted, the following rules can be used to construct the MO diagram for the distorted molecule:
1. Determine within the point group symmetry of the distorted molecule the irreducible representation according to which the MO's of the undistorted molecule transform.
2. Determine the additional interaction between MO's that become symmetry allowed as a result of the molecular distortion (i.e, lowering in symmetry).
3. The distortion becomes energetically favorable if $\sum_{i}^{occ} \Delta\varepsilon_i < 0$, where $\Delta\varepsilon_i$ is the change in orbital energy of MO ψ_i caused by distorting the molecule.

In this set of rules we have discussed the position of the molecular energy levels relative to the energy levels of the isolated atoms. When the molecule is formed, the Hamiltonian changes from the sum of atomic Hamiltonians to the molecular Hamiltonian. This means for example, that the position of the energy level of a nonbonding orbital changes even though the atomic orbital is identical to the molecular orbital. The orbital energies of bonding and antibonding molecular orbitals are also affected by this change in Hamiltonian. It is generally assumed, however, that all molecular levels exhibit approximately the *same* shift and this shift is usually not included in the MO diagram. This is the reason for using the same orbital energy in rule E.

Textbooks that give sufficient background for solving the problems in this section

Ballhausen and Gray (1965). DeKock and Gray (1980). Gray (1965). Karplus and Porter (1970). La Paglia (1971). Levine (1974). Lowe (1978). McWeeny (1979). Murrell, Kettle, and Tedder (1970, 1978). Pilar (1968).

Textbook that gives a more elaborate treatment of the topics

Gimarc (1979).

Problem Description

E.1 Construction of qualitative molecular orbitals from atomic orbitals

E.2 The bonding-antibonding splitting in the formation of molecular orbitals

E.3 Dissociation products of FH and FH^+ in their ground states

E.4 Potential energy curves of the O_2 molecule in its ground and excited states

E.5 Stability of the O_2 dimer in a frontier orbital description

E.6 Photoelectron spectrum of C_2H_2 and C_4H_2

E.7 Photoelectron spectrum of N_2

E.8 Photoelectron spectrum of HCl

E.9 Photoelectron spectrum and electron affinity of CO_2

E.10 Photoelectron spectrum and spin magnetic moment of CO_3^{2-}

E.11 Excited states of the azide anion, N_3^-

E.12 Orbital correlation diagram and Jahn-Teller distortion of CH_4 and CH_4^+

E.13 Classification of states of H_2 formed from H + H in ^2S and ^2P states

E.14 Classification of states of VH formed from vanadium and hydrogen in their ground states

E.15 Excited states of CH^+ in a molecular orbital description and in a model where C is perturbed by H^+

E.16 The ground and lowest excited states of O_2 in the presence of an external perturbation

PROBLEMS

E.1 In this problem we will consider the qualitative change in the orbital energies that occurs when a bonding and an antibonding MO of a diatomic molecule AB is formed from an atomic orbital ϕ_A on atom A and an atomic orbital ϕ_B on atom B. This change in orbital energy is very fundamental when constructing an MO diagram (see, for example, rules E-H in the introduction to this set). The bonding and antibonding MO's are thus of the form

$$\psi = c_A \phi_A + c_B \phi_B$$

where the coefficients c_A, c_B are determined from a variational calculation. Let us assume that

$$S_{KL} = \langle \phi_K | \phi_L \rangle = \delta_{KL}$$

and

$$H_{KL} = \langle \phi_K | \hat{H} | \phi_L \rangle, \qquad L,K = A,B$$

where $\hat{H}$ is the molecular Hamiltonian. Assume further that the atomic orbital energies ε_A and ε_B are approximately equal to H_{AA} and H_{BB}, respectively. Assume initially that

(1) $\varepsilon_B > \varepsilon_A$ (E.1,1)

(2) $\gamma_{AB} = H_{AB} | H_{BB} - H_{AA} |^{-1} \ll 1$ (E.1,2)

1. Calculate through the lowest nonvanishing order in γ_{AB} the differences between the atomic and molecular orbital energies. Sketch graphically the effect of the molecular formation on the orbital energies; that is, draw an MO diagram.

Assume now that Eqs. (E.1,1-2) are replaced by

(1') $\varepsilon_B \approx \varepsilon_A$

(2') $\gamma_{AB} \gg 1$

2. Calculate through the lowest nonvanishing order in γ_{AB}^{-1} the differeneces between the atomic and the molecular orbital energies and draw an MO diagram.

When MO's are formed, the energy gain and energy loss in both the limiting cases of γ_{AB} of questions 1 and 2 are proportional to $|H_{AB}|$. Thus, in the case in which $|H_{AB}| = 0$, there will be no splitting of the atomic energy levels.

3. Which condition must ϕ_A and ϕ_B fulfill in order for $|H_{AB}|$ to be equal to zero? Which condition must ϕ_A and ϕ_B fulfill if $|H_{AB}|$ is to be large and small, respectively?

E.2 In this problem we will prove the validity of rule H in the introduction to this chapter for the case where two overlapping atomic orbitals, ϕ_A and ϕ_B, form a bonding and antibonding MO.

We will thus show that the energy gain, $\Delta\varepsilon$, obtained when forming the bonding MO is smaller than the energy loss, $\Delta\varepsilon^*$, resulting from forming an antibonding MO; that is,

$$\Delta\varepsilon^* - \Delta\varepsilon \geq 0$$ (E.2,1)

$\Delta\varepsilon^*$ and $\Delta\varepsilon$ are calculated relative to the largest and smallest of H_{AA} and H_{BB}, respectively, where

$$H_{ij} = \langle \phi_i | \hat{H} | \phi_j \rangle, \qquad i, j = A, B$$

and $\hat{H}$ is the *molecular* Hamiltonian.

The atomic overlap is different from zero and will be denoted

$$S = \langle \phi_A | \phi_B \rangle$$

where ϕ_A and ϕ_B are normalized to one.

1.   Show that the MO energies for the bonding and antibonding MO are

$$\varepsilon_{\pm} = [H_{AA} + H_{BB} - 2SH_{AB}$$

$$\pm [(H_{AA} - H_{BB})^2 + 4H_{AB}^2 - 4S(H_{AA} + H_{BB})H_{AB} + 4S^2 H_{AA} H_{BB}]^{1/2}]$$

$$\times [2(1 - S^2)]^{-1} \tag{E.2.2}$$

2.   Determine $\Delta\varepsilon^*$ and $\Delta\varepsilon$ in terms of H_{ij} and S.

In order to obtain a bonding (and antibonding) MO we must of course require that $\Delta\varepsilon$ (and $\Delta\varepsilon^*$) be greater than zero.

3.   Show that $\Delta\varepsilon > 0$ and $\Delta\varepsilon^* > 0$ implies that

$$H_{AB}^2 > SH_{AA} (2H_{AB} - SH_{AA}) \tag{E.2.3}$$

$$H_{AB}^2 > SH_{BB}(2H_{AB} - SH_{BB}) \tag{E.2.4}$$

4.   Show that

$$\Delta\varepsilon^* - \Delta\varepsilon = \frac{S}{1 - S^2}[S(H_{AA} + H_{BB}) - 2H_{AB}] \tag{E.2.5}$$

5.   Show that Eqs. (E.2,3) and (E.2,4) imply that $\Delta\varepsilon^* - \Delta\varepsilon$ is positive, that is, Eq. (E.2,1) is fulfilled. (Note that all integrals $H_{ij} < 0$.)

6.   In which circumstances would $\Delta\varepsilon = \Delta\varepsilon^*$, that is, the "bonding" energy gain be equal to the "antibonding" energy loss?

E.3 In this problem we will determine the dissociation products of the FH and FH^+ molecules in their ground states by using simple MO considerations. We will show that a Hartree-Fock calculation gives the wrong dissociation products for FH but the correct dissociation products for FH^+. We include the $1s$, $2s$, and $2p$ orbitals on F and the $1s$ orbital on H in the MO description. The $1s$, $2s$, and $2p$ fluorine orbital energies are -717.89 eV, -42.79 eV, and -19.86 eV, respectively (Clementi and Roetti 1974).

1.   Sketch an MO diagram for FH.
2.   Construct an *orbital* correlation diagram for the process FH → F + H.
3.   Determine the electronic configuration and the corresponding molecular term symbols for the ground and two lowest singly excited electronic configurations of FH and FH^+.

Dissociation of FH^+ may give $F^+ + H$ or $F + H^+$, while dissociation of FH may give F + H or $F^- + H^+$. The ionization potential and electron affinity of F are 17.34 eV and 3.45 eV, respectively.

4. Use this information to determine the dissociation products of the FH and FH$^+$ molecules in their ground state.

5. Determine the atomic term symbols for the dissociation products F$^+$, F, F$^-$, and H. Indicate the molecular term symbols for the molecular states that are formed when binding occurs between the following dissociation products: F + H, F$^-$ + H$^+$, F + H$^+$, and F$^+$ + H.

6. Construct a *configuration* correlation diagram for the dissociation of FH and FH$^+$. The diagram must include all the dissociation products that were discussed in question 5 and the electronic configurations determined in question 3.

7. Are there any avoided crossings in the configuration correlation diagrams of question 6? If so, which state symmetries do they correspond to?

8. How would the ground state of FH and of FH$^+$ dissociate in a restricted Hartree-Fock calculation? How would they dissociate in a CI calculation?

Bender and Davidson (1968) have performed a CI calculation of the potential energy curves for the ground and lowest excited states of FH. The potential energy curves are given in Fig. E.3,1. The bonding and/or repulsive characters of the potential energy curves can often be explained by considering the bonding and/or antibonding character of the dominant electronic configuration of the state.

9. Why does the potential energy curve for the ground state of FH have a large potential energy minimum, and why do not any of the excited states have a minimum? Can excited states have potential energy minima?

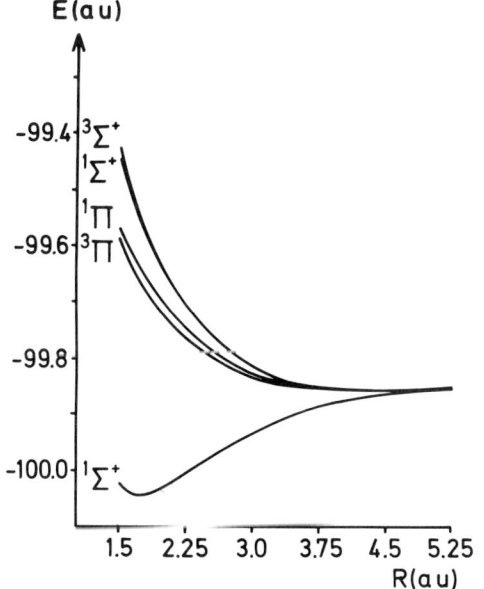

Fig. E.3,1. Potential energy curves for the ground and lowest excited states of FH. (Reprinted from J. Chem. Phys. by permission of the American Institute of Physics and the authors)

E.4 In the following we will analyze the form of the potential energy curves of the oxygen molecule for the ground state and for some excited states using simple MO arguments.

1. Draw an MO diagram for the oxygen molecule. Include in the diagram the $1s$, $2s$, and $2p$ orbitals of the oxygen atoms. The orbital energies of the $1s$, $2s$, and $2p$ orbitals are -562.41 eV, -33.86 eV, and -17.20 eV, respectively (Clementi and Roetti 1974).

2. Sketch an orbital correlation diagram for the O_2 system with special indication of the united atom, the equilibrium distance, and the separated atom limit. Include in the diagram the orbitals that correlate with the $1s$, $2s$, $2p$, $3s$, $3p$, and $3d$ atomic orbitals of the united atom limit.

3. Determine the ground-state electronic configuration of the oxygen molecule and indicate the molecular terms that originate from this configuration. Indicate the molecular term symbol for the ground state of O_2.

Some of the lowest states of u ("ungerade") symmetry in O_2 may be obtained from the ground-state electronic configuration by exciting an electron from the $1\pi_u$ to the $1\pi_g$ orbital ($1\pi_u \rightarrow 1\pi_g$).

4. Indicate the electronic configuration and the molecular terms which result from this excited electronic configuration.

The next lowest state of $^3\Sigma_u^-$ symmetry at equilibrium distance is a Rydberg state that may be obtained from the ground-state electronic configuration by exciting an electron from the $1\pi_g$ to the $2\pi_u$ orbital.

5. Show, by using the orbital correlation diagram of question 2, that the Rydberg state of $^3\Sigma_u^-$ symmetry (Ry) and the valence $^3\Sigma_u^-$ state (V_π) of question 4 have an avoided crossing.

The result of a CI calculation of the curve crossing discussed in question 5 is given in Fig. E.4,1 (Buenker and Peyerimhoff 1975). The ordinate in Fig. E.4,1 is the total energy relative to the total energy at the minimum of the ground state, $X\,^3\Sigma_g^-$.

6. Indicate the dominant electronic configuration of the potential energy curves in Fig. E.4,1 at various internuclear separations.

The lowest and next lowest state of $^3\Delta_u$ symmetry have an avoided crossing at about the same internuclear distance as the crossings for the states of $^3\Sigma_u^-$ symmetry.

7. How would you explain that the avoided crossing of $^3\Delta_u$ symmetry occurs at approximately the same internuclear separation?

In Fig. E.4,2 (Buenker and Peyerimhoff 1975) the potential energy curves of Fig.E.4,1 are displayed together with the potential energy curve for the ground state, $X\,^3\Sigma_g^-$, and for the $^3\Delta_u$ states of question 7. Based on the bonding and/or antibonding character of the dominant configuration, qualitative MO arguments can be used to explain the shape and the structure of the individual potential energy curves of Fig.E.4,2.

8. a Why do all the potential energy curves increase in energy at an internuclear distance smaller than about 2 a.u.?

   b Why does the $X\,^3\Sigma_g^-$ state have a potential energy minimum?

   c Why do the lowest states of $^3\Sigma_u^-$ and $^3\Delta_u$ symmetry have a very shallow minimum at a large internuclear distance?

d   Why are there bumps in the potential energy curves for the lowest states of $^3\Sigma_u^-$ and $^3\Delta$ symmetry at about 2.2-2.3 a.u.?

e   Why do the next lowest states of $^3\Sigma_u^-$ and $^3\Delta_u$ both have large potential energy minima?

9.   Indicate which states in Fig. E.4,2 may be reached in a dipole allowed transition from the $X\,^3\Sigma_g^-$ ground state. Sketch qualitatively as a function of the internuclear distance the shape of the transition moments for the dipole allowed transitions.

E.5 In this problem we will investigate the geometry and stability of the hypothetical O_2 dimers by requiring maximum overlap between the frontier orbitals of the two oxygen molecules. Consider initially the oxygen molecule.

1.   Sketch graphically, with indication of nodes, the frontier orbital of the O_2 molecule.

The oxygen molecules may have a tendency to form an O_2 dimer.

2.   Determine the point group symmetry of the O_2 dimer such that the overlap between the frontier MO's of the two O_2 molecules is as large as possible.

The frontier orbitals of the two O_2 molecules will form bonding and antibonding MO's when the O_2 dimer is formed.

3.   Determine the irreducible representations spanned by these MO's in the point group of the O_2 dimer that was determined in question 2.

4.   What is the energy ordering of these four MO's of the O_2 dimer? Indicate the occupation of the MO's and whether the O_2 dimer is expected to be stable relative to its dissociation products.

No experimental evidence exists of a *covalent* O_2 dimer. A van der Waals structure with large $O_2 - O_2$ equilibrium distances ($\simeq 4$ A) are known. Recent

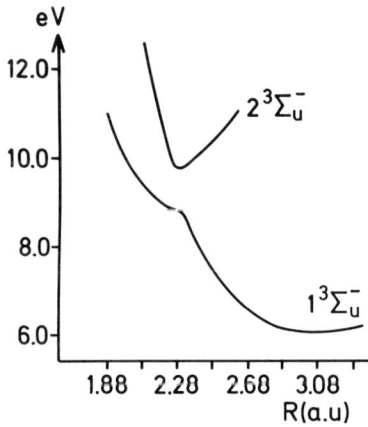

Fig. E.4,1. Potential energy curves of O_2 for the two lowest states of $^3\Sigma_u^-$ symmetry in the region of the avoided crossing. (Reprinted from Chem. Phys. Lett. by permission of North-Holland Publishing Company and the authors).

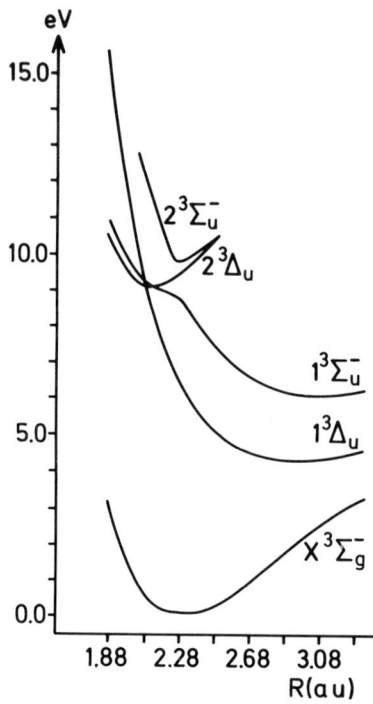

Fig. E.4,2. Potential energy curves for the ground state $(X\,^3\Sigma_g^-)$ and some of the lowest excited states of O_2 (Reprinted from Chem. Phys. Lett. by permission of North-Holland Publishing Company and the authors).

ab initio SCF – CI calculations by Adamantides *et al.* (1980) indicate the existence of a metastable covalent O_2 dimer but with a different geometry than that determined by means of the simple MO arguments of the present problem.

E.6 In the following we will analyze the electronic structure of the acetylene $(C_2 H_2)$ and the diacetylene $(C_4 H_2)$ molecules. We assume that both molecules have $D_{\infty h}$ point group symmetry, and we include in the MO description the $1s$, $2s$, and $2p$ orbitals on the carbon atoms and the $1s$ orbital on the hydrogen atoms. The $1s$, $2s$, and $2p$ orbital energies of carbon are -308.18 eV, -19.20 eV, and -11.79 eV, respectively (Clementi and Roetti 1974).

1.   Draw an MO diagram for the $C_2 H_2$ molecule.
2.   Determine the electronic configuration and the molecular term symbol for the ground state of $C_2 H_2$.
3.   Determine the electronic configuration and the molecular term symbol for the three electronic states of the $C_2 H_2^+$ ions that are formed by ionizing one electron out of the three highest occupied MO's in the ground state of $C_2 H_2$.

The photoelectron spectrum of C_2H_2 in the region between 0 eV and 20 eV has been assigned to three electronic states of $C_2H_2^+$. The ionization potentials are 11.40 eV, 16.36 eV, and 18.38 eV, respectively (Turner *et al.* 1970).

4. Perform a symmetry assignment of the photoelectron spectrum based on the MO diagram.

5. Use the MO diagram of question 1 to predict the position of the photoelectron peaks that correspond to ionizing an electron out of the occupied orbitals of C_2H_2 which were not considered in question 3.

The photoelectron spectrum of C_4H_2 in the region between 0 and 20 eV has been assigned to four electronic states of $C_4H_2^+$. The ionization potentials are 10.17 eV, 12.62 eV, 16.61 eV, and 19.8 eV, respectively (Turner *et al.* 1970).

6. Use simple MO considerations to explain why the lowest ionization potential of C_2H_2 is located halfway between the lowest and next lowest ionization potential of C_4H_2 and why the second ionization potential of C_2H_2 is not located halfway between the third and fourth ionization potential of C_4H_2. Perform a symmetry assignment of the ionization potentials of C_4H_2 at 10.17 eV and 12.62 eV.

E.7 In this problem we will perform an assignment of the photoelectron spectrum of N_2 based on an MO description. The orbital energies of $1s$, $2s$, and $2p$ of the N atom are -425.28 eV, -25.72 eV, and -15.44 eV, respectively (Clementi and Roetti 1974).

1. Sketch an MO diagram for N_2 that includes the $1s$, $2s$, and $2p$ orbitals of N.

2. Indicate the electronic configuration and the term symbol for the ground state of N_2.

3. Indicate the electronic configurations and the term symbols for the states of N_2^+ that correspond to ionizing one electron out of the four highest occupied MO's of N_2.

The photoelectron spectrum of N_2 in the 15- to 20-eV region (Turner *et al.* 1970) is displayed in Fig. E.7,1. The spectrum shows three major peaks, which correspond to ionization out of the three highest occupied MO's of N_2. The transitions all have vibrational structure. The ionization potentials corresponding to the three vertical ionizations are marked with arrows.

4. Perform an assignment of the photoelectron spectrum of N_2 in Fig. E.7,1 based on the MO diagram.

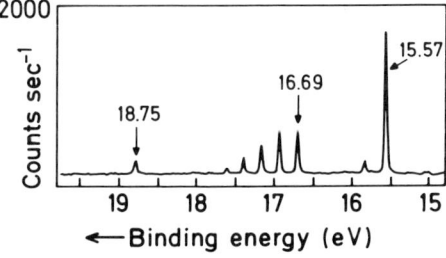

Fig. E.7,1 The photoelectron spectrum of N_2 (Reprinted from Turner *et al.* (1970) by permission of John Wiley and Sons, Ltd).

At the Hartree-Fock limit, an SCF calculation gives the orbital energies (Cade *et al.* 1966) $\varepsilon_{1\sigma_g} = -426.61$ eV, $\varepsilon_{1\sigma_u} = -426.61$ eV, $\varepsilon_{2\sigma_g} = -40.10$ eV, $\varepsilon_{2\sigma_u} = -21.17$ eV, $\varepsilon_{3\sigma_g} = -17.28$ eV, and $\varepsilon_{1\pi_u} = -16.73$ eV .

5.   Perform an assignment of the photoelectron spectrum of N_2 based on the Hartree-Fock calculation. Is this assignment in agreement with the one in question 4?

The three peaks in the photoelectron spectrum of N_2 are experimentally (in order of increasing binding energy) assigned to ionization out of the $3\sigma_g$, $1\pi_u$, and $2\sigma_u$ orbitals, respectively.

6.   Compare the experimental assignment with the one based on the Hartree-Fock calculation. Indicate the discrepancy between the experimental assignment and the assignment based on the Hartree-Fock calculation. Discuss how you might improve the calculation in order to obtain agreement between the experimental and the theoretical assignments.

E.8 In this problem we will perform an assignment of the photoelectron spectrum of HCl based on an MO diagram in which only the $1s$ orbital on H and the $3p$ orbital on Cl are considered. The ionization potential for Cl is 13.01 eV (Moore 1949).

1.   Sketch an MO diagram for HCl.
2.   Determine the molecular electronic configuration and term symbol of the ground state of HCl.
3.   Determine the molecular electronic configuration and term symbol for the two electronic states of HCl^+ that are formed by ionizing one electron out of the two highest occupied MO's in the ground state of HCl.

A crude photoelectron spectrum of HCl is displayed in Fig. E.8,1 (Turner *et al.* 1970).

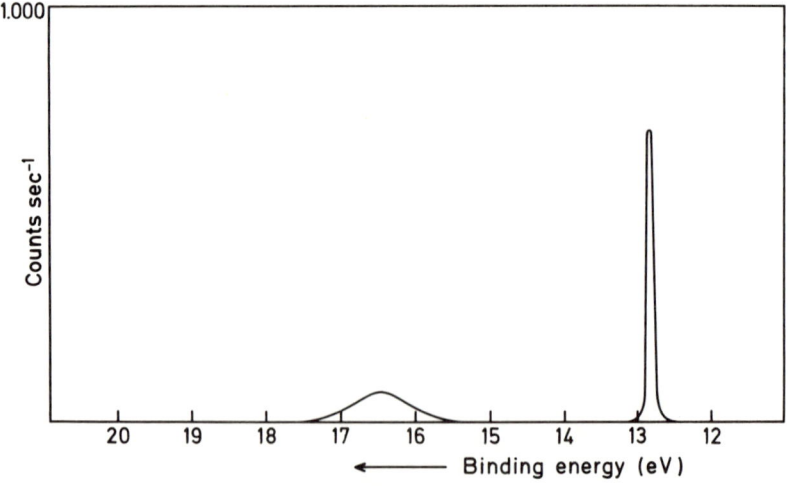

Fig. E.8,1. A crude photoelectron spectrum of HCl (Reprinted from Turner *et al.* (1970) by permission of John Wiley and Sons, Ltd).

4. Perform an assignment of the photoelectron spectrum based on the MO diagram.

The peak at 12.9-12.7 eV may be resolved into two peaks (see Fig. E.8.2), the splitting of which is caused by the spin-orbit interaction.

5. Perform an assignment of the two peaks in Fig. E.8,2.

E.9 In the following we will analyze the electronic structure of the CO_2^+, CO_2, and CO_2^- molecules. We assume initially that all the molecules have $D_{\infty h}$ point group symmetry and we include in the MO description only the $2s$ and $2p$ orbitals on oxygen and carbon. The $2s$ and $2p$ orbital energies of oxygen are -33.86 eV and -17.20 eV, respectively, while the $2s$ and $2p$ orbital energies of carbon are -19.20 eV ad -11.79 eV, respectively (Clementi and Roetti 1974).

1. Draw an MO diagram for CO_2.

2. Determine the molecular electronic configuration and the term symbol for the ground state of CO_2.

3. Determine the molecular electronic configuration and the term symbol for the four electronic states of CO_2^+ that correspond to ionizing one electron out of one of the four highest occupied orbitals in the ground state of CO_2.

The photoelectron spectrum of CO_2 in the region between 0 eV and 20 eV has been assigned to four electronic states of CO_2^+ with the ionization potentials 13.8 eV, 17.6 eV, 18.1 eV, and 19.4 eV, respectively. The peaks at 13.8 eV and 17.6 eV both show a splitting of 0.02 eV which has been attributed to the spin-orbit coupling (Turner *et al.* 1970).

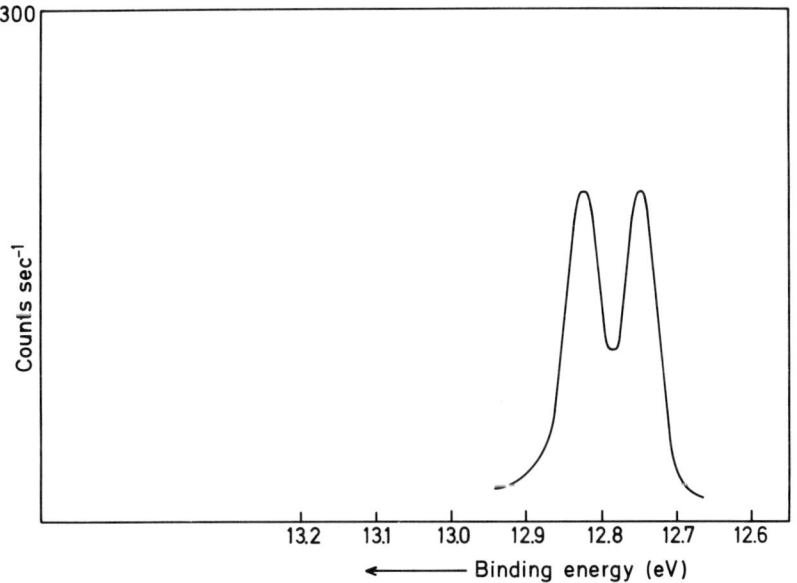

Fig. E.8,2. A high resolution of the first peak of the photoelectron spectrum of HCl. The splitting of the peak into two peaks is caused by the spin-orbit interaction (Reprinted from Turner *et al.* (1970) by permission of John Wiley and Sons, Ltd).

4.   Perform an assignment of the photoelectron spectrum based on the MO diagram and indicate the spectroscopic term symbol for the two spin-orbit levels which arise from the peaks at 13.8 eV and 17.6 eV.

5.   Use the MO diagram of question 1 to predict the location of the next two peaks in the photoelectron spectrum.

In an SCF calculation on CO_2 the following orbital energies have been obtained (McLean and Yoshimine 1968):

$$\varepsilon_{1\sigma_u} = -561.67 \text{ eV} \qquad \varepsilon_{3\sigma_u} = -20.34 \text{ eV}$$

$$\varepsilon_{1\sigma_g} = -561.67 \text{ eV} \qquad \varepsilon_{1\pi_u} = -19.77 \text{ eV}$$

$$\varepsilon_{2\sigma_g} = -311.41 \text{ eV} \qquad \varepsilon_{1\pi_g} = -14.85 \text{ eV}$$

$$\varepsilon_{3\sigma_g} = -42.03 \text{ eV} \qquad \varepsilon_{2\pi_u} = 6.14 \text{ eV}$$

$$\varepsilon_{2\sigma_u} = -40.59 \text{ eV} \qquad \varepsilon_{5\sigma_g} = 6.91 \text{ eV}$$

$$\varepsilon_{4\sigma_g} = -21.68 \text{ eV} \qquad \varepsilon_{4\sigma_u} = 9.22 \text{ eV}$$

6.   Perform an assignment of the photoelectron spectrum based on the Hartree-Fock calculation. Is this assignment in agreement with the one in question 5? If not, discuss the reason for the disagreement.

Accurate calculations (England et al. 1976) on CO_2^- show that both the ground and lowest excited state in CO_2^- have C_{2v} symmetry. The molecular term symbol for the ground state is $^2A_1$ while the term symbol for the lowest excited state is $^2B_1$. Assume that the distortion of the geometry of CO_2^- from $D_{\infty h}$ to C_{2v} is so small that the electronic configuration of CO_2^- can be described by the ground-state electronic configuration of CO_2 and an additional electron in an unoccupied CO_2 orbital.

7.   In which unoccupied orbital of the MO diagram in question 1 should the electron be added in order to give the symmetry of the two states of CO_2^- as found by England et al. (1976)?

8.   According to the Hartree-Fock calculation on CO_2, what would the molecular term symbol be for the ground, first, second, and third excited states of CO_2^- ?

It is observed experimentally (Cooper and Compton 1972) that CO_2 has a positive electron affinity.

9.   Is this in agreement with the result of the Hartree-Fock calculation on CO_2 ?

E.10

In this problem we will use qualitative MO theory to perform a gross interpretation of the photoelectron spectrum of the carbonate ion, CO_3^{2-}, and to estimate the magnitude of the spin magnetic moment of ionic $MnCO_3$ crystals. We initially examine the transformation properties of the s and p orbitals on carbon and oxygen in the point group of CO_3^{2-} (D_{3h}).

1.   Determine the irreducible representations spanned by the following atomic orbitals: $2s$ (carbon), $2p$ (carbon), $2s$ on the three oxygens, and $2p$ on the three oxygens.

The orbital energies of the valence atomic orbitals are -33.86 eV for $2s$ (oxygen), -19.20 eV for $2s$ (carbon), -17.20 eV for $2p$ (oxygen), and -11.79 eV for $2p$ (carbon) (Clementi and Roetti 1974).

2. Construct an MO diagram for CO_3^{2-}.

3. Determine the ground-state electronic configuration and term symbol for CO_3^{2-}.

Conner *et al.* (1972) have recorded the photoelectron spectrum of Li_2CO_3 displayed in Fig. E.10,1. This spectrum shows three main peaks I, II + III, and IV, each of which represents ionization out of more than one MO of CO_3^{2-}.

4. Perform an assignment of the three peaks I, II + III, and IV on the basis of the MO diagram in question 2.

We will now use the MO diagram of CO_3^{2-} to calculate the spin magnetic moment of crystalline $MnCO_3$, which is assumed to have the ionic structure $Mn^{2+}CO_3^{2-}$.

5. Determine within the *LS* coupling scheme the spin quantum number, S, for atomic Mn^{2+} in its ground state.

6. Use the MO diagram from question 2 to calculate the magnitude (in units of e/mc) of the spin magnetic moment of one molecule in crystalline $Mn^{2+}CO_3^{2-}$.

E.11

In this problem we will use qualitative MO theory to analyze the geometry and the electronic structure of the azide anion, N_3^-. Initially, we assume that N_3^- is linear with equal N – N bond lengths. We assume that the bond formation can be described by considering only the $2s$ and $2p$ atomic orbitals on nitrogen.

1. Construct an MO diagram for N_3^-. Use the two end nitrogen atoms and the central N atom as the two fragments from which the MO diagram is constructed (cf. rules B and C of the introduction to this set). (*Hint*: the lowest unoccupied molecular orbital (LUMO) has π_u symmetry.)

2. Determine the ground-state electronic configuration and term symbol for N_3^-.

Consider the three singly excited electronic configurations that are formed by exciting one electron from one of the three highest occupied molecular orbitals (HOMO's) and into the LUMO.

3. Determine the electronic configurations and the term symbol for the three singly excited configurations.

The lowest excitation energies of N_3^- have in an *ab initio* SCF calculation been calculated to be (Rossi and Bartram 1979): $^3\Sigma_u^+$ (2.92), $^3\Delta_u$ (3.55), $^1\Sigma_u^-$ (4.16),

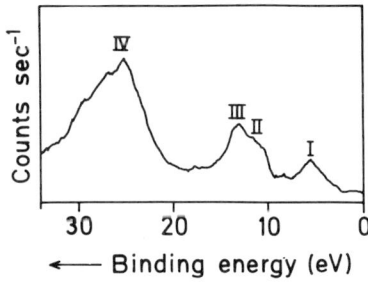

Fig. E.10,1. The photoelectron spectrum of Li_2CO_3 (Reprinted from Mol. Phys. by permission of Taylor and Francis Ltd. and the authors).

$^3\Sigma_u^-$(4.16), $^1\Delta_u$(4.56), $^3\Pi_g$(9.93), $^1\Sigma_u^+$(11.19), $^1\Pi_g$(11.29), $^3\Pi_u$(11.97), and
$^1\Pi_u$(14.26) where the numbers in parentheses give the energy (in eV) of the
states relative to the ground state at the ground-state equilibrium geometry.

4.   Comment on the qualitative agreement (disagreement) between the energy
     ordering of the excited states as obtained from the SCF calculation and
     from the MO diagram of question 1. Use the energy ordering of the excited
     states to estimate the energy ordering of the three highest occupied
     molecular orbitals of N_3^-.

We will now investigate the stability of the bent azide anion relative to the linear
configuration. We assume that the two N – N bond lengths are the same in the
linear and in the bent configuration. Furthermore, we assume that it is sufficient
to consider how the *frontier* orbital energies change with bending angle in order
to determine the geometry of N_3^- in its ground and lowest excited state. The
frontier orbitals are both doubly degenerate.

5.   Determine the irreducible representations spanned by the frontier orbitals
     of linear N_3^- in the point group of the bent molecule.

When bending N_3^- around the middle N atom, we will assume that the increase
(decrease) in orbital energy is proportional to the decrease (increase) in electron
density between the three nitrogen atoms (the so-called bonding density).

6.   Indicate qualitatively how the orbital energies of the HOMO and the
     LUMO change for $90° \leq \theta \leq 180°$ where θ is the angle between the two
     N – N bonds in N_3^-.

7.   Would you, from the results of question 6, expect N_3^- to be bent or linear
     in its ground state?

Consider the lowest excited state of N_3^-, which corresponds to exciting one
electron from the HOMO to the LUMO of the linear molecule.

8.   Would you, from the result of question 6, expect N_3^- to be bent or linear in
     this excited state? What are the possible molecular term symbols for this
     state?

E.12

Four hydrogen atoms initially at infinite nuclear separation approach
adiabatically a carbon atom, thus forming an Ne atom in the united atom limit.
We assume that T_d symmetry is conserved during this process. The Hartree-Fock
orbital energies for atomic carbon and neon are $\varepsilon_{1sC} = -308.18$ eV,
$\varepsilon_{2sC} = -19.20$ eV, $\varepsilon_{2pC} = -11.79$ eV, $\varepsilon_{1sNe} = -891.76$ eV,
$\varepsilon_{2sNe} = -52.53$ eV, and $\varepsilon_{2pNe} = -23.14$ eV (Clementi and Roetti 1974).

1.   Draw an orbital correlation diagram that includes all orbitals that are
     occupied in the separated atom limit. Indicate the irreducible
     representations spanned by the orbitals in the separated atom limit, at the
     CH_4 equilibrium geometry, and in the united atom limit.

2.   Determine the electronic configuration and the term symbols of the ground
     state and the first excited state of CH_4.

3.   Would you expect a Jahn-Teller distortion of CH_4 in either of the states
     that were determined in question 2?

4.   Determine the electronic configuration and the term symbol of the ground
     state of the CH_4^+ ion.

The ground state of the CH_4^+ ion is subject to a Jahn-Teller distortion from T_d
symmetry. Possible geometries of the distorted CH_4^+ ion are C_{3v}, C_{2v}, or D_{2d}.

5. In how many *equivalent* ways is it possible to obtain distorted CH_4^+ molecules with C_{3v} symmetry? Answer the same question for C_{2v} and D_{2d} symmetry.

No experimental information about the geometry of the CH_4^+ ion in its ground state is available, but a theoretical calculation (Lathan *et al.* 1974) predicts D_{2d} geometry.

6. What is the spectroscopic term symbol of the ground state of CH_4^+ if we assume a distortion of D_{2d} symmetry?

E.13

Two hydrogen atoms, initially at infinite separation, approach adiabatically until they reach the united atom limit, where a helium atom is formed.

1. Determine the term symbols of the electronic states of H_2 that are formed if both hydrogen atoms were initially in the lowest ^2S state.
2. Draw a state correlation diagram indicating term symbols for both the separated atoms, the H_2 molecule, and the united atom limit.

Let us now consider the case where the two hydrogen atoms are initially both in the ^2P state of lowest energy.

3. Construct a microstate diagram for the H_2 molecular states that are formed in the molecular limit.
4. Indicate the spectroscopic term symbol for the resulting molecular states.
5. How many of the molecular states from question 4 may correlate with atomic states that are formed from the $2p^2$ configuration of He? Give an example of higher-lying atomic states of He that may correlate with the rest of the molecular states of question 4.

E.14

In this problem we will determine the spectroscopic term symbols for the states of VH, which correlate with the ground state of vanadium and hydrogen in the separated atom limit. The ground state of the vanadium atom has the electron configuration $1s^2 2s^2 2p^6 3s^2 3p^6 3d^3 4s^2$ and will be treated within the Russell-Saunders (LS) coupling scheme.

1. Show by constructing a microstate diagram that the ground-state electron configuration of V gives rise to the following atomic terms: ^2H, ^2G, ^4F, ^2F, ^2D, ^2D, ^4P, and ^2P. Which of these terms correspond to the ground state of V?

A linear combination of Slater determinants constructed from the set of orthogonal orbitals $1s$, $2s$, $2p$, ... is an approximate wave function for the individual M_S and M_L components of a given term.

2. Determine the simplest determinantal wave function for the $M_S = \frac{3}{2}$ and $M_L = 0$ component of the ground-state term.

We will now assume that a VH molecule is formed by the adiabatic approach of a vanadium to a hydrogen atom. Both atoms are in their ground states.

3. Determine the Λ and S quantum numbers for the molecular states of VH, which is formed during this process.

Two of the states will be Σ states ($\Lambda = 0$).

4. Write a wave function for the largest M_S component of these states expressed as an antisymmetric product of determinantal wave functions for vanadium and hydrogen.
5. Determine the term symbols for all the states of question 3.

E.15

In this problem we will estimate the lowest excitation energies of the CH^+ ion by means of two simple physical models. In questions 1-3 we perform an MO description, and in questions 4-7 the CH^+ ion will be described as a carbon atom that is perturbed by a proton. In question 8 the excitation energies obtained in the two models will be compared with the experimental excitation energies. Questions 1-3 and 4-7 can be solved independently of each other. We first consider the MO description of the CH^+ ion.

1. Draw an MO diagram for the CH^+ ion. Include the $2s$ and $2p$ orbitals on carbon and the $1s$ orbital on hydrogen in the diagram. The $2s$ and $2p$ carbon orbital energies are -19.20 eV and -11.79 eV, respectively (Clementi and Roetti 1974).

2. What is the molecular term symbol for the ground state for the CH^+ ion? Indicate the term symbol for the states that originate from the two lowest singly excited electronic configurations.

None of the excited states determined in question 2 have Δ symmetry.

3. How would you expect the lowest electronic state of Δ symmetry to be described within an MO picture?

The CH^+ ion will now be described within a model where the carbon atom is perturbed by a proton. The corresponding perturbation operator will be denoted by $\hat{V}$. Let $\hat{V}_{EE}$ denote the electronic repulsion operator of the carbon atom. The ground-state electronic configuration of carbon gives rise to three states, ^3P, ^1D, and ^1S, which are degenerate in the absence of $\hat{V}_{EE}$ and $\hat{V}$.

4. Determine the components of ^3P, ^1D, and ^1S that interact in the presence of $\hat{V}$. (*Hint*: Determine the molecular term symbols for the atomic terms in the $C_{\infty v}$ point group.)

We will now carry out a first-order perturbation calculation in $\hat{V}_{EE} + \hat{V}$ to determine the splitting of the ground-state electronic configuration of carbon. $\hat{V}_{EE}$ splits the carbon ground-state electronic configuration $1s^2 2s^2 2p^2$ into the atomic terms $^3P(0.0\text{ eV})$, $^1D(1.3\text{ eV})$, and $^1S(2.7\text{ eV})$ where the numbers in parentheses are the relative energies of the three states (Moore 1949).

5. Sketch the splitting of the ground-state electronic configuration of carbon that is obtained by carrying out the first-order perturbation calculation. The result should be expressed as a function of a parameter k, where k is the value of the nondiagonal matrix element of $\hat{V}$ (see question 4). All diagonal matrix elements of $\hat{V}$ are assumed to be equal to a constant.

The ground state for the CH^+ ion is a $^1\Sigma^+$ state (Botterud *et al.* 1973).

6. Which values of k will, according to the analysis in question 4, give a ground state of $^1\Sigma^+$ symmetry?

The lowest electronic singlet excitation energy of Π symmetry, $\Delta E(^1\Sigma^+ \rightarrow {}^1\Pi)$, is experimentally determined to be 3.0 eV (Botterud *et al.* 1973).

7. Use $\Delta E(^1\Sigma^+ \rightarrow {}^1\Pi)$ to determine k and compute the excitation energies for the other singlet and triplet excitation from the ground state of the CH^+ ion.

Experimentally, the lowest singlet excitation energy of Δ symmetry is found to be 6.0 eV (Carré and Dufay 1968), and the lowest triplet excitation energy of Π symmetry is calculated by Green *et al.* (1972) to be 1.0 eV, while the lowest singlet excitation energy of Π symmetry, as mentioned earlier, is 3.0 eV. The lowest singlet excitation energy of Σ^+ symmetry is not yet known.

8. Compare the experimental excitation spectrum with the ones obtained in the two previously described models. (Approximate excitation energies in the MO model are MO energy differences. To obtain comparable excitation spectra in the MO and in the perturbation model requires also that the lowest $^1\Pi$ excitation energy be 3.0 eV in the MO model.) Give an estimate of the magnitude of the lowest singlet excitation energy of Σ^+ symmetry.

E.16

In this problem we will consider the energy splitting of the valence electronic configuration $(\pi_g)^2$ of O_2, caused by both the electronic repulsion, $\hat{V}_{EE}$, and by an external perturbation, $\hat{V}_{ext}$. The $(\pi_g)^2$ electronic configuration gives rise to three electronic states, $X^3\Sigma_g^-$, $a^1\Delta_g$, and $b^1\Sigma_g^+$, which are degenerate if we disregard $\hat{V}_{EE}$ and $\hat{V}_{Ext}$.

1. Write the wave functions for the $X^3\Sigma_g^-$, $a^1\Delta_g$, and $b^1\Sigma_g^+$ states as linear combinations of Slater determinants constructed from the set of complex π_g orbitals: π_g^+ and π_g^-.

The effect of the spin-independent external one-electron perturbation, $\hat{V}_{ext}$, is described through the matrix elements of $\hat{V}_{ext}$ in the valence orbital basis:

$$\langle \pi_g^x | \hat{V}_{ext} | \pi_g^x \rangle = -\frac{a}{2} \text{ eV} \qquad (E.16,1)$$

$$\langle \pi_g^y | \hat{V}_{ext} | \pi_g^y \rangle = +\frac{a}{2} \text{ eV} \qquad (E.16,2)$$

$$\langle \pi_g^x | \hat{V}_{ext} | \pi_g^y \rangle = 0 \qquad (E.16,3)$$

where π_g^x and π_g^y refer to the set of real π_g orbitals, that is,

$$\pi_g^x = \frac{1}{\sqrt{2}} (\pi_g^+ + \pi_g^-) \qquad (E.16,4)$$

and

$$\pi_g^y = \frac{i}{\sqrt{2}} (\pi_g^- - \pi_g^+) \qquad (E.16,5)$$

2. Determine the matrix elements of $\hat{V}_{ext}$ between the wave functions obtained in question 1.

The splitting of the ground-state electronic configuration of the oxygen molecule which results from $\hat{V}_{EE}$ and $\hat{V}_{ext}$ may be determined by carrying out a first-order perturbation calculation in $\hat{V}_{EE} + \hat{V}_{ext}$. $\hat{V}_{EE}$ splits the ground-state electronic configuration into the following states: $X^3\Sigma_g^-$ (0.0 eV), $a^1\Delta_g$ (1.0 eV), and $b^1\Sigma_g^+$ (1.7 eV) where the numbers in parentheses indicate the relative ordering of the states at the ground-state equilibrium distance (Huber and Herzberg 1979).

3. Sketch the splitting of the ground-state electronic configuration as a function of a.

The perturbation $\hat{V}_{ext}$ is caused by external charges of δ located at R and $-R$ in the y-direction and charges of $-\delta$ located at R and $-R$ in the x-direction. All charges are further placed in the σ_h reflection plane for the O_2 molecule, which is perpendicular to the z-axis. R is an arbitrary distance.

4.   Determine within the point group of the perturbed oxygen molecule the term symbols for the three lowest states of O_2. Explain why symmetry considerations alone could have told us that we should have approximately the changes in state energies as a function of the parameter a that were computed in the previous question.

The perturbation $\hat{V}_{ext}$ may change the symmetry of the ground state.

5.   Determine the value $a = a_0$ for which this change occurs and indicate the ground-state term symbol for $a > a_0$ and $a < a_0$.

6.   Do the physical properties of the ground state change at $a = a_0$?

SOLUTIONS

E.1 1. The MO energies, $\varepsilon_\pm$, are obtained from the secular determinant

$$\begin{vmatrix} H_{AA} - \varepsilon & H_{AB} \\ H_{BA} & H_{BB} - \varepsilon \end{vmatrix} = 0$$

or

$$\begin{aligned}
\varepsilon_\pm &= \frac{H_{AA} + H_{BB}}{2} \pm \left[\frac{1}{4}(H_{BB} - H_{AA})^2 + |H_{AB}|^2 \right]^{1/2} \\
&= \frac{H_{AA} + H_{BB}}{2} \pm \frac{1}{2}(H_{BB} - H_{AA})[1 + 4\gamma_{AB}^2]^{1/2} \qquad \text{(E.1,I)} \\
&\approx \frac{H_{AA} + H_{BB}}{2} \pm \frac{1}{2}(H_{BB} - H_{AA})[1 + 2\gamma_{AB}^2]
\end{aligned}$$

$$= \begin{cases} H_{BB} + \gamma_{AB}^2(H_{BB} - H_{AA}) \\ \\ H_{AA} - \gamma_{AB}^2(H_{BB} - H_{AA}) \end{cases} \approx \begin{cases} \varepsilon_B + \dfrac{|H_{AB}|^2}{H_{BB} - H_{AA}} \\ \\ \varepsilon_A - \dfrac{|H_{AB}|^2}{H_{BB} - H_{AA}} \end{cases} \qquad \text{(E1,II)}$$

The orbital energy of the bonding MO is slightly lower than ε_A, and the orbital energy of the antibonding MO is increased relative to ε_B by the *same* amount. This seems to be in contradiction with rule H in the introduction. However, differences between the energy loss and energy gain when forming MO's first show up when the atomic overlap is included explicitly in the calculation (see problem H.2). The MO diagram for $\gamma_{AB} \ll 1$ is given in Fig. E.1,I

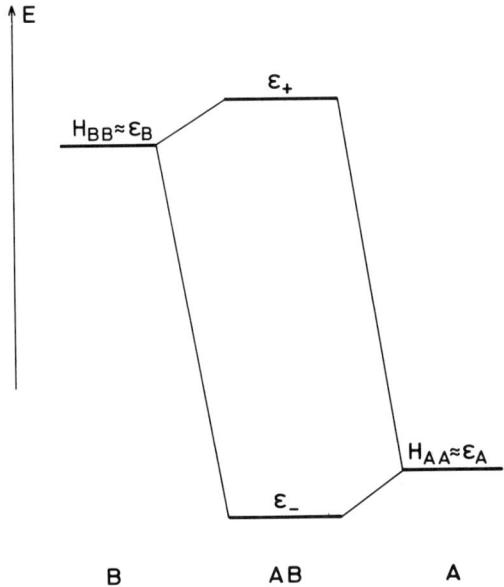

Fig. E.1,I. MO diagram illustrating the bonding situation that occurs when atomic orbital ϕ_A with orbital energy ε_A interacts with atomic orbital ϕ_B with orbital energy ε_B and $\varepsilon_B \gg \varepsilon_A$.

2. Using Eq. (E.1,I) we find

$$\varepsilon_{\pm} = \frac{H_{AA} + H_{BB}}{2} \pm |H_{AB}| \left[1 + \frac{1}{4}\gamma_{AB}^{-2} \right]^{1/2}$$

$$\approx \varepsilon_A \pm \left[|H_{AB}| + \frac{1}{8}|H_{AB}|\gamma_{AB}^{-2} \right]$$

$$= \begin{cases} \varepsilon_A + |H_{AB}| + \dfrac{|H_{BB} - H_{AA}|^2}{8|H_{AB}|} \\[2mm] \varepsilon_B - |H_{AB}| - \dfrac{|H_{BB} - H_{AA}|^2}{8|H_{AB}|} \end{cases} \qquad \text{(E.1,III)}$$

where $|H_{BB} - H_{AA}|^2/8|H_{AB}|$ is positive and small compared with $|H_{AB}|$. The splitting is determined almost solely by $|H_{AB}|$ and the MO diagram is displayed in Fig. E.1,II for a case with a rather large value of $|H_{AB}|$ (cf. the discussion in question 1).

3. Equations (E.1,II) and (E.1,III) show that $|H_{AB}|$ determines the magnitude of the splitting.

Fig. E.1,II. MO diagram illustrating the bonding situation that occurs when atomic orbital ϕ_A with orbital energy ε_A interacts with atomic orbital ϕ_B with orbital energy ε_B and $\varepsilon_A \approx \varepsilon_B$.

If ϕ_A and ϕ_B transform according to different irreducible representations of the molecular point group then $H_{AB} = 0$ and ϕ_A and ϕ_B will not form bonding orbitals, (rule E in the introduction).

If ϕ_A and ϕ_B have amplitudes in the same region of space, H_{AB} will be relatively large, whereas small values for H_{AB} are obtained for orthogonal or nearly orthogonal atomic orbitals. When H_{AB} is large, a large splitting occurs in the MO diagram and when H_{AB} is small, a small splitting is observed in the MO diagram.

E.2 1. The MO energies are roots of the secular determinant

$$\begin{vmatrix} H_{AA} - \varepsilon & H_{AB} - \varepsilon S \\ H_{AB} - \varepsilon S & H_{BB} - \varepsilon \end{vmatrix} = 0$$

or

$$\varepsilon^2(1 - S^2) - \varepsilon(H_{AA} + H_{BB} - 2SH_{AB}) + H_{AA}H_{BB} - H_{AB}^2 = 0$$

which readily yields Eq. (E.2,2).

2. Assume, for example, that $H_{BB} > H_{AA}$. Then

$$\Delta\varepsilon = H_{AA} - \varepsilon_- = [H_{AA} - H_{BB} + 2S(H_{AB} - SH_{AA}) + C] \tag{E.2,I}$$
$$\times [2(1 - S^2)]^{-1}$$

$$\Delta\varepsilon^* = \varepsilon_+ - H_{BB} = [H_{AA} - H_{BB} - 2S(H_{AB} - SH_{BB}) + C] \tag{E.2,II}$$
$$\times [2(1 - S^2)]^{-1}$$

where

$$C = [(H_{AA} - H_{BB})^2 + 4H_{AB}^2 \tag{E.2,III}$$
$$- 4S(H_{AA} + H_{BB})H_{AB} + 4S^2 H_{AA} H_{BB}]^{1/2}$$

3. Using Eq. (E.2,I) we see that $\Delta\varepsilon > 0$ implies that

$$C^2 > (H_{BB} - H_{AA})^2 + 4S^2(H_{AB} - SH_{BB})^2$$
$$+ 4S(H_{AB} - SH_{AA})(H_{AA} - H_{BB})$$

which straightforwardly yields Eq. (E.2,3). Requiring $\Delta\varepsilon^* > 0$ gives Eq. (E.2,4).

4. Equation (E.2,5) follows from Eqs. (E.2,I-II).

5. Adding Eqs. (E.2,3-4), we obtain

$$2H_{AB}^2 > (2H_{AB} - S)S(H_{AA} + H_{BB})$$

Since $2H_{AB} - S < 0$ we have the inequality

$$\frac{2H_{AB}^2}{2H_{AB} - S} < S(H_{AA} + H_{BB})$$

Thus

$$S(H_{AA} + H_{BB}) - 2H_{AB} > 2H_{AB}\frac{S - H_{AB}}{2H_{AB} - S} > 0$$

and it follows from Eq. (E.2,5) that $\Delta\varepsilon^* - \Delta\varepsilon > 0$ Q.E.D.

    6. $\Delta \varepsilon^* = \Delta \varepsilon$ for $S = 0$, which was also found in problem E.1.

E.3 1. See Fig. E.3,I.
   2. See Fig. E.3,II.
   3. FH: Ground-state configuration: $1\sigma^2 2\sigma^2 3\sigma^2 1\pi^4$; $^1\Sigma^+$
              Lowest excited configuration: $1\sigma^2 2\sigma^2 3\sigma^2 1\pi^3 4\sigma^1$; $^1\Pi$, $^3\Pi$
              Next lowest excited configuration: $1\sigma^2 2\sigma^2 3\sigma^1 1\pi^4 4\sigma^1$; $^1\Sigma^+$, $^3\Sigma^+$

       FH$^+$: Ground-state configuration: $1\sigma^2 2\sigma^2 3\sigma^2 1\pi^3$; $^2\Pi$
              Lowest excited configuration: $1\sigma^2 2\sigma^2 3\sigma^1 1\pi^4$; $^2\Sigma^+$
              Next lowest excited configuration:

$$1\sigma^2 2\sigma^2 3\sigma^2 1\pi^2 4\sigma^1; \;\; ^2\Sigma^+, \;\; ^2\Delta, \;\; ^2\Sigma^-, \text{ and } ^4\Sigma^-, \quad \text{(E.3,I)}$$

since $1\pi^2$ gives the states

$$^1\Sigma^+, \;\; ^1\Delta, \text{ and } ^3\Sigma^-. \quad \text{(E.3,II)}$$

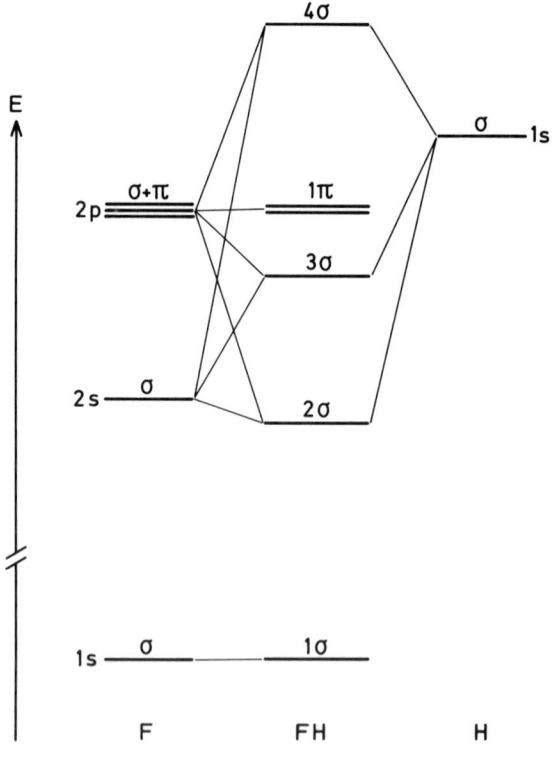

Fig. E.3,I. MO diagram for FH.

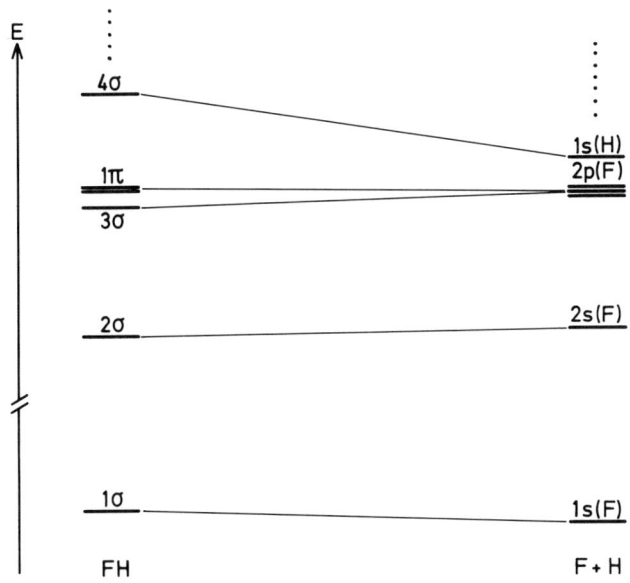

Fig. E.3,II. Orbital correlation diagram for the dissociation of FH.

4.

$$[E(F) + E(H)] - [E(F^-) + E(H^+)] = \text{EA (F)} - \text{IP (H)}$$
$$= -10.16 \text{ eV}$$

$$[E(F) + E(H^+)] - [E(F^+) + E(H)] = \text{IP (H)} - \text{IP (F)}$$
$$= -3.63 \text{ eV}$$

where IP is the ionization potential and EA the electron affinity. The experimental dissociation product of FH is thus F ┼ H and of FH^+ it is $\text{F} + \text{H}^+$.

5.

$$\text{F } (2p^6); \; {}^1S$$
$$\text{F}^- (2p^5); \; {}^2P$$
$$\text{F}^+ (2p^4); \; {}^3P, \; {}^1D, \; {}^1S$$
$$\text{H } (1s^1); \; {}^2S$$

When binding occurs,

$$F\ (2p^5) + H\ (1s^1) \rightarrow {}^1\Sigma^+,\ {}^3\Sigma^+,\ {}^1\Pi,\ {}^3\Pi$$

$$F^-\ (2p^6) + H^+ \rightarrow {}^1\Sigma^+$$

$$F\ (2p^5) + H^+ \rightarrow {}^2\Sigma^+,\ {}^2\Pi$$

$$F^+\ (2p^4) + H\ (1s^1) \rightarrow {}^2\Sigma^+(2),\ {}^2\Sigma^-,\ {}^4\Sigma^-,\ {}^2\Pi(2),\ {}^4\Pi,\ {}^2\Delta \quad (E.3,III)$$

In deriving Eq. (E.3,III), we have used that p^2 and p^4 give the same atomic terms. The atomic electronic configuration p^2 corresponds to the molecular electronic configurations and terms:

$$\pi^2 + \sigma^1 \pi^1 + \sigma^2 \rightarrow {}^1\Sigma^+,\ {}^1\Delta,\ {}^3\Sigma^- + {}^1\Pi,\ {}^3\Pi + {}^1\Sigma^+ \quad (E.3,IV)$$

Coupling of the molecular terms of Eq. (E.3,IV) with the ${}^2\Sigma^+$ term of H gives Eq. (E.3,III). The number in parentheses indicates the number of molecular terms of the same symmetry.

6. See Figs E.3,III and E.3,IV.
7. It follows from the answer to question 5 that both the dissociation products F^- (p^6) and F (p^5) + H (s^1) correlate with molecular states of symmetry ${}^1\Sigma^+$. There is thus an avoided crossing in Fig. E.3,III between the curves labeled 1 and 2, both of which have ${}^1\Sigma^+$ symmetry.
8. In a restricted Hartree-Fock calculation FH would dissociate incorrectly into F^- + H^+ since the $2p$ atomic orbital energy of fluorine is lower than the hydrogen $1s$ orbital energy. FH^+ would in a restricted Hartree-Fock calculation dissociate correctly into F + H^+. In CI calculations where appropriate configurations are included, both FH and FH^+ would, of course, dissociate correctly.

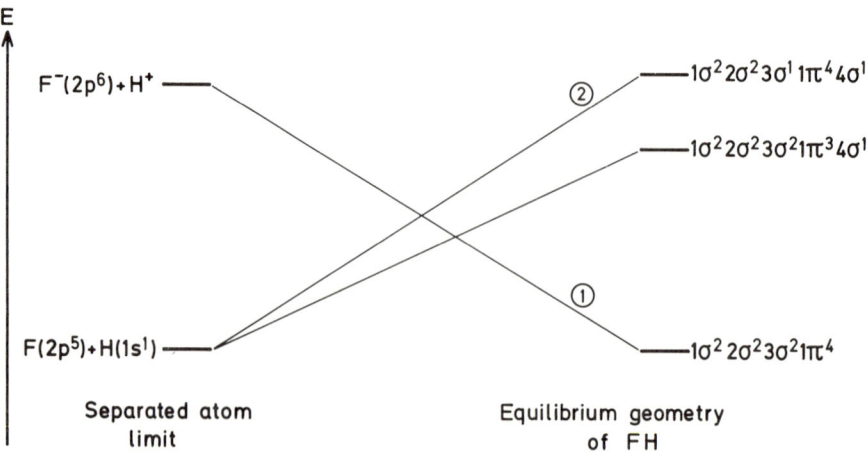

Fig. E.3,III. Configuration correlation diagram for the dissociation of FH.

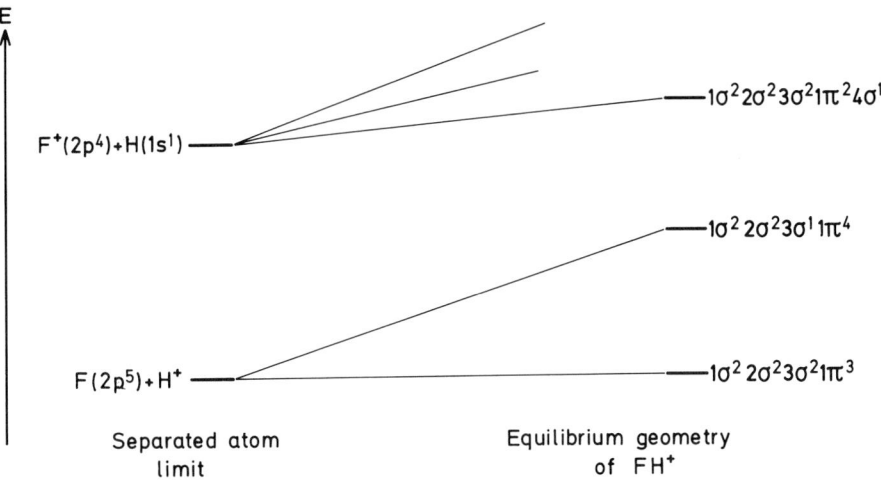

Fig. E.3,IV. Configuration correlation diagram for the dissociation of FH^+.

9. In the ground state of FH all electrons occupy bonding and nonbonding orbitals (see Fig. E.3,I) and a large potential energy minimum is therefore observed. The excited configurations are obtained from the ground configuration by exciting an electron from the (non-) bonding $(1\pi)3\sigma$ to the strongly antibonding 4σ orbital. The state therefore gets predominantly antibonding character and becomes repulsive.

   Excited states can of course have a potential energy minimum and be bound states if the dominant configuration has bonding character.

E.4 1. See Fig. E.4,I.
   2. See Fig. E.4,II.
   3. core $3\sigma_g^2\,1\pi_u^4\,1\pi_g^2$; $^3\Sigma_g^-$, $^1\Delta_g$, $^1\Sigma_g^+$ where the core, $1\sigma_g^2\,1\sigma_u^2\,2\sigma_g^2\,2\sigma_u^2$, will be referred to as "core" in the rest of this solution. The ground state is of $^3\Sigma_g^-$ symmetry (Hund's rule).
   4. core $3\sigma_g^2\,1\pi_u^3\,1\pi_g^3$; $^1\Sigma_u^+$, $^3\Sigma_u^+$, $^1\Sigma_u^-$, $^3\Sigma_u^-$, $^1\Delta_u$, $^3\Delta_u$.
   5.

		Equilibrium distance	United atom
V_π:	core	$3\sigma_g^2\,1\pi_u^3\,1\pi_g^3$	$1s^2 2s^2 2p^5 3s^2 3p^2 3d^3$
Ry:	core	$3\sigma_g^2\,1\pi_u^4\,1\pi_g^1\,2\pi_u^1$	$1s^2 2s^2 2p^6 3s^2 3p^3 3d^1$

The energy of the electronic configuration of the Ry configuration is much lower than the energy for the V_π configuration in the united atom limit and an avoided crossing therefore has to occur between

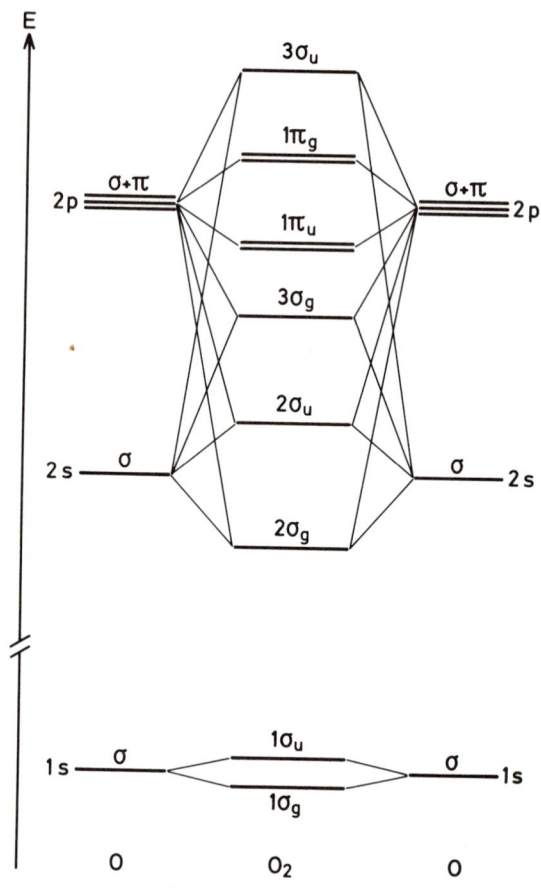

Fig. E.4,I. MO diagram of O_2.

the equilibrium distance and the united atom limit as indicated in the configuration correlation diagram of Fig. E.4,III.

6. The dominant electronic configurations are indicated in Fig. E.4,IV.

7. The $^3\Delta_u$ states originate from the same electronic configurations as those used to describe the avoided crossing for the $^3\Sigma_u^-$ states. Since a very small molecular term splitting is expected for Rydberg configurations (diffuse configuration) and only a moderate-sized molecular term splitting normally occurs for valence configurations, the location of the $^3\Delta_u$ and $^3\Sigma_u^-$ avoided crossing should be at about the same internuclear distance.

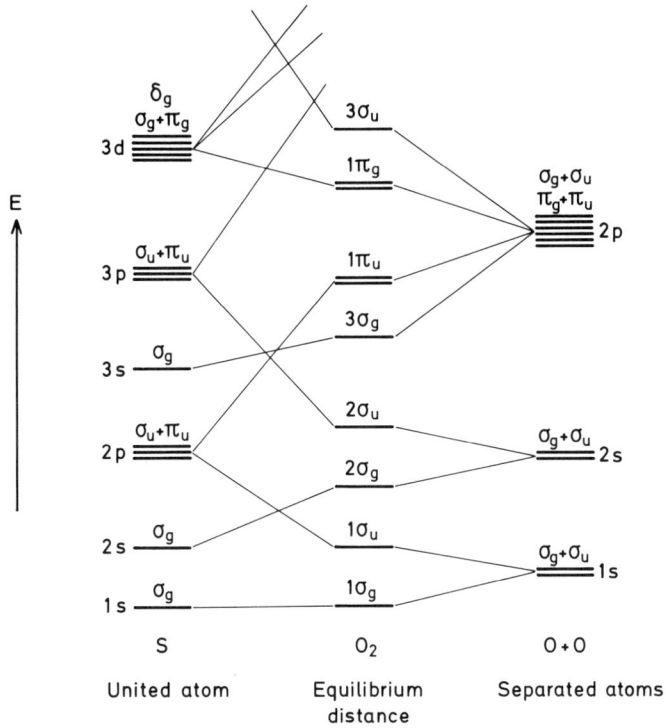

Fig. E.4,II. Orbital correlation diagram for O_2 with indication of the united atom limit, the equilibrium distance, and the separated atom limit.

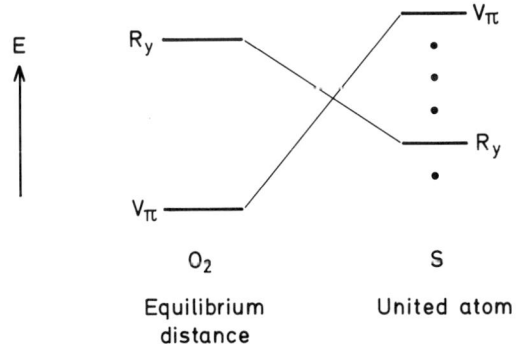

Fig. E.4,III. Configuration correlation diagram for the lowest two states of $^3\Sigma_u^-$ symmetry in O_2.

8. (a) Nuclear repulsion dominates at small internuclear distances.
(b) The $X^3\Sigma_g^-$ state has two electrons in the weakly antibonding orbital $1\pi_g$. Four electrons are in the weakly bonding orbital $1\pi_u$. The bonding and antibonding characters about balance for the residual electrons. The $X^3\Sigma_g^-$ state therefore has totally bonding characters and shows a potential energy minimum.
(c) The dominant configurations of the lowest states of $^3\Sigma_g^-$ and $^3\Delta_u$ symmetry are obtained from the ground-state electronic configuration by removal of one electron from the weakly bonding $1\pi_u$ orbital to the weakly antibonding $1\pi_g$ orbital. The lowest $^3\Sigma_u^-$ and $^3\Delta_u$ states therefore have less bonding character than the ground state.
(d) The bumps occur because of the avoided crossings in this region.
(e) The next lowest states of $^3\Sigma_u^-$ and $^3\Delta_u$ symmetry all have dominant Rydberg character. The Rydberg configurations differ from the ground-state electronic configuration by removal of an electron from the weakly antibonding $1\pi_g$ orbital to a basically nonbonding Rydberg orbital ($2\pi_u$ in $^3\Sigma_u^-$ and $^3\Delta_u$). The next lowest states of $^3\Sigma_u^-$ and $^3\Delta_u$ are thus more bonding than the $X^3\Sigma_g^-$ ground state and a deeper and more pronounced potential energy minimum is observed.

9. The z- and (x,y)-coordinates transform as Σ_u^+ and Π_u, respectively. The states that may be reached in a dipole allowed transition from the $X^3\Sigma_g^-$ state therefore must have either $^3\Sigma_u^-$ or $^3\Pi_u$ symmetry.

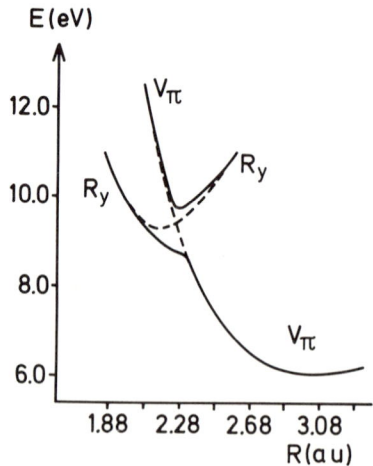

Fig. E.4,IV. The dominant electronic configuration for the various parts of the potential energy curves in Fig. E.4,1 (Reprinted from Chem. Phys. Lett. by permission of North-Holland Publishing Company and the authors).

Thus only the $^3\Sigma_u^-$ states in Fig. E.4,2 are reached in a dipole allowed transition.

The transition moments from the ground state to the part of the potential energy curves where Rydberg configurations dominate will be very small (the Rydberg state and the ground-state occupy very different parts of the space). The transition moments from the ground state to the part of the potential energy curves where the valence configurations dominate must be reasonably large (both states are valence configurations). A very rough estimate of the dependence of the transition moment M of the internuclear distance is given in Fig. E.4,V. Accurate transition moments have been evaluated by Buenker *et al.* (1976).

E.5   1.   The electronic configuration of O_2 is (see, e.g., Fig. E.4,I)

$$1\sigma_g^2 \, 1\sigma_u^2 \, 2\sigma_g^2 \, 2\sigma_u^2 \, 3\sigma_g^2 \, 1\pi_u^4 \, 1\pi_g^2$$

where the frontier orbital $1\pi_g$ is the antibonding linear combination of the $2p\pi$ orbitals on the oxygen atoms (see Fig. E.4,I):

$$1\pi_g^x = N_g(2px_A - 2px_B) \qquad (E.5,I)$$

$$1\pi_g^y = N_g(2py_A - 2py_B) \qquad (E.5,II)$$

and $N_g = (2 - 2S)^{1/2}$, $S = \langle 2px_A | 2px_B \rangle = \langle 2py_A | 2py_B \rangle$.

The $2p\pi$ orbitals ($2px_A, 2py_B$, etc.) are centered on the two oxygen atoms, A and B.

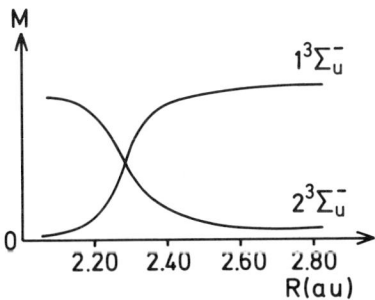

Fig. E.4,V. A plot (schematic) of the transition moments as a function of the internuclear distance for the transition from the ground state $(X\,^3\Sigma_g^-)$ to the lowest two excited states of $^3\Sigma_u^-$ symmetry in O_2.

A cut of the electronic charge distribution of $1\pi_g^x$ in the x-z plane is given in Fig. E.5,I. A cut of the electronic charge distribution of $1\pi_g^y$ in the y-z plane is identical to Fig. E.5,I.

2. In Fig. E.5,II we have sketched the bonding situation of the $1\pi_{g1}^x$ and $1\pi_{g2}^x$ oxygen MO's. Maximum overlap is obtained when the O_2 dimer is formed as indicated in Fig. E.5,II. In the situation of Fig. E.5,II both the $1\pi_g^x$ and $1\pi_g^y$ overlap. Nearly σ-type overlap occurs between the $1\pi_g^x$ orbitals (strictly speaking, the overlap is more like overlaps between d orbitals) and π-type overlap occurs between the $1\pi_g^y$ orbitals. Thus, the O_2 dimer is rectangular with point group symmetry D_{2h}.

3. Bonding MO's:

$$\psi_1 = N_1(1\pi_{g1}^x - 1\pi_{g2}^x) = N_1(2px_{A1} - 2px_{B1} - 2px_{A2} + 2px_{B2})$$

$$(E.5,III)$$

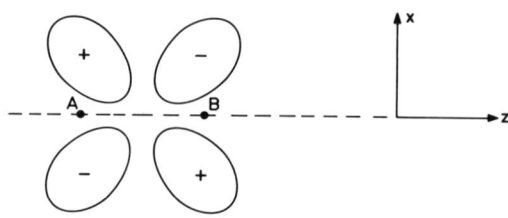

Fig. E.5,I. A contour plot of the electronic charge distribution of the $1\pi_g^x$ orbital.

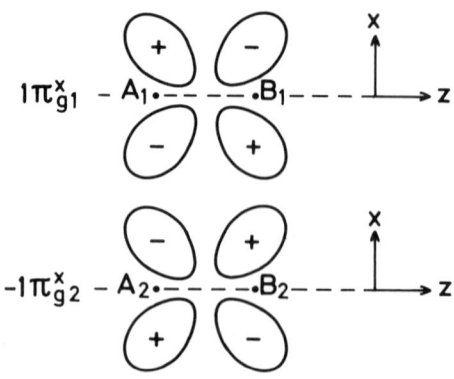

Fig. E.5,II. A contour plot that illustrates the bonding situation that occurs when $1\pi_g^x$ orbitals of two oxygen molecules approach each other.

Table E.5,I
The Transformation Properties of the MO's of the O_2 Dimer in the D_{2h} Point Group

E	$C_2(z)$	$C_2(y)$	$C_2(x)$	i	$\sigma(xy)$	$\sigma(xz)$	$\sigma(yz)$	IR
ψ_1	ψ_1	$-\psi_1$	$-\psi_1$	$-\psi_1$	$-\psi_1$	ψ_1	ψ_1	b_{1u}
ψ_2	$-\psi_2$	$-\psi_2$	ψ_2	ψ_2	$-\psi_2$	$-\psi_2$	ψ_2	b_{3g}
ψ_3	$-\psi_3$	ψ_3	$-\psi_3$	ψ_3	$-\psi_3$	ψ_3	$-\psi_3$	b_{2g}
ψ_4	ψ_4	ψ_4	ψ_4	$-\psi_4$	$-\psi_4$	$-\psi_4$	$-\psi_4$	a_u

$$\psi_2 = N_2(1\pi_{g1}^y + 1\pi_{g2}^y) = N_2(2py_{A1} - 2py_{B1} + 2py_{A2} - 2py_{B2})$$

$$(E.5,IV)$$

Antibonding MO's:

$$\psi_3 = N_3(1\pi_{g1}^x + 1\pi_{g2}^x) = N_3(2px_{A1} - 2px_{B1} + 2px_{A2} - 2px_{B2})$$

$$(E.5,V)$$

$$\psi_4 = N_4(1\pi_{g1}^y - 1\pi_{g2}^y) = N_4(2px_{A1} - 2py_{B1} - 2py_{A2} + 2py_{B2})$$

$$(E.5,VI)$$

Using the coordinate system in Fig. E.5,III we obtain from Eqs.
(E.5,III-VI) the transformation properties of ψ_i given in Table E.5,I.

4. When the O_2 molecules approach each other as indicated in Fig.
E.5,III the $1\pi_g^x$ orbitals overlap more strongly than the $1\pi_g^y$ orbitals.
As discussed in question 2 we may classify the MO's of the O_2
dimer as $1b_{1u} \sim \sigma$, $1b_{3g} \sim \pi$, $1a_u \sim \pi^*$, and $1b_{2g} \sim \sigma^*$. Since the
bonding-antibonding splitting is proportional to the overlap, rule F,
we obtain the MO diagram of Fig. E.5,IV.

Both the bonding orbitals, $1b_{1u}$ and $1b_{3g}$, contain two electrons,
and a frontier orbital description would thus predict the O_2 dimer to
be stable relative to two isolated O_2 molecules on the basis of rule K.3
of the introduction to this set.

E.6 1. See Fig. E.6,I.
2. $1\sigma_g^2 1\sigma_u^2 2\sigma_g^2 2\sigma_u^2 3\sigma_g^2 1\pi_u^4$; $^1\Sigma_g^+$
3. $1\sigma_g^2 1\sigma_u^2 2\sigma_g^2 2\sigma_u^2 3\sigma_g^2 1\pi_u^3$; $^2\Pi_u$
    $1\sigma_g^2 1\sigma_u^2 2\sigma_g^2 2\sigma_u^2 3\sigma_g^1 1\pi_u^4$; $^2\Sigma_g^+$
    $1\sigma_g^2 1\sigma_u^2 2\sigma_g^2 2\sigma_u^1 3\sigma_g^2 1\pi_u^4$; $^2\Sigma_u^+$
4. 11.40 eV $\sim$ $^2\Pi_u$ $(1\pi_u)$
    16.36 eV $\sim$ $^2\Sigma_g^+$ $(3\sigma_g)$
    18.38 eV $\sim$ $^2\Sigma_u^+$ $(2\sigma_u)$
5. The $2\sigma_g$ orbital is predominantly a bonding combination of the $2s$
    orbitals on the carbon atoms. The $2\sigma_g$ orbital energy is therefore

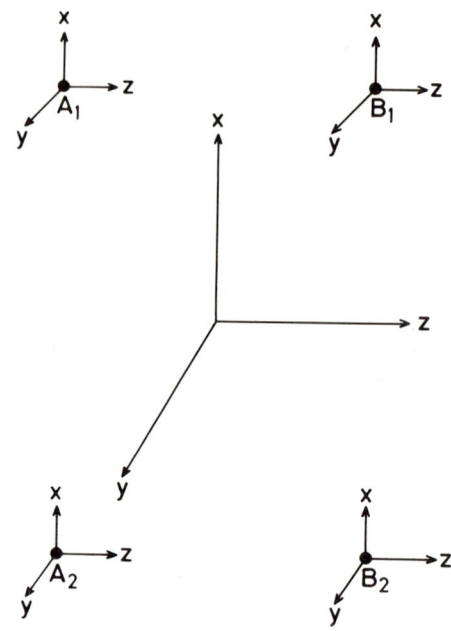

Fig. E.5,III. The coordinate system used in the description of the formation of a O_2 dimer.

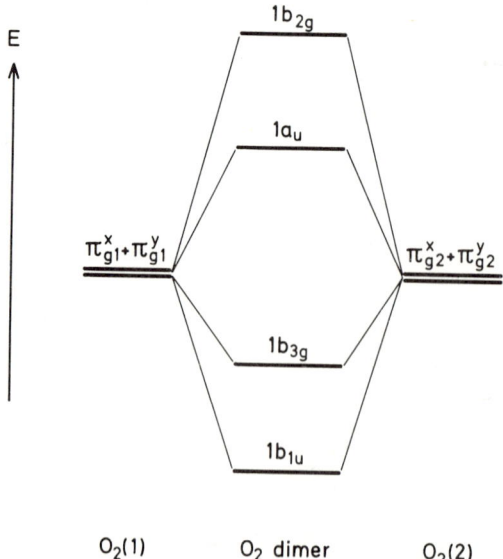

Fig. E.5,IV. MO diagram displaying the MO's of the O_2 dimer formed from $1\pi_g$ orbitals of O_2.

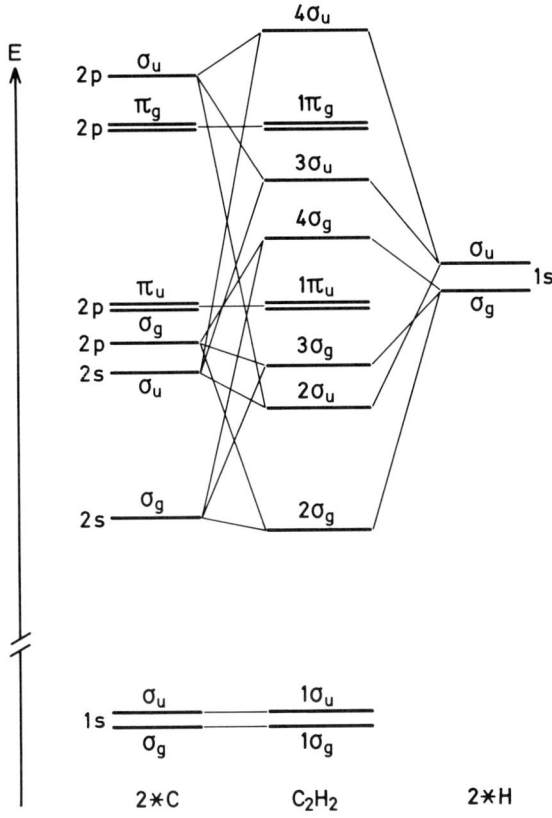

Fig. E.6,I. MO diagram for acetylene.

substantially lower than the $2s$ orbital energy of the carbon atom. A reasonable estimate of its orbital energy is -25 to -30 eV. The $1\sigma_u$ and $1\sigma_g$ orbitals represent a very weak bonding and antibonding combination of the $1s$ orbitals on carbon. A reasonable estimate of their orbital energies is -311 to -309 eV and -309 to -307 eV for the $1\sigma_g$ and $1\sigma_u$ orbitals, respectively. Correlation effects may change the absolute energies of the $1\sigma_g$ and $1\sigma_u$ orbitals but not their relative positions.

6. Diacetylene may be thought of as "formed" by adding two acetylene molecules such that one C-H bond on each acetylene molecule is broken and one new sigma C-C bond is formed. The $1\pi_u$ orbital in acetylene is the only occupied orbital of π symmetry in acetylene. When diacetylene is formed, a bonding and antibonding combination of the $1\pi_u$ orbitals is formed. Thus, the ionization

potential at 10.17 eV represents the bonding combination of the $1\pi_u$ orbitals of acetylene and must be assigned to the state $^2\Pi_u(1\pi_u)$ of $C_4H_2^+$. The ionization potential at 12.62 eV represents the antibonding combination and must be assigned to the $^2\Pi_g(1\pi_g)$ state of $C_4H_2^+$.

Two σ bonds are broken and a new one formed in diacetylene relative to two acetylene molecules. Nothing simple can therefore be predicted concerning the position of the σ orbitals in diacetylene relative to those of acetylene, that is, concerning the relative ordering of the rest of the ionization potentials of C_4H_2.

E.7 1. The ordering of the $3\sigma_g$ and $1\pi_u$ orbitals is difficult to decide from simple MO arguments and may be the opposite of that in Fig. E.7,I.

2. $1\sigma_g^2\,1\sigma_u^2\,2\sigma_g^2\,2\sigma_u^2\,3\sigma_g^2\,1\pi_u^4;\ ^1\Sigma_g^+$

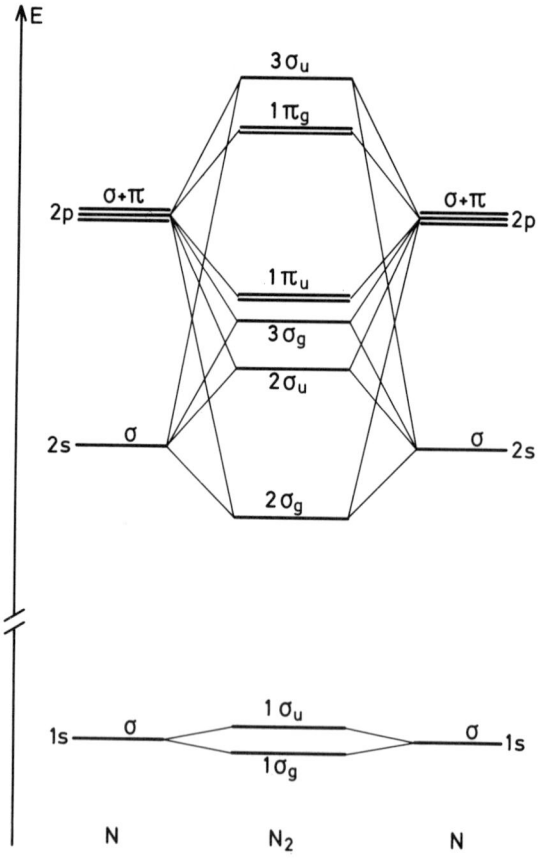

Fig. E.7,I. MO diagram for N_2.

3. $1\sigma_g^2 \, 1\sigma_u^2 \, 2\sigma_g^2 \, 2\sigma_u^2 \, 3\sigma_g^2 \, 1\pi_u^3 \, ; \, ^2\Pi_u$

 $1\sigma_g^2 \, 1\sigma_u^2 \, 2\sigma_g^2 \, 2\sigma_u^2 \, 3\sigma_g^1 \, 1\pi_u^4 \, ; \, ^2\Sigma_g^+$

 $1\sigma_g^2 \, 1\sigma_u^2 \, 2\sigma_g^2 \, 2\sigma_u^1 \, 3\sigma_g^2 \, 1\pi_u^4 \, ; \, ^2\Sigma_u^+$

 $1\sigma_g^2 \, 1\sigma_u^2 \, 2\sigma_g^1 \, 2\sigma_u^2 \, 3\sigma_g^2 \, 1\pi_u^4 \, ; \, ^2\Sigma_g^+$

4. $15.57 \text{ eV} \sim {}^2\Pi_u \quad (1\pi_u)$

 $16.69 \text{ eV} \sim {}^2\Sigma_g^+ \quad (3\sigma_g)$

 $18.75 \text{ eV} \sim {}^2\Sigma_u^+ \quad (2\sigma_u)$

5. The Hartree-Fock calculation gives the same assignment (Koopmans' theorem) as in question 4.

6. The assignment based on the Hartree-Fock calculation does not give the correct ordering of the $3\sigma_g$ and $1\pi_u$ orbital. Koopmans' theorem uses the orbitals of the parent molecule to determine the total energy of the ionized molecule. The orbital relaxation that is not included in Koopmans' theorem is important, and it can be evaluated if separate SCF calculations are performed on the parent molecule and the ionized species. The ionization potential obtained in this way is referred to in the literature as the ΔSCF value. Cade *et al.* (1966) have computed ΔSCF values for the N_2 ionization potentials. They still obtain the incorrect assignment. The correct assignment is obtained when correlation effects are included explicitly in both calculations on the N_2 and N_2^+ molecules. The correct ordering of the $^2\Pi_u$ and $^2\Sigma_g^+$ states has been obtained by Cederbaum and von Niessen (1975) and Purvis and Öhrn (1974) using Green's function methods.

E.8 1. See Fig. E.8,I.

 2. We will label 1σ to 4σ and 1π as core orbitals (core).

 core $5\sigma^2 2\pi^4 \, ; \, {}^1\Sigma^+$

 3. core $5\sigma^2 2\pi^3 \, ; \, {}^2\Pi$

 core $5\sigma^1 2\pi^4 \, ; \, {}^2\Sigma^+$

 4. $12.8 \text{ eV} \sim {}^2\Pi \quad (2\pi)$

 $16.5 \text{ eV} \sim {}^2\Sigma^+ \quad (5\sigma)$

 5. The spin-orbit coupling couples the orbital angular momentum and spin angular momentum. The component of the total angular momentum along the internuclear axes, J_z, is preserved. The numerical value of the quantum number for J_z is

$$\Omega = |\Lambda + M_S|$$

where Λ denotes the projection of the orbital angular momentum on the internuclear axes and M_S is the projection of the spin angular momentum on the internuclear axes. Thus we obtain the states $^2\Pi_{1/2}$

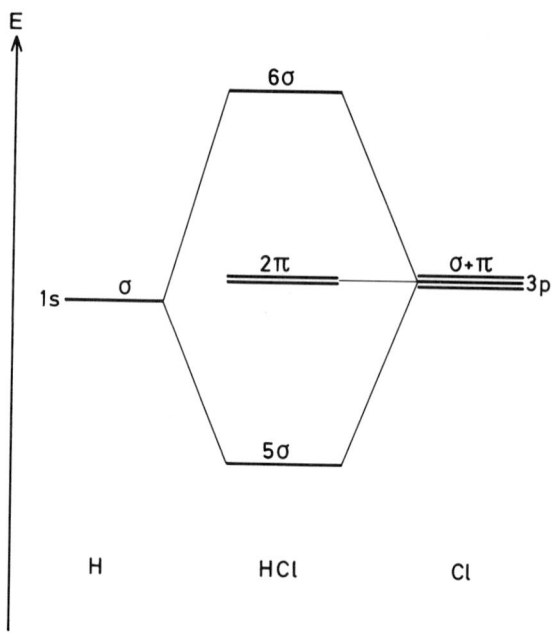

Fig. E.8,I. MO diagram of HCl.

and $^2\Pi_{3/2}$ where the $^2\Pi_{3/2}$ represents the peak at 12.7 eV, assuming an inverted multiplet splitting.

E.9 1. The MO diagram is given in Fig. E.9,I.
    2. We have labeled $1\sigma_g$, $2\sigma_g$, and $1\sigma_u$ as core orbitals.

$$\text{core } 3\sigma_g^2\, 2\sigma_u^2\, 4\sigma_g^2\, 1\pi_u^4\, 3\sigma_u^2\, 1\pi_g^4\,;\; {}^1\Sigma_g^+$$

3.

$$\text{core } 3\sigma_g^2\, 2\sigma_u^2\, 4\sigma_g^2\, 1\pi_u^4\, 3\sigma_u^2\, 1\pi_g^3\,;\; {}^2\Pi_g$$

$$\text{core } 3\sigma_g^2\, 2\sigma_u^2\, 4\sigma_g^2\, 1\pi_u^4\, 3\sigma_u^1\, 1\pi_g^4\,;\; {}^2\Sigma_u^+$$

$$\text{core } 3\sigma_g^2\, 2\sigma_u^2\, 4\sigma_g^2\, 1\pi_u^3\, 3\sigma_u^2\, 1\pi_g^4\,;\; {}^2\Pi_u$$

$$\text{core } 3\sigma_g^2\, 2\sigma_u^2\, 4\sigma_g^1\, 1\pi_u^4\, 3\sigma_u^2\, 1\pi_g^4\,;\; {}^2\Sigma_g^+$$

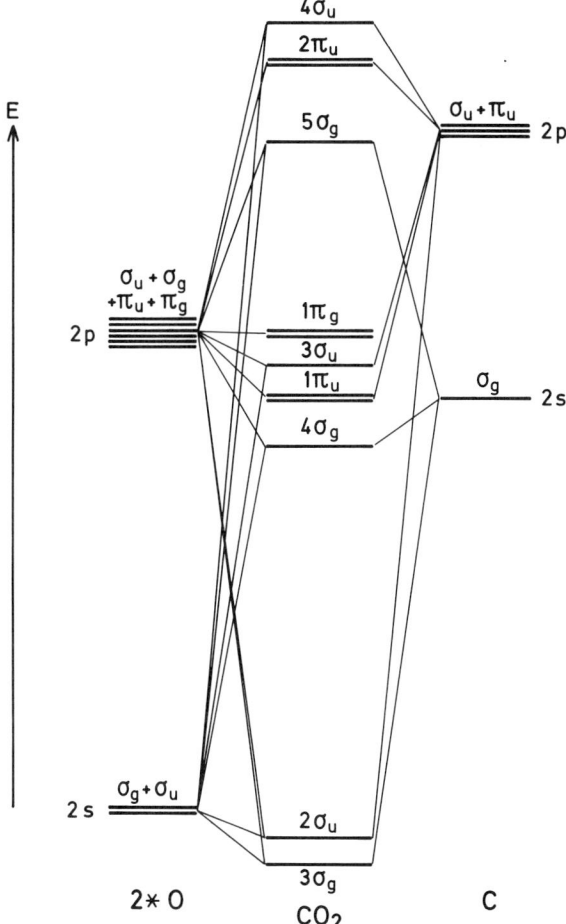

Fig. E.9,I. MO diagram of CO_2.

4. Only the $^2\Pi$ states split as a result of the spin-orbit coupling, and we therefore have the assignment:

$$13.8 \text{ eV} \sim {}^2\Pi_g \quad (1\pi_g)$$
$$17.6 \text{ eV} \sim {}^2\Pi_u \quad (1\pi_u)$$
$$18.1 \text{ eV} \sim {}^2\Sigma_u^+ \quad (3\sigma_u)$$
$$19.4 \text{ eV} \sim {}^2\Sigma_g^+ \quad (4\sigma_g)$$

The spin-orbit coupling splits the molecular terms into the levels (see solution to question 5 of problem E.8).

$$^2\Pi_g \rightarrow {}^2\Pi_{(3/2)u} \text{ and } {}^2\Pi_{(1/2)g}$$
$$^2\Pi_u \rightarrow {}^2\Pi_{(3/2)u} \text{ and } {}^2\Pi_{(1/2)u}$$

5. We expect a $^2\Sigma_u^+$ and a $^2\Sigma_g^+$ state to be located at about 35 eV (the ionization potential of the oxygen $2s$ orbital) with the $^2\Sigma_u^+$ state having the numerically smallest ionization potential.

6. The same assignment as in question 4. Notice that the sequence of the orbital energies is in agreement with the actual assignment. In the MO diagram it is difficult to establish a definite sequence of the orbital energies $3\sigma_u$ and $1\pi_u$. Very accurate calculations are required in order to obtain a reliable sequence of these orbital energies.

7. The choice of coordinate systems is indicated in Fig. E.9,II. It then follows from Table E.9,I that the ground state of CO_2^- may be obtained by adding an electron in the $2\pi_u$ or in the $5\sigma_g$ orbital. The lowest excited state is obtained by adding an electron in the $2\pi_u$ orbital.

8. According to Table E.9,I the ground and lowest excited state is obtained by adding an electron in the $2\pi_u$ Hartree-Fock orbital ($^2A_1$ and $^2B_1$). The second excited state is obtained by adding an electron in the $5\sigma_g$ orbital ($^2A_1$) and the third excited state by adding an electron in $4\sigma_u(^2B_2)$.

9. Koopmans' theorem gives a negative electron affinity since $\varepsilon_{2\pi_u}$ is positive, in disagreement with the experimental data.

Table E.9.I
Correspondence between Coordinate Axis[a] and IR's when Symmetry is Reduced from $D_{\infty h}$ to C_{2v}

C_{2v}	z	y	x	$a_1(y^2)$	$b_2(y)$	$a_1(x) + b_1(z)$
$D_{\infty h}$	y	z	x	$\sigma_g(z^2)$	$\sigma_u(z)$	$\pi(x,y)$

[a] The function in parentheses denotes the coordinate according to which the orbital transforms.

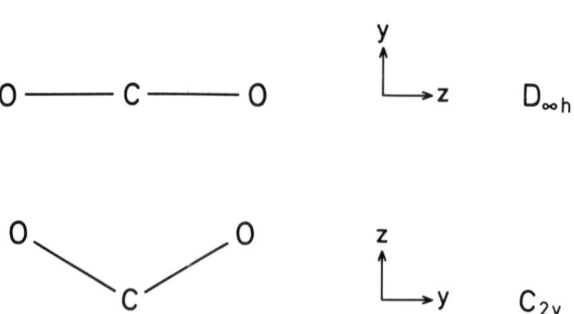

Fig. E.9,II. The coordinate system that is used for CO_2^- in the linear and in the bent conformation.

E.10 1. If we choose the coordinate system as indicated in Fig. E.10,I, the
transformation properties of the valence AO's are as given in Table
E.10,I. The p orbitals on the oxygens are divided into three $p\sigma$
orbitals directed along the bonds and six $p\pi$ orbitals perpendicular
to the bond [three in the plane of the paper ($2p\pi_{\parallel}$), three
perpendicular to the plane of the paper ($2p\pi_{\perp}$)].

2. The $1e''$ and $1a_2'$ are nonbonding MO's.
The e' symmetry orbital from the $2p\pi_{\parallel}$ (O) set will only have a minor
overlap with the $2p_x$ and $2p_y$ orbitals on C. The three a_1' and the three
e' MO's are placed in the MO diagram, in Fig. E.10,II, according to
rule G taking into account that the a_1' overlap is much larger than the
e' overlap. The a_2'' orbitals have π-type overlap and thus show a
rather small splitting.

3. core $1a_1'^2 1e'^4 2a_1'^4 1a_2''^2 2e'^4 3e'^4 1e''^4 1a_2'^2 ; {}^1A_1'$

4. Since the MO energies fall energywise into three rather distinct
groups (see Fig. E.10,II), we can make the following assignment:

I corresponds to ionization out of the orbitals $1a_2'$, $1e''$, and $3e'$.
II, III correspond to ionization out of the orbitals $2a_1'$, $1a_2''$, and $2e'$.
IV corresponds to ionization out of the orbitals $1a_1'$ and $1e'$.

Table E.10,I
Characters for the Symmetry Equivalent Sets of CO_3^{2-} Orbitals in D_{3h}

Orbitals	E	$2C_3$	$3C_2$	σ_h	$2S_3$	$3\sigma_v$	IR in D_{3h}
$2s(C)$	1	1	1	1	1	1	a_1'
$2p_z(C)$	1	1	-1	-1	-1	1	a_2''
$\{2p_x, 2p_y\}(C)$	2	-1	0	2	-1	0	e'
$2s(O)$	3	0	1	3	0	1	$a_1' + e'$
$2p\sigma(O)$	3	0	1	3	0	1	$a_1' + e'$
$2p\pi_{\parallel}(O)$	3	0	-1	3	0	-1	$e' + a_2'$
$2p\pi_{\perp}(O)$	3	0	-1	-3	0	1	$e'' + a_2''$

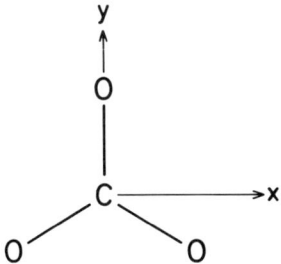

Fig. E.10,I. The geometry and coordinate system for CO_3^{2-}.

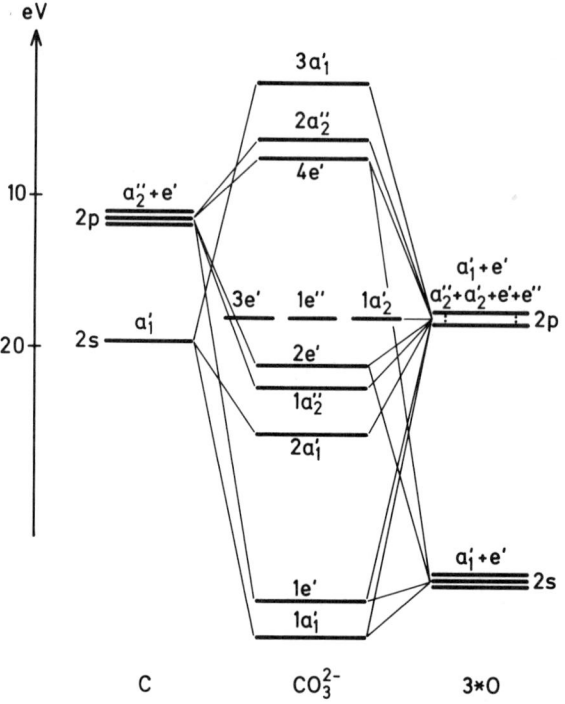

Fig. E.10,II. MO diagram for CO_3^{2-}.

5. Mn^{2+} in its ground state has the electronic configuration $[Ar]3d^5$ and the ground-state term is ^6S.

6. All electrons are paired in the MO diagram in Fig. E.10,II; that is,

$$S(CO_3^{2-}) = 0 \qquad \text{or} \qquad S(Mn^{2+}CO_3^{2-}) = \frac{5}{2}$$

The magnitude of the spin magnetic moment is

$$\mu = \frac{e}{mc}\sqrt{S(S+1)}$$

and thus

$$\mu(Mn^{2+}CO_3^{2-}) = \frac{e}{mc}\frac{\sqrt{35}}{2}$$

E.11 1. Let us label the three nitrogen atoms N(1), N(2), and N(3) with N(2) at the center. The point group symmetry is $D_{\infty h}$. The orbitals on N(1) + N(3) form equivalent sets. Due to the large internuclear distance between N(1) and N(3) the orbital energies of the symmetry orbitals are nearly the orbital energies of the atomic nitrogen orbitals. The MO orbital diagram is displayed in Fig. E.11,I. In constructing this diagram we have used that the overlap between the N(2) and N(1) + N(3) orbitals increases in the order $\sigma_u < \pi_u < \sigma_g$.

On the basis of the simple rules E-H, it is difficult to estimate the relative ordering of the $1\pi_u$, $2\sigma_u$, $2\sigma_g$ and of the $2\pi_u$, $3\sigma_u$, $3\sigma_g$ orbitals. Rule G has been used to place the $2\sigma_u$ and $2\sigma_g$ orbitals.

2. core $1\sigma_g^2 \, 1\sigma_u^2 \, 1\pi_u^4 \, 2\sigma_g^2 \, 2\sigma_u^2 \, 1\pi_g^4$; $^1\Sigma_g^+$

3. $\cdots 2\sigma_g^2 \, 2\sigma_u^2 \, 1\pi_g^3 \, 2\pi_u^1$; $^1\Sigma_u^+$, $^3\Sigma_u^+$, $^1\Sigma_u^-$, $^3\Sigma_u^-$, $^1\Delta_u$, $^3\Delta_u$

   $\cdots 2\sigma_g^2 \, 2\sigma_u^1 \, 1\pi_g^4 \, 2\pi_u^1$; $^1\Pi_g$, $^3\Pi_g$

   $\cdots 2\sigma_g^1 \, 2\sigma_u^2 \, 1\pi_g^4 \, 2\pi_u^1$; $^1\Pi_u$, $^3\Pi_u$

4. We see from Fig. E.11,II that the excited states fall into three groups. Except for the $^1\Sigma_u^+$ states, this grouping corresponds to states originating from the three singly excited electronic

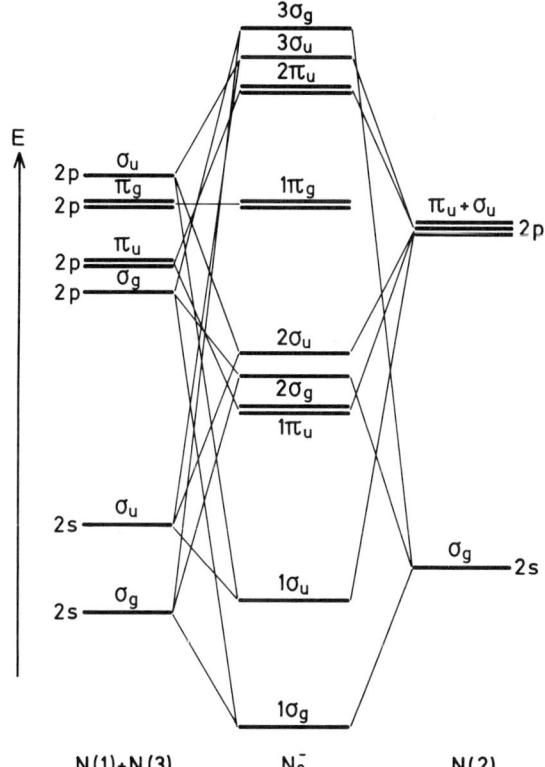

Fig. E.11,I. MO diagram for N_3^- .

configurations examined in question 3, and in a simple independent-particle model we thus expect the relative energies of the three HOMO's to be

$$\varepsilon(1\pi_g) > \varepsilon(2\sigma_u) \gtrsim \varepsilon(2\sigma_g)$$

with the energy difference between $2\sigma_u$ and $2\sigma_g$ being rather small.

5. Let us choose the coordinate system of the bent molecule (C_{2v}) such that the z-axis is pointing along the C_2 axis and the molecule is bending in the y-z plane. Then the HOMO $(1\pi_g)$ and LUMO $(2\pi_u)$ can be written as

$$1\pi_g^x = c(px_1 - px_3) \tag{E.11,I}$$

Fig. E.11,II. Energy diagram for the lowest states of N_3^-.

$$1\pi_g^z = c(pz_1 - pz_3) \qquad \text{(E.11,II)}$$

$$2\pi_u^x = b(px_1 + a(px_2) + px_3) \qquad \text{(E.11,III)}$$

$$2\pi_u^z = b(pz_a + a(pz_2) + pz_3) \qquad \text{(E.11,IV)}$$

where b and c are normalization constants and a is a negative LCAO coefficient since $2\pi_u$ is an antibonding orbital.

Table E.11,I
Transformation Properties of the Orbitals in Eqs. (E.11, I–IV) in the C_{2v} Point Group (see Fig. E.11, III)

Orbital	E	$C_2(z)$	$\sigma_v(xz)$	$\sigma_v'(yz)$	IR
$1\pi_g^x$	1	1	-1	-1	a_2
$1\pi_g^z$	1	-1	-1	1	b_2
$2\pi_u^x$	1	-1	1	-1	b_1
$2\pi_u^z$	1	1	1	1	a_1

6. Figure E.11,III shows how the geometry change will effect the HOMO and LUMO of the linear molecule. Since $d_\Theta < d$, that is, the two terminal atoms of N_3^- are closer to each other, the "antibonding" density in $1\pi_g^x$ and $1\pi_g^z$ will increase slightly when going from $D_{\infty h}$ to C_{2v} symmetry, resulting in a slight increase of the orbital energies of the a_2 and b_2 orbitals. Because of the σ-type antibonding character of the π_g^z orbitals we would expect that b_2 will increase more than a_2.

   Based on the same type of argument, $\varepsilon(2\pi_u^z) = \varepsilon(a_1)$ and $\varepsilon(2\pi_u^x) = \varepsilon(b_1)$ must decrease (larger bonding density). The $2\pi_u^z$ bonding overlap is of σ type and the $2\pi_u^x$ bonding overlap is of the π type and we thus have the Θ dependence indicated in Fig. E.11,IV.

7. Linear.

8. On the basis of Fig. E.11,IV it is not possible to determine the geometry of N_3^- in its first excited state. If a_1 decreases more than the sum of b_2 and a_2 increases, N_3^- is bent and the first excited state is

$$\cdots a_2^2 b_2^1 a_1^1 ; \; {}^3B_2$$

If that is not the case, N_3^- is linear and the first excited state is

$$\cdots 1\pi_g^3 2\pi_u^1 ; \; {}^3\Sigma_u^+ \text{ or } {}^3\Sigma_u^- \text{ or } {}^3\Delta_u$$

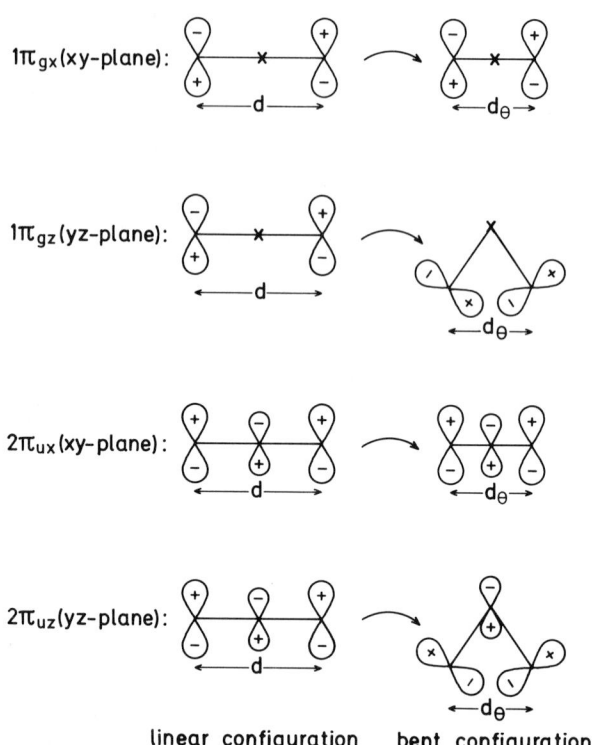

$1\pi_{gx}$(xy-plane):

$1\pi_{gz}$(yz-plane):

$2\pi_{ux}$(xy-plane):

$2\pi_{uz}$(yz-plane):

linear configuration   bent configuration

Fig. E.11,III. The change in "bonding densities" of the $1\pi_g$ and $2\pi_u$ orbitals of the linear N_3^- when the N_3^- molecule is bent.

E.12 1. The four $1s$ orbitals on H form an equivalent set in T_d which transforms according to the $a_1 + t_2$ irreducible representation. The $2p$ orbitals on carbon transform according to a t_2 representation. The orbital correlation diagram is displayed in Fig. E.12,I.

2. Ground state: $1a_1^2\,2a_1^2\,1t_2^6$; $^1A_1$
First excited state: $1a_1^2\,2a_1^2\,1t_2^5\,3a_1^1$; $^1T_2 + {}^3T_2$

3. The ground state, $^1A_1$, is spatially nondegenerate, and no Jahn-Teller distortion would be expected. The excited states $^1T_2$ and $^3T_2$ are spatially degenerate, and a Jahn-Teller distortion is expected.

4. $1a_1^2\,2a_1^2\,1t_2^5$; $^2T_2$

5. C_{3v}: Eight C_3 axes in T_d. The distortion can thus be performed in eight equivalent ways.
C_{2v}: Extension (contraction) of an H-H bond gives C_{2v} symmetry, thus six equivalent ways.
D_{2d}: pulling an H-H bond such that it is still orthogonal to the other

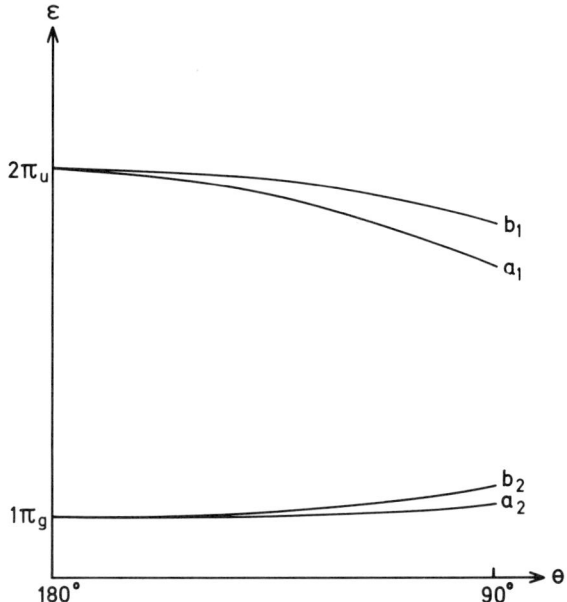

Fig. E.11,IV. The orbital energies of the $1\pi_g$ and $2\pi_u$ orbitals of the linear N_3^- molecule as a function of the N-N-N angle Θ.

H-H bond gives D_{2d} symmetry. Six different H-H bonds may be pulled, and D_{2d} symmetry may therefore be obtained in six different ways.

6.

When $T_d \rightarrow D_{2d}$ then $T_2 \rightarrow B_2 + E$

The ground-state term is $^2B_2$, since a ^2E state would be Jahn-Teller distorted, thereby leading to a lower symmetry than D_{2d}.

E.13 1. Let s_K denote an s orbital on hydrogen K, $(K = A, B)$. In the molecular region the symmetry of the states is determined from the microstate diagram in Table E.13,I.

(a)Since s orbitals are symmetric with respect to reflection in a plane containing the internuclear axis, the molecular states have Σ^+ symmetry.

Table E.13,I
Microstate Diagram for H_2 Correlating with Two
H(^2S) in the Separated Atom Limit

$M_L\backslash M_S$	1	0
0	$\lvert s_A s_B\rvert (^3\Sigma)$	$\lvert s_A \bar{s}_B\rvert\lvert\bar{s}_A s_B\rvert\ (^3\Sigma, {}^1\Sigma)$

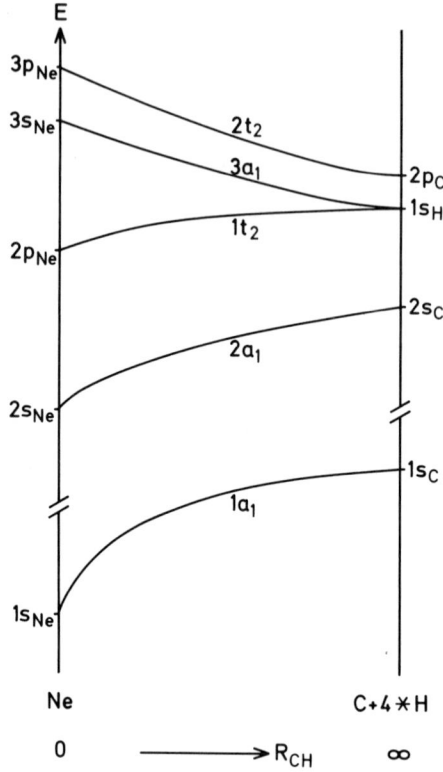

Fig. E.12,I. Orbital correlation diagram for CH_4 with indication of the separated atom and the united atom limit when T_d symmetry is conserved.

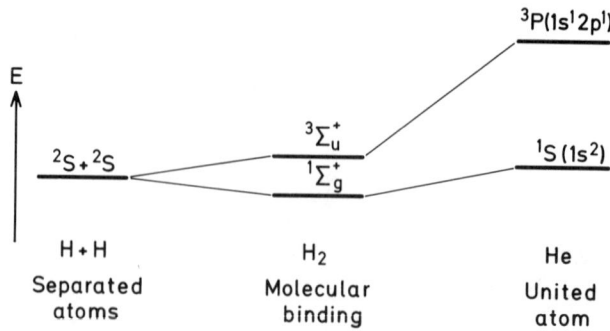

Fig. E.13,I. State correlation diagram for two hydrogen atoms, both in ^2S states.

(b)Let $\hat{i}$ denote the inversion operator; then

$$\hat{i}(1)s_A(1) = s_B(1)$$

$$\hat{i}|s_A s_B| = -|s_A s_B| \qquad \text{where} \quad \hat{i} = \hat{i}(1)\hat{i}(2)$$

Thus the triplet state is a $^3\Sigma_u^+$ state.

(c)The $M_L = 0$, $M_S = 0$ block contains a $^3\Sigma_u^+$ and a $^1\Sigma^+$ state. Since $\langle |s_A \bar{s}_B| |\hat{i}| |s_A \bar{s}_B| \rangle + \langle |\bar{s}_A s_B| |\hat{i}| |\bar{s}_A s_B| \rangle = 0$, the $^1\Sigma^+$ state must be of $^1\Sigma_g^+$ symmetry (the trace of the inversion operator $\hat{i}$ is independent of the representation used).

2. The lowest states of He that correlate with the $^1\Sigma_g^+$ and $^3\Sigma_u^+$ states of H_2 are $^1S(1s^2)$ and $^3P(1s^1 2p^1)$, respectively, and we obtain the correlation diagram in Fig. E.13,I.

3. **Table E.13,II**
Microstate diagram^a for H_2 Correlating with Two $H(^2P)$ in the Separated Atom Limit

$M_L \backslash M_S$	1	0
2	$\|p_{+1}^A p_{+1}^B\|(^3\Delta)$	$\|\bar{p}_{+1}^A p_{+1}^B\| \,\|p_{+1}^A \bar{p}_{+1}^B\|(^3\Delta,\,^1\Delta)$
1	$\|p_{+1}^A p_0^B\|(^3\Pi)$	$\|\bar{p}_{+1}^A p_0^B\| \,\|p_{+1}^A \bar{p}_0^B\|(^3\Pi,\,^1\Pi)$
	$\|p_0^A p_{+1}^B\|(^3\Pi)$	$\|\bar{p}_0^A p_{+1}^B\| \,\|p_0^A \bar{p}_{+1}^B\|(^3\Pi,\,^1\Pi)$
0	$\|p_0^A p_0^B\|(^3\Sigma)$	$\|\bar{p}_0^A p_0^B\| \,\|p_0^A \bar{p}_0^B\|(^3\Sigma,\,^1\Sigma)$
	$\|p_{+1}^A p_{-1}^B\|(^3\Sigma)$	$\|\bar{p}_{+1}^A p_{-1}^B\| \,\|p_{+1}^A \bar{p}_{-1}^B\|(^3\Sigma,\,^1\Sigma)$
	$\|p_{-1}^A p_{+1}^B\|(^3\Sigma)$	$\|\bar{p}_{-1}^A p_{+1}^B\| \,\|p_{-1}^A \bar{p}_{+1}^B\|(^3\Sigma,\,^1\Sigma)$

^{a} p_{+1}^A denotes the $m_l = +1$ component of the p state of hydrogen atom A, etc.

4. Let $\hat{i}$ be the inversion operator defined in question 1. Then

$$\hat{i}p_{+1}^A = \hat{i}\left[\frac{1}{\sqrt{2}}(p_x^A + ip_y^A)\right] = \frac{1}{\sqrt{2}}(-p_x^B - ip_y^B) = -p_{+1}^B$$

$$\hat{i}p_{-1}^A = -p_{-1}^B, \qquad \hat{i}p_0^A = -p_0^B$$

(a)Ad $^3\Delta$ *term*:
Consider the $^3\Delta(M_S = 1)$ term:

$$\hat{i}|p_{+1}^A p_{+1}^B| = -|p_{+1}^A p_{+1}^B|$$

Thus

$$\Psi(^3\Delta_u, M_S = 1) = |p_{+1}^A p_{+1}^B|$$

(b)Ad $^1\Delta$ *term*:
$\Psi(^3\Delta_u, M_S = 0)$ is obtained as $\hat{S}_- \Psi(^3\Delta_u, M_S = 1)$; that is,

$$S_- |p_{+1}^A p_{+1}^B| = |\bar{p}_{+1}^A p_{+1}^B| + |p_{+1}^A \bar{p}_{+1}^B| \text{ where } \hat{S}_- = \hat{S}_-(1) + \hat{S}_-(2)$$

or

$$\Psi(^3\Delta_u, M_S = 0) = \frac{1}{\sqrt{2}}(|\bar{p}_{+1}^A p_{+1}^B| + |p_{+1}^A \bar{p}_{+1}^B|)$$

The $^1\Delta$ wave function must be orthogonal to $\Psi(^3\Delta_u, M_S = 0)$. Since

$$\hat{\imath}\left[\frac{1}{\sqrt{2}}(|\bar{p}_{+1}^A p_{+1}^B| - |p_{+1}^A \bar{p}_{+1}^B|)\right] = \frac{1}{\sqrt{2}}(|\bar{p}_{+1}^A p_{+1}^B| - |p_{+1}^A \bar{p}_{+1}^B|)$$

we obtain

$$\Psi(^1\Delta_g) = \frac{1}{\sqrt{2}}(|\bar{p}_{+1}^A p_{+1}^B| - |p_{+1}^A \bar{p}_{+1}^B|)$$

(c)the Π states can be analyzed similarly, and are found to be of $^3\Pi_g$, $^3\Pi_u$, $^1\Pi_g$, and $^1\Pi_u$, symmetry.

(d)The determinants, which are constructed from p_0 orbitals alone, must be of Σ^+ symmetry. Application of the $\hat{\imath}$ operator gives that the triplet determinant is *ungerade* and the singlet determinant is *gerade*; that is, the states are of $^3\Sigma_u^+$ and $^1\Sigma_g^+$ symmetry. The wave functions are

$$\Psi(^3\Sigma_u^+, M_S = 1) = |p_0^A p_0^B| \qquad \text{(E.13,I)}$$

$$\Psi(^3\Sigma_u^+, M_s = 0) = \frac{1}{\sqrt{2}}(|p_0^A \bar{p}_0^B| + |\bar{p}_0^A p_0^B|) \qquad \text{(E.13,II)}$$

$$\Psi(^1\Sigma_g^+) = \frac{1}{\sqrt{2}}(|p_0^A \bar{p}_0^B| - |\bar{p}_0^A p_0^B|) \qquad \text{(E.13,III)}$$

(e)*The other Σ states:*
Let $\hat{\sigma}_v$ denote a reflection in the plane containing the internuclear axis for which

$$\hat{\sigma}_v p_{+1} = p_{-1} \qquad \text{and} \qquad \hat{\sigma}_v p_{-1} = p_{+1}$$

Consider the $^3\Sigma$ state with $M_S = 1$ which is a linear combination of two determinants (see Table E.13,II). The $^3\Sigma$ state is an eigenfunction for $\frac{1}{2}(1 + \hat{\sigma}_v)$ (see the Appendix). Since

$$\frac{1}{2}(1 + \hat{\sigma}_v)(|p^A_{+1}p^B_{-1}| + |p^A_{-1}p^B_{+1}|) = |p^A_{+1}p^B_{-1}| + |p^A_{-1}p^B_{+1}|$$

this wave function represents a $^3\Sigma^+$ state. The inversion symmetry of this state is *ungerade*, that is,

$$\Psi(^3\Sigma_u^+, M_S = 1) = \frac{1}{\sqrt{2}}(|p^A_{+1}p^B_{-1}| + |p^A_{-1}p^B_{+1}|) \quad (E.13,IV)$$

Consider now the $\pm$ symmetry of the antisymmetric linear combination of Eq. (E.13,IV)

$$\frac{1}{2}(1 - \hat{\sigma}_v)(|p^A_{+1}p^B_{-1}| - |p^A_{-1}p^B_{+1}|) = (|p^A_{+1}p^B_{-1}| - |p^A_{-1}p^B_{+1}|)$$

This wave function thus represents a $^3\Sigma^-$ state which is *gerade* with respect to inversion; that is,

$$\Psi(^3\Sigma_g^-, M_S = 1) = \frac{1}{\sqrt{2}}(|p^A_{+1}p^B_{-1}| - |p^A_{-1}p^B_{+1}|) \quad (E.13,V)$$

Using similar arguments the two $^1\Sigma$ states can be determined to be of $^1\Sigma_g^+$ and $^1\Sigma_u^-$ symmetry. The wave functions are

$$\Psi(^1\Sigma_g^+) = \frac{1}{2}(|\bar{p}^A_{+1}p^B_{-1}| - |p^A_{+1}\bar{p}^B_{-1}| + |\bar{p}^A_{-1}p^B_{+1}| - |p^A_{-1}\bar{p}^B_{+1}|)$$
$$(E.13,VI)$$

$$\Psi(^1\Sigma_u^-) = \frac{1}{2}(|\bar{p}^A_{+1}p^B_{-1}| - |p^A_{+1}\bar{p}^B_{-1}| - |\bar{p}^A_{-1}p^B_{+1}| + |p^A_{-1}\bar{p}^B_{+1}|)$$
$$(E.13,VII)$$

5. The p^2 configuration of He gives the terms ^3P, ^1D, and ^1S. Since a p^2 configuration has even parity, all molecular states must be *gerade* and the correlation between these and the molecular states is

United atom	Molecular binding
$^1D(p^2)$	$^1\Delta_g$, $^1\Pi_g$, $^1\Sigma_g^+$
$^3P(p^2)$	$^3\Pi_g$, $^3\Sigma_g^-$
$^1S(p^2)$	$^1\Sigma_g^+$

The molecular terms of *ungerade* symmetry vanish as a result of the Pauli principle in the united atom limit when A = B. We see, for example, from Eq. (E.13,IV) that this wave function becomes identically zero, whereas $\Psi(^3\Sigma_g^-, M_S = 1)$ in Eq. (E.13,V) is nonvanishing.

Thus, ^3P must correlate with the $^3\Sigma_g^-$ molecular state. Similarly, we see that $\Psi(^1\Sigma_u^-)$ in Eq. (E.13,VII) and $\Psi(^3\Sigma_u^+)$ in Eqs. (E.13,I-II) vanish for A = B.

The rest of the molecular states are all *ungerade*. The lowest possible excited electronic configurations of He with which they may correlate are given in Fig. E.13,II (arbitrary energy scale).

E.14 1. From Table E.14,I we obtain the terms

$$^2H, \ ^2G, \ ^4F, \ ^2F, \ ^2D, \ ^2D, \ ^4P, \ ^2P$$

According to Hund's rule ^4F represents the ground state.

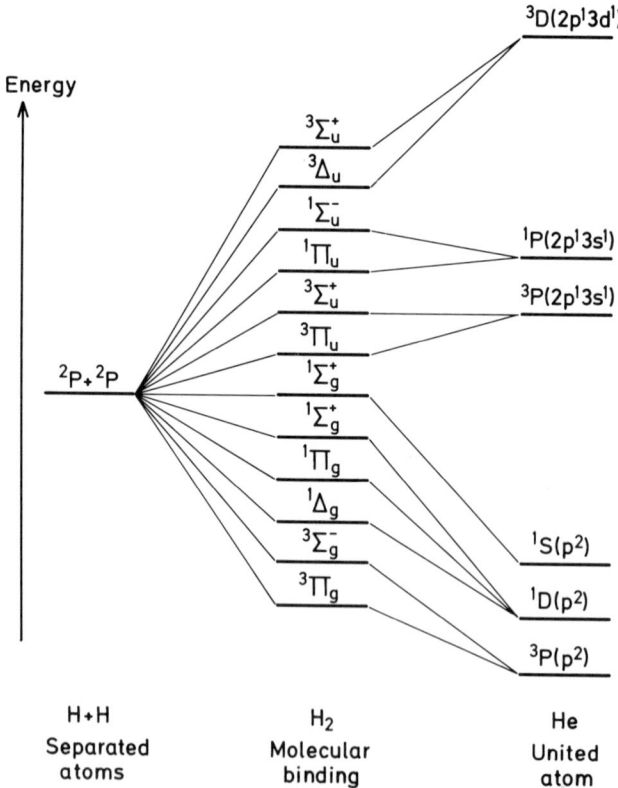

Fig. E.13,II. State correlation diagram for two hydrogen atoms, both in ^2P states.

Table E.14,I
Microstate diagram for a d^3 Configuration[a]

$M_L \backslash M_S$	3/2	1/2
5		$\|2\ \bar{2}\ 1\|\ (^2H)$
4		$\|2\ 1\ \bar{0}\|\|2\ 1\ \bar{1}\|\ (^2H,^2G)$
3	$\|2\ 1\ 0\|\ (^4F)$	$\|2\ 1\ \bar{0}\|\|2\ \bar{1}\ 0\|\|\bar{2}\ 1\ 0\|\|2\ \bar{2}\ -1\|\ (^2H,^4F,^2G,^2F)$
2	$\|2\ 1\ -1\|\ (^4F)$	$\|2\ 1\ -\bar{1}\|\|2\ 0\ \bar{0}\|\|1\ \bar{1}\ 0\|\|\bar{2}\ 1\ -1\|\|2\ \bar{2}\ -2\|\|2\ \bar{1}\ -1\|$
		$(^2H,^4F,^2G,^2F,^2D(2))$
1	$\|2\ 1\ -2\|$	$\|\bar{2}\ 1\ -2\|\|2\ \bar{1}\ -2\|\|2\ 1\ -\bar{2}\|\|\bar{2}\ 0\ -1\|\|2\ \bar{0}\ -1\|$
	$\|2\ 0\ -1\|\ (^4F,^4P)$	$\|2\ 0\ -\bar{1}\|\|0\ \bar{0}\ 1\|\|1\ \bar{1}\ -1\|\ (^2H,^4F,^2G,^2F,^2D(2),^4P,^2P)$
0	$\|2\ 0\ -2\|$	$\|\bar{2}\ 0\ -2\|\|2\ \bar{0}\ -1\|\|2\ 0\ -\bar{2}\|\|\bar{1}\ 0\ -1\|\|1\ \bar{0}\ -1\|$
	$\|1\ 0\ -1\|\ (^4F,^4P)$	$\|1\ 0\ -\bar{1}\|\|2\ -1\ -\bar{1}\|\|1\ \bar{1}\ -2\|\ (^2H,^4F,^2G,^2F,$
		$^2D(2),^4P,^2P)$

[a] Only the m_l quantum numbers are indicated in the determinants.

2. The determinant $\|2\quad 1\ -1\|$ is the wave function for the component
    of $\|^4F\rangle$ having $M_S = \dfrac{3}{2}$ and $M_L = 2$. Using the step operator

$$\hat{L}_- = \sum_{i=1}^{3} \hat{L}_-(i)$$

we obtain

$$L_-^2 \|V,^4F, M_L = 2, M_S = \tfrac{3}{2}\rangle = L_-^2 \|2\quad 1\ -1\|$$

$$= \sqrt{3}\,(2\|1\quad 0\ -1\|+4\|2\quad 0\ -2\|)$$

The normalized wave function with $M_S = \dfrac{3}{2}$ and $M_L = 0$ is thus

$$\|V,^4F, M_l = 0, M_S = \tfrac{3}{2}\rangle = \frac{1}{\sqrt{5}}(\|1\quad 0\ -1\|+2\|2\quad 0\ -2\|) \quad (\text{E.14,I})$$

3. $V\ (^4F) + H\ (^2S) \rightarrow VH$
    $\Lambda = 3, 2, 1, 0$
    $S = 2, 1$
4. The Σ states are $^5\Sigma$ and $^3\Sigma$.

$$\|VH,^5\Sigma, M_S = 2\rangle = \Big|\|V,^4F, M_L = 0, M_S = \tfrac{3}{2}\rangle\ H,^2S, M_s = \tfrac{1}{2}\rangle\Big|$$

$$(\text{E.14,II})$$

where

$$|\mathrm{H},\,{}^2S,\,M_S = \tfrac{1}{2}\rangle = 1s_\mathrm{H}$$

Using Eq. (E.14,I) we obtain

$$S_-|\mathrm{VH},\,{}^5\Sigma,\,M_S = 2\rangle = N|\,\mathrm{VH},\,{}^5\Sigma,\,M_S = 1\rangle$$

$$= \frac{1}{\sqrt{5}}[|\bar{1}\quad 0 - 1\ \ 1s_\mathrm{H}| + |1\quad \bar{0} - 1\ \ 1s_\mathrm{H}| + |1\quad 0 - \bar{1}\ \ 1s_\mathrm{H}|$$

$$+\ 2|\bar{2}\quad 0 - 2\ \ 1s_\mathrm{H}| + 2|2\quad \bar{0} - 2\ \ 1s_\mathrm{H}|$$

$$+\ 2|2\quad 0 - \bar{2}\ \ 1s_\mathrm{H}| + |1\quad 0 - 1\ \ \overline{1s_\mathrm{H}}| + 2|2\quad 0 - 2\ \ \overline{1s_\mathrm{H}}|]$$

The normalized ${}^3\Sigma$, $M_S = 1$ wave function must be orthogonal to this wave function, that is,

$$|\mathrm{VH},\,{}^3\Sigma,\,M_s = 1\rangle = \frac{1}{\sqrt{20}}[|\bar{1}\quad 0 - 1\ \ 1s_\mathrm{H}| - |1\quad \bar{0} - 1\ \ 1s_\mathrm{H}| + |1\quad -0\bar{1}\ \ 1s_\mathrm{H}|$$

$$+\ 2|\bar{2}\quad 0 - 2\ \ 1s_\mathrm{H}| - 2|2\quad \bar{0} - 2\ \ 1s_\mathrm{H}| + 2|2\quad 0 - \bar{2}\ \ 1s_\mathrm{H}|$$

$$-\ |1\quad 0 - 1\ \ \overline{1s_\mathrm{H}}| - 2|2\quad 0 - 2\ \ \overline{1s_\mathrm{H}}|] \qquad \text{(E.14,III)}$$

5. $\Lambda = 1:\ {}^3\Pi,\,{}^5\Pi$
   $\Lambda = 2:\ {}^3\Delta,\,{}^5\Delta$
   $\Lambda = 3:\ {}^3\phi,\,{}^5\phi$
   $\Lambda = 0$: The behavior of the Σ states under a reflection in a plane containing the internuclear axis is determined from the effect of the operator $\frac{1}{2}(1 - \hat{\sigma}_v)$ (see the Appendix).

$$\frac{1}{2}(1 - \hat{\sigma}_v)|\mathrm{VH},\,{}^5\Sigma,\,M_S = 2\rangle = -|\,\mathrm{VH},\,{}^5\Sigma,\,M_S = 2\rangle$$

$$\frac{1}{2}(1 - \hat{\sigma}_v)|\mathrm{VH},\,{}^3\Sigma,\,M_S = 1\rangle = -|\,\mathrm{VH},\,{}^3\Sigma,\,M_S = 1\rangle$$

The Σ states are thus ${}^5\Sigma^-$ and ${}^3\Sigma^-$.

E.15 1. The MO diagram is given in Fig. E.15,I.
    2. Ground state: core $2\sigma^2 3\sigma^2$; ${}^1\Sigma^+$
    The two lowest singly excited states:

$$\text{core } 2\sigma^2 3\sigma^1 1\pi^1;\ {}^3\Pi,\,{}^1\Pi$$

$$\text{core } 2\sigma^2 3\sigma^1 4\sigma^1;\ {}^1\Sigma^+,\,{}^3\Sigma^+$$

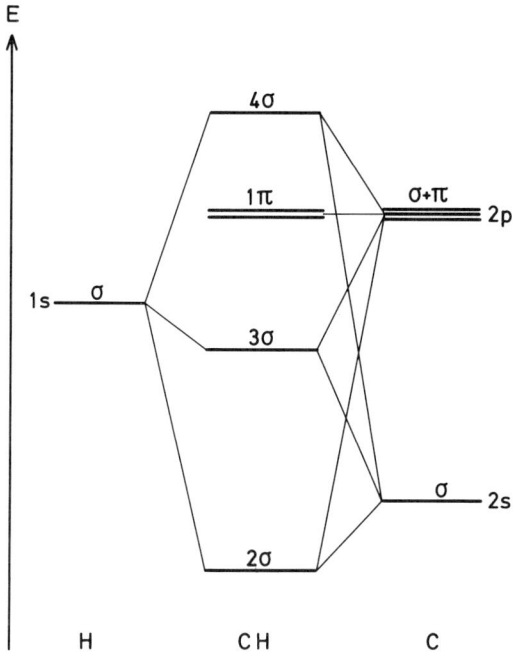

Fig. E.15,I. MO diagram for CH.

3. The lowest state of $^1\Delta$ symmetry does most likely originate from the doubly excited electronic configuration

$$\text{core } 2\sigma^2 1\pi^2; \ ^1\Sigma^+, \ ^3\Sigma^-, \ ^1\Delta$$

If d functions had been included in the basis set, Δ states could have shown up when an electron was excited from a σ orbital to a δ orbital.

4. The ground-state electronic configuration of carbon splits into [see Eq. (E.3,IV)]

$$^3P \rightarrow \ ^3\Sigma^-, \ ^3\Pi; \ ^1D \rightarrow \ ^1\Delta, \ ^1\Pi, \ ^1\Sigma^+; \ ^1S \rightarrow \ ^1\Sigma^+ \qquad \text{(E.15,I)}$$

Since the perturbation operator must transform as the totally symmetric IR of the point group $C_{\infty v}$, interaction occurs between states of the same symmetry, that is, between $^1D(^1\Sigma^+)$ and $^1S(^1\Sigma^+)$.

5. The zero-order states are given in Eq. (E.15,I) and first-order perturbation theory then gives the secular equation

$$|\langle \underline{\Psi}|\hat{V}_{EE} + \hat{V} - E|\underline{\Psi}\rangle|$$

$$= \begin{vmatrix} 0-E & 0 & 0 & 0 & 0 & 0 \\ 0 & 0-E & 0 & 0 & 0 & 0 \\ 0 & 0 & 1.3-E & 0 & 0 & 0 \\ 0 & 0 & 0 & 1.3-E & 0 & 0 \\ 0 & 0 & 0 & 0 & 1.3-E & k \\ 0 & 0 & 0 & 0 & k & 2.7-E \end{vmatrix}$$

$$= 0$$

where $\underline{\Psi}$ is a row vector that has the elements

$$\underline{\Psi} = \{|^3\Sigma^-\rangle, |^3\Pi\rangle, |^1\Delta\rangle, |^1\Pi\rangle, |^1\Sigma^+(^1D)\rangle, |^1\Sigma^+(^1S)\rangle\}$$

We have set all the diagonal matrix elements of $\hat{V}$ equal to zero. The $^1\Sigma^+$ block gives

$$\begin{vmatrix} 1.3-E & k \\ k & 2.7-E \end{vmatrix} = 0 \quad \text{or} \quad E_{\pm} = 2 \pm \sqrt{0.49 + k^2} \text{ eV}$$

and the relative energies of the states are plotted as a function of k in Fig. E.15,II.

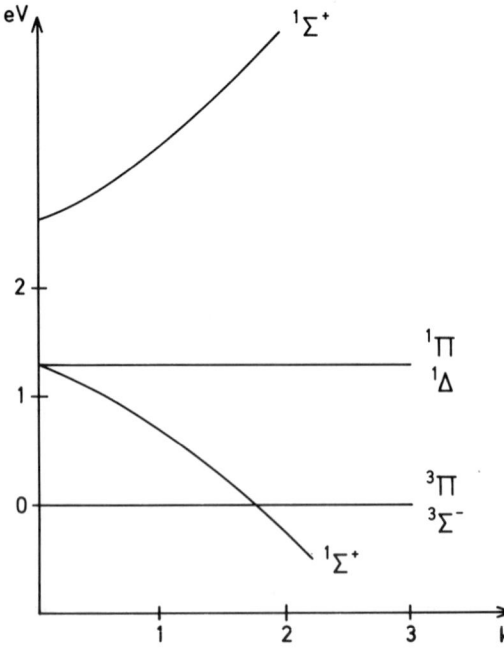

Fig. E.15,II. The total energies of the states of CH^+ as obtained by carrying out a first-order perturbation calculation on C.

6. $E_- = 0$ for $k = 1.87$ eV.
    For $k < 1.87$ eV the ground state has $^3\Sigma^-$ symmetry and for
    $k > 1.87$ eV the ground state has $^1\Sigma^+$ symmetry.
7. $\Delta E(^1\Sigma^+ \to {}^1\Pi) = E(^1\Pi) - E_- = 3.0$ eV implies that $k = 2.63$ eV.
    The other triplet and singlet excitation energies become
    $\Delta E(^1\Sigma^+ \to {}^3\Sigma^-) = \Delta E(^1\Sigma^+ \to {}^3\Pi) = 1.7$ eV;
    $\Delta E(^1\Sigma^+ \to {}^1\Delta) = 3.0$ eV; $\Delta E(^1\Sigma^+ \to {}^1\Sigma^+) = 7.4$ eV.
8. The excitation energies are estimated in Table E.15,I.
    $\Delta E(x \quad , {}^1\Sigma^+ \to {}^1\Pi) = 3.0$ eV implies the orbital energy difference
    $\varepsilon_{1\pi} - \varepsilon_{3\sigma}$ is 3.0 eV. The simple treatment given in question 3
    determines $\Delta E(X\,{}^1\Sigma^+ \to {}^1\Delta)$ to be twice the energy difference
    $\varepsilon_{1\pi} - \varepsilon_{3\sigma}$. The MO diagram is used to estimate the magnitude of the
    last MO excitation energy which is listed in Table E.15,I.

Table E.15,I
Excitation Energies of CH^+ in an MO and a Perturbative
Calculation

Excitation	Best estimate	Perterbative calc.	MO
$X\,{}^1\Sigma^+ \to {}^3\Pi$	1.0 eV	1.7 eV	3.0 eV
$X\,{}^1\Sigma^+ \to {}^1\Pi$	3.0 eV	3.0 eV	3.0 eV
$X\,{}^1\Sigma^+ \to {}^1\Delta$	6.0 eV	3.0 eV	6.0 eV
$X\,{}^1\Sigma^+ \to {}^1\Sigma^+$	?	7.4 eV	~ 5.0 eV

E.16 1. From Table E.16, I we see that the wave functions for the $a\,{}^1\Delta_g$ and
    $X\,{}^3\Sigma_g^- (M_S = \pm 1)$ states are

$$\Psi_1(^1\Delta_g, M_L = 2) = |\pi_g^+ \bar{\pi}_g^+| \qquad (\text{E.16,I})$$

$$\Psi_2(^1\Delta_g, M_L = -2) = |\pi_g^- \bar{\pi}_g^-| \qquad (\text{E.16,II})$$

$$\Psi_3(^3\Sigma_g^-, M_S = 1) = |\pi_g^+ \pi_g^-| \qquad (\text{E.16,III})$$

$$\Psi_4(^3\Sigma_g^-, M_S = -1) = |\bar{\pi}_g^+ \bar{\pi}_g^-| \qquad (\text{E.16,IV})$$

The two last wave functions are combinations of the determinants with $M_S = M_L = 0$. The spatial part of the triplet wave
function must be antisymmetric under interchange of the electrons
and the singlet space wave function must be symmetric:

$$\Psi_5(^3\Sigma_g^-, M_S = 0) = \frac{1}{\sqrt{2}}[|\pi_g^+ \bar{\pi}_g^-| + |\bar{\pi}_g^+ \pi_g^-|] \qquad (\text{E.16,V})$$

$$\Psi_6(^1\Sigma_g^+, M_S = 0) = \frac{1}{\sqrt{2}}[|\pi_g^+ \bar{\pi}_g^-| - |\bar{\pi}_g^+ \pi_g^-|] \qquad (\text{E.16,VI})$$

Table E.16,I
Microstate diagram for a π_g^2 configuration

$M_L\backslash M_S$	1	0	-1
2		$\|\pi_g^+ \bar{\pi}_g^+\| (^1\Delta_g)$	
0	$\|\pi_g^+ \pi_g^-\| (^3\Sigma_g)$	$\|\pi_g^+ \bar{\pi}_g^-\| \|\bar{\pi}_g^+ \pi_g^-\| (^3\Sigma_g, {}^1\Sigma_g)$	$\|\bar{\pi}_g^+ \bar{\pi}_g^-\| (^3\Sigma_g)$
-2		$\|\pi_g^- \bar{\pi}_g^-\| (^1\Delta_g)$	

The $\pm$ symmetry of the Σ states is as indicated, which may be checked, using the projection operators $\hat{O}_+$ and $\hat{O}_-$ (see the Appendix).

2. Using Eqs. (E.15,1-5) and the Slater-Condon rules for evaluating matrix elements of one-electron operators, we obtain

$$\langle\Psi_1({}^1\Delta_g, M_L = 2)|\hat{V}_{ext}|\Psi_6({}^1\Sigma_g^+, M_S = 0)\rangle = -\frac{a}{\sqrt{2}} \quad \text{(E.16,VII)}$$

$$\langle\Psi_2({}^1\Delta_g, M_L = -2)|\hat{V}_{ext}|\Psi_6({}^1\Sigma_g^+, M_S = 0)\rangle = -\frac{a}{\sqrt{2}} \quad \text{(E.16,VIII)}$$

where

$$\hat{V}_{ext} = \sum_{i=1}^{2} \hat{V}_{ext}(i)$$

The rest of the matrix elements are zero.

3. The $X\,{}^3\Sigma_g^-$ state does not mix with the other states. However, the three other states with $M_S = 0$ interact, and degenerate first-order perturbation theory in the perturbation $\hat{V} = \hat{V}_{ext} + \hat{V}_{EE}$ leads to a 3×3 secular problem

$$\begin{vmatrix} E_1 - E & 0 & -\dfrac{a}{\sqrt{2}} \\ 0 & E_1 - E & -\dfrac{a}{\sqrt{2}} \\ -\dfrac{a}{\sqrt{2}} & -\dfrac{a}{\sqrt{2}} & E_6 - E \end{vmatrix} = 0 \quad \text{(E.16,IX)}$$

where we have used Eqs. (E.16,VII and VIII) and where

$$E_1 = \langle\Psi_1|\hat{V}|\Psi_1\rangle = \langle\Psi_2|\hat{V}|\Psi_2\rangle = 1.0 \text{ eV}$$

$$E_6 = \langle\Psi_6|\hat{V}|\Psi_6\rangle = 1.7 \text{ eV}$$

Equation (E.16,IX) yields

$$E = E_1$$

$$E = \frac{E_1 + E_6}{2} \pm \frac{1}{2}\sqrt{(E_1 - E_6)^2 + 4a^2}$$

which is sketched in Fig. E.16,I.

4. The point group symmetry is D_{2h} as seen from Fig. E.16,II. The relation between IR's in the perturbed and unperturbed cases is given in Table E.16,I. The $b\,{}^1\Sigma_g^+$ state and one component of the $a\,{}^1\Delta_g$ state will have the same symmetry after introducing $\hat{V}_{ext}$ and will thus form bonding and antibonding combinations as observed in Fig. E.16,I.

5. The symmetry changes from ${}^3B_{1g}$ to ${}^1A_g$ at a_0, which is determined as

$$a_0^2 = E_1 E_6 \qquad \text{or} \qquad a_0 = 1.3 \text{ eV}.$$

6. Yes, from paramagnetic to diamagnetic. We observe a so-called spin crossover.

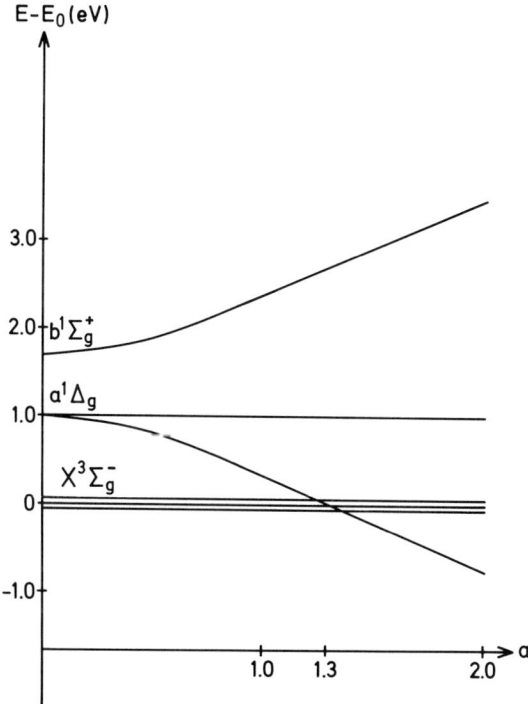

Fig. E.16,I. The total energies of the states of the perturbed O_2 molecule as obtained by carrying out a first-order perturbation calculation.

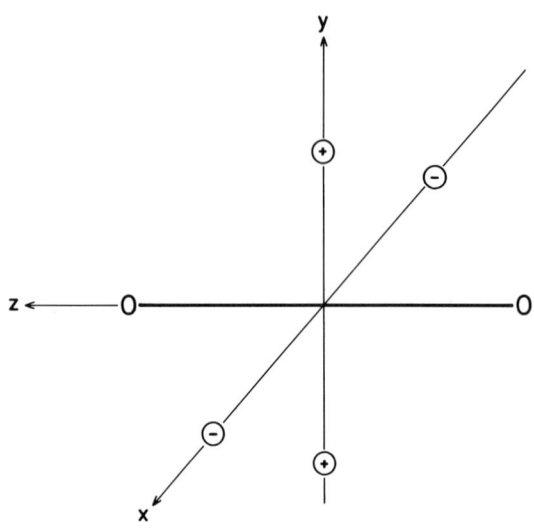

Fig. E.16,II. The location of the external charges of the perturbed O_2 molecule.

Table E.16,I
Relations between Irreducible
Representations of $D_{\infty h}$ and D_{2h}

$D_{\infty h}$	D_{2h}
$^1\Sigma_g^+ \sim x^2 + y^2, z^2$	$^1A_g \sim x^2, y^2, z^2$
$^3\Sigma_g^- \sim R_z$	$^3B_{1g} \sim R_z, xy$
$^1\Delta_g \sim (x^2 - y^2, xy)$	$^1B_{1g} \sim xy$
	$^1A_g \sim x^2, y^2, z^2$

π-ELECTRON SYSTEMS

INTRODUCTION

The problems in this set use the π-electron approximation, that is, the approximation in which the π electrons are treated separately and the effect of the σ electrons is included through the choice of semiempirical parameters.

Most of the problems use standard Hückel MO theory (F.1–F.12). The general Hückel solutions for monocyclic and linear conjugated polyenes are derived in problems F.1 and F.2, respectively. The geometry and stability of various planar conjugated molecules in their ground and excited states are examined in problems F.3, F.7, F.8, F.9, F.11, and F.12, while spectral properties are calculated in F.4, F.5, F.6, F.7, and F.10. The standard Hückel MO theory as presented in any elementary textbook on quantum chemistry is the only background needed for solving these problems.

In problems F.13–F.15, we introduce electronic repulsion through the use of the Pariser–Parr–Pople (PPP) method. The PPP equations are often not included in quantum chemistry texts, and we will therefore give a brief summary of the PPP equations here.

The PPP method is a self-consistent field method (see the introduction to section B) for the π-electron system. The matrix representation of the Fock operator ($\hat{F}$) is given in Eq. (B.0, 9). In the *atomic* orbital representation $\{\phi_\mu\}$ (i.e., the pπ orbitals of the conjugated system) the overlap integral and the matrix elements of the Fock operator are (Murrell et al. 1970)

$$S_{\mu\nu} = \langle \phi_\mu | \phi_\nu \rangle = \delta_{\mu\nu} \qquad (\text{F.0, 1})$$

$$F_{\mu\mu} = \langle \phi_\mu | \hat{F} | \phi_\mu \rangle = I_\mu + \tfrac{1}{2} q_\mu \gamma_{\mu\mu} + \sum_{\rho \neq \mu} (q_\rho - Z_\rho)\gamma_{\mu\rho} \qquad (\text{F.0, 2})$$

$$F_{\mu\nu} = \langle \phi_\mu | \hat{F} | \phi_\nu \rangle = \beta_{\mu\nu} - \tfrac{1}{2} p_{\mu\nu} \gamma_{\mu\nu}, \qquad \mu \neq \nu \qquad (\text{F.0, 3})$$

where q_μ is the charge order, $p_{\mu\nu}$ represents the bond order matrix elements [Eq. (B.0,10)], and Z_μ, I_μ, $\beta_{\mu\nu}$, and $\gamma_{\mu\nu}$ are parameters that are normally determined by comparing calculated spectra of a few selected compounds with measured spectra. I_μ and $\beta_{\mu\nu}$ are parametrized one-electron integrals, while

$$\gamma_{\mu\nu} = (\phi_\mu \phi_\mu | \phi_\nu \phi_\nu) \qquad\qquad \text{(F.0, 4)}$$

are parametrized two-electron integrals. The $\beta_{\mu\nu}$ parameter is chosen as

$$\beta_{\mu\nu} = \begin{cases} \beta^c & \mu \text{ and } \nu \text{ are neighbors} \\ 0 & \mu \text{ and } \nu \text{ are nonneighbors} \end{cases} \qquad \text{(F.0, 5)}$$

Z_μ is the net charge of nucleus μ as seen by a valence electron; that is, the charge of the "core" when the π electron has been removed. Z_μ is always equal to 1.0 for a conjugated carbon atom. The actual values of I_μ, β^c, and $\gamma_{\mu\nu}$ will be given in the text when necessary for solving the problems.

The PPP molecular orbitals and orbital energies are obtained as the eigenvectors and eigenvalues for the Fock $\{F_{\mu\nu}\}$ matrix [Eq. (B.0, 7)]. From an initial guess on the MO's (often the Hückel MO's) we can determine the (zeroth) order charge and bond order matrix and thus get a first estimate of the PPP Fock matrix from Eqs. (F.0, 2–3). By diagonalizing this matrix, we obtain a new (first-order) set of MO's and thus a new charge–bond order matrix. This iterative procedure continues until self-consistency. The self-consistency requirement is normally a limit put on the variation of the total energy or the charge–bond order matrix in two consecutive iterations.

Textbooks that give sufficient background for solving the problems in this section

Atkins (1970). Avery (1972). Chandra (1974). George (1972). Hanna (1969). La Paglia (1971). Levine (1974). Lowe (1978). McWeeny (1979). Murrell et al. (1970, 1978). Pilar (1968). Streitwieser (1961).

Textbooks that give a more elaborate treatment of the topics

Dewar (1969). Murrell (1963). Murrell and Harget (1972). Parr (1963). Salem (1966).

Problem Description

F.1 Hückel solution for a monocyclic conjugated polyene, $C_N H_N$
F.2 Hückel solution for a linear conjugated polyene, $C_N H_{N+2}$
F.3 Aromaticity of trimethylene cyclopropane
F.4 Photoelectron spectrum of furan, σ–π separability.
F.5 Ultraviolet absorption spectrum of $C_4 O_4^{2-}$
F.6 Electron spin resonance spectrum of styrene
F.7 Electronic structure of NO_2^+, NO_2, and NO_2^-
F.8 Geometry of substituted allene in its ground and excited states

F.9 Cis-trans isomerization of 1, 3-butadiene

F.10 Singlet excited states of benzene and singly substituted benzenes (aniline, phenol, toluene, and chlorobenzene)

F.11 Geometry of the cyclopentadienyl radical, Jahn–Teller distortion

F.12 Geometry of cyclobutadiene in its ground and lowest excited state, pseudo-Jahn–Teller effect

F.13 Pariser–Parr–Pople calculation on the allyl cation

F.14 Pariser–Parr–Pople calculation of the lowest triplet–triplet excitation energy of cyclobutadiene

F.15 The π–π* transition of ethylene: excitation energy and oscillator strength

PROBLEMS

F.1 In this problem we will determine the analytic solution to the Hückel secular problem for a monocyclic conjugated polyene

$$C_N H_N, \qquad N = 3, 4, 5, \ldots \tag{F.1, 1}$$

Let the molecular orbitals be

$$\psi_p = \sum_{\mu=1}^{N} c_{p\mu} \phi_\mu \tag{F.1, 2}$$

where ϕ_μ is a $2p\pi$ atomic orbital on atom μ.

1. Show that the cyclic boundary condition is satisfied by the two sets of (normalized) coefficients

$$c_{p\mu} = \sqrt{\frac{2}{N}} \sin \frac{2\pi p\mu}{N}; \qquad c_{p\mu} = \left(\frac{2 - \delta_{p0} - \delta_{p^{N/2}}}{N} \right)^{1/2} \cos \frac{2\pi p\mu}{N} \tag{F.1, 3}$$

where

$$p = 0, 1, \ldots, \frac{N}{2} \qquad (N \text{ even})$$
$$p = 0, 1, \ldots, \frac{N-1}{2} \qquad (N \text{ odd}) \tag{F.1, 4}$$

2. Show that the corresponding Hückel orbital energies are

$$\varepsilon_p = \alpha + 2\beta \cos \frac{2\pi p}{N} \tag{F.1, 5}$$

where p satisfies the condition in (F.1, 4). Discuss the degeneracy of ε_p. Consider now the cyclic polyenes for which $N = 4n + 2$ and n is a positive integer (i.e., the so-called annulenes).

3. Show that the π-electron delocalization energy per carbon atom is

$$E_D = \frac{4\beta}{(4n + 2) \sin \dfrac{\pi}{4n + 2}} - \beta \tag{F.1, 6}$$

4. Determine E_D for $n \to \infty$.

5. From consideration of the π-electron system, would you expect that the stability of the annulenes increases or decreases as n increases when n is large?

We will now calculate the bond order, $p_{\mu\nu}$, for the annulenes.

6. Show that

$$p_{\mu\nu} = \frac{2 \sin \dfrac{\pi(\mu - \nu)}{2}}{(4n + 2) \sin \dfrac{\pi(\mu - \nu)}{4n + 2}} \qquad (\text{F.1, 7})$$

7. Determine $p_{\mu\nu}$ for $n \to \infty$.

F.2 In this problem, we will determine the analytic solution to the Hückel secular problem for a linear conjugated polyene

$$C_N H_{N+2}, \qquad N = 2, 3, 4, \ldots$$

We assume that we have N carbon atoms in the chain and that the MO's are

$$\psi_p = \sum_{\mu=1}^{N} c_{p\mu} \phi_\mu \qquad (\text{F.2, 1})$$

where ϕ_μ is a $2p\pi$ atomic orbital on atom μ. As boundary conditions for the MO coefficients we use that $c_{p0} = c_{pN+1} = 0$, where c_{p0} and c_{pN+1} denote the MO coefficients on the zeroth and $(N + 1)$ st atomic site, respectively.

1. Show that the coefficients

$$c_{p\mu} = \sqrt{\frac{2}{N+1}} \sin \frac{p\pi\mu}{N+1}, \qquad p = 1, 2, \ldots, N \qquad (\text{F.2, 2})$$

satisfy the boundary condition and yield normalized molecular orbitals.

2. Show that the corresponding Hückel orbital energies are

$$\varepsilon_p = \alpha + 2\beta \cos \frac{p\pi}{N+1}, \qquad p = 1, 2, \ldots, N \qquad (\text{F.2, 3})$$

3. Show that the π-electron delocalization energy per carbon atom is

$$E_D = \beta \left(\frac{2}{N \sin \dfrac{\pi}{2N+2}} - \frac{N+2}{N} \right) \qquad (\text{F.2, 4})$$

4. Show that the limit of E_D for $N \to \infty$ is equal to limiting value of E_D for the annulenes (question 4 of problem F.1).

F.3 In this problem we examine the aromaticity of the trimethylene cyclopropane molecule given in Fig. F.3, 1 by carrying out a Hückel π-electron calculation. We compare the stability of trimethylene cyclopropane and the iso-electronic molecule benzene.

1. Which point group does the trimethylene cyclopropane molecule belong to?

2. Which subgroup of the molecular point group does the π-electron system belong to?

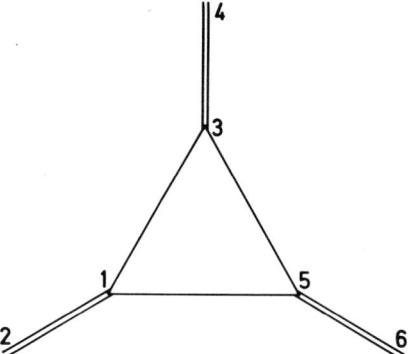

Fig. F.3, 1. The geometry of trimethylene cyclopropane.

We will carry out the Hückel calculation assuming

$$\alpha_1 = \alpha_2 = \alpha_3 = \alpha_4 = \alpha_5 = \alpha_6 = \alpha$$

$$\beta_{12} = \beta_{34} = \beta_{56} = \beta$$

$$\beta_{13} = \beta_{35} = \beta_{15} = \beta'$$

3. Determine the Hückel MO energies and their relative energy ordering. Determine the ground-state total energy in terms of α, β, and β'.

4. Determine the bond orders p_{12} and p_{13} as a function of β and β'. Determine the numerical value for the bond orders p_{12} and p_{13} for $\beta = \beta'$.

5. Compute the delocalization energy of trimethylene cyclopropane for $\beta = \beta'$. Trimethylene cyclopropane is iso-electronic with benzene.

6. Compare, on the basis of the magnitude of the delocalization energy and the bond orders, the relative aromaticity of trimethylene cyclopropane and benzene (the delocalization energy of benzene is 2.0β).

The analysis of the aromaticity of trimethylene cyclopropane in the preceding questions is based on the assumption that $\beta = \beta'$. However, it is known that the $C_1 - C_3$, $C_3 - C_5$, and the $C_1 - C_5$ bonds are substantially longer than the $C_1 - C_2$, $C_3 - C_4$, and $C_5 - C_6$ bonds.

7. What does this imply for the relative magnitudes of β and β' and thus for the ratio $k = \beta'/\beta$?

8. Show that $p_{12} - p_{13}$ decreases with increasing values of $k (k > 0)$.

9. Does the result of question 8 imply that the aromaticity of trimethylene cyclopropane increases or decreases relative to what was found in question 5 for $\beta = \beta'$?

F.4 In this problem we will perform an assignment of the photoelectron spectrum of furan, C_4H_4O, using Hückel MO theory. We will assume that the oxygen Hückel parameters are

$$\alpha_O = \alpha + 2\beta, \qquad \beta_{OC} = 0.8\,\beta$$

where α, β are the usual carbon Hückel parameters.

1. Determine the π symmetry orbitals of furan.
2. Determine the Hückel MO energies and the total π-electron energy.

The experimental photoelectron spectrum of furan (Turner *et al.* 1970) is given in Fig. F.4, 1. The numerical values of the ionization potentials and the assignment given by Derrick *et al.* (1971) are also indicated in Fig. F.4, 1. Accurate calculations show that the peaks at 8.88 eV and 10.31 eV and the "shoulder" at 15.6 eV can be assigned to ionization out of the three π MO's of furan (von Niessen *et al.* 1976). The rest of the peaks have been assigned to ionization out of σ orbitals.

3. Indicate the orbital labeling of the three π ionization potentials and discuss whether the assignment of von Niessen *et al.* (1976) could have been obtained from the Hückel calculation.
4. Which of the three π ionization potentials corresponds mainly to ionization out of the $p\pi$ orbital on oxygen?

F.5 We will use a Hückel MO calculation to analyze the ultraviolet absorption spectrum of the divalent anion of squaric acid, $C_4 O_4^{2-}$. Squaric acid is planar and its geometry is displayed in Fig. F.5, 1.

1. Determine the normalized π symmetry orbitals.

We will assume that

$$\alpha_O = \alpha + \beta, \qquad \beta_{CO} = \beta$$

where α, β are the usual carbon Hückel parameters.

2. Determine the Hückel MO energies.
3. Indicate the π-electron configuration and term symbol for the ground state and lowest excited singlet state.
4. Is the transition from the ground state to the lowest excited singlet state electric dipole allowed?

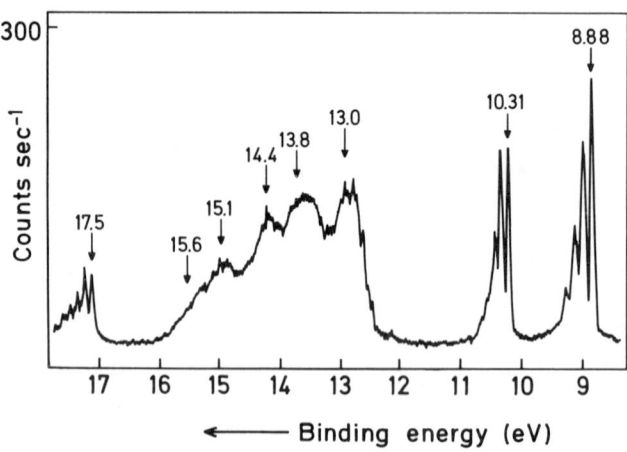

Fig. F.4, 1. Photoelectron spectrum of furan (Reprinted from Turner *et al.* (1970) by permission of John Wiley and Sons, Ltd).

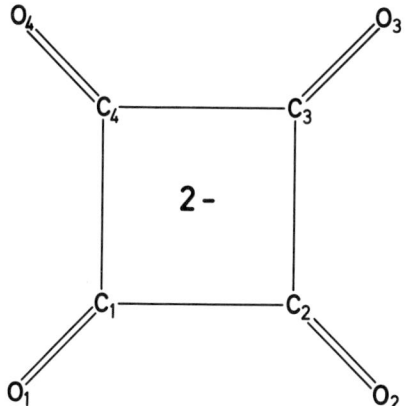

Fig. F.5, 1. The geometry of the divalent anion of squaric acid, $C_4O_4^{2-}$.

The ultraviolet absorption spectrum for the transition discussed in question 4 is given in Fig. F.5, 2 (Takahashi *et al.* 1978).

5. How would you explain that the spectrum in Fig. F.5, 2 exhibits two close-lying absorption lines (marked with arrows)?

F.6 In this problem we use Hückel MO theory to calculate the hyperfine structure of the electron spin resonance (ESR) spectrum of the styrene molecule. The $2p\pi$ orbital on carbon atom μ will be denoted ϕ_μ and the carbon atoms will be labeled as indicated in Fig. F.6, 1. The bonding Hückel MO's are

$$\psi_1 = 0.513\,\phi_1 + 0.394\,\phi_2 + 0.329\,\phi_3 + 0.308\,\phi_4 + 0.329\,\phi_5 + 0.394\,\phi_6$$
$$+ 0.308\,\phi_7 + 0.144\,\phi_8$$

$$\psi_2 = -0.354\,\phi_1 + 0.354\,\phi_3 + 0.500\,\phi_4 + 0.354\,\phi_5 - 0.500\,\phi_7 - 0.354\,\phi_8$$

$$\psi_3 = 0.500\,\phi_2 + 0.500\,\phi_3 - 0.500\,\phi_5 - 0.500\,\phi_6$$

$$\psi_4 = -0.334\,\phi_1 - 0.308\,\phi_2 + 0.130\,\phi_3 + 0.394\,\phi_4 + 0.130\,\phi_5 - 0.308\,\phi_6$$
$$+ 0.394\,\phi_7 + 0.595\,\phi_8$$

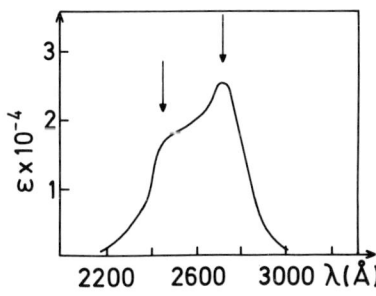

Fig. F.5, 2. Ultraviolet absorption spectrum of the divalent anion of squaric acid.

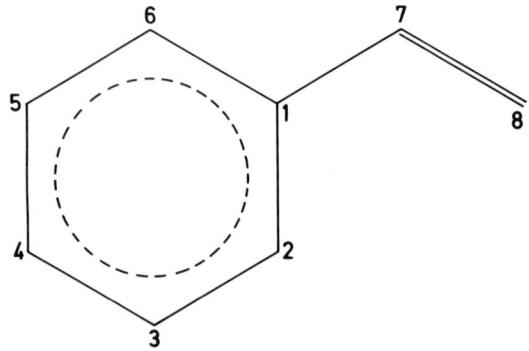

Fig. F.6, 1. The geometry of styrene.

and the corresponding orbital energies are

$$\varepsilon_1 = \alpha + 2.136 \beta$$

$$\varepsilon_2 = \alpha + 1.414 \beta$$

$$\varepsilon_3 = \alpha + 1.000 \beta$$

$$\varepsilon_4 = \alpha + 0.662 \beta$$

1.   Determine the antibonding Hückel MO's and their orbital energies.
2.   Determine the simplest determinantal wave function for the ground state and for the spin component of the first excited triplet state that has $M_S = 1$.
3.   Determine the charge orders, q_μ, of the carbon atoms in the ground state and in the lowest excited triplet state.

The spin density on a carbon atom μ in a triplet state is defined as the difference in the π charge density on atom μ originating from orbitals with $m_s = \frac{1}{2}$ and the density originating from orbitals with $m_s = -\frac{1}{2}$ for the component of the wave function with $M_S = \sum m_s = \pm 1$.

4.   Calculate the spin density on all carbon atoms in the lowest excited triplet state.

Transitions between (Zeeman) levels with different values of M_S for $S \neq 0$ are measured in ESR spectroscopy. The separations between the (Zeeman) levels are influenced by the presence of magnetic nuclei in the molecules. McConnell (1956) has shown that the splitting of the Zeeman levels of aromatic hydrocarbons caused by the protons is given approximately by the relation

$$A_H = Q_{CH} \, \rho_{\mu \to H} \tag{F.6, 1}$$

where Q_{CH} is a constant and $\rho_{\mu \to H}$ is the spin density on the carbon atom to which the proton is bonded. Experimental studies of ESR spectra of electronic systems similar to that of styrene indicate that the proton bonded to carbon atom 4 has the largest effect on the splitting of the Zeeman levels when only protons which are part of the benzene ring are considered.

5.   Is the experimental interpretation of the ESR experiment in agreement with McConnell's relation?

F.7 The NO_2 radical is bent and will in this problem be treated within the π-electron approximation where the $2p\pi$ orbitals on nitrogen and oxygen are used as basis functions.

1.   Determine the Hückel MO's and orbital energies for NO_2 when
$$\alpha_N = \alpha_O = -7.0 \text{ eV}, \beta_{ON} = -7.0/\sqrt{2} \text{ eV, and } \beta_{OO} = 0.0 \text{ eV} .$$

2.   Determine from the Hückel calculation the electron affinity, the ionization potential, and the lowest excitation energy for NO_2 .

3.   Calculate the charge and bond order matrix for NO_2, NO_2^-, and NO_2^+ . NO_2^- forms an alkali salt with K^+ .

4.   How would you expect the K^+ ion to be oriented relative to the NO_2^- ion? Hydrogen reacts with NO_2 and forms the acid HNO_2 .

5.   Which atom would you expect the hydrogen atom to attack in the NO_2 radical?

NO_2 has an electronic excited state which dissociates to a NO radical and an oxygen atom $(NO_2 \rightarrow NO + O)$.

6.   Determine, by calculating the charge and bond order matrix of the singly excited states of NO_2, the excited state of NO_2 that is most likely to dissociate into NO plus O.

7.   Would you expect the singly excited states in NO_2 that do not dissociate to NO plus O to have a stronger or weaker NO bond strength than the NO bond strength in the ground state of NO_2 ?

8.   Which electronic excited states can be reached from the ground state via allowed electric dipole transitions?

9.   Which polarization direction of the incident light is most likely to cause a dissociation of NO_2 into NO + O ?

F.8 In this problem we will examine the geometry of the ground and excited states of substituted allene molecules using the Hückel π-electron approximation. Consider the dialkyl-substituted allene molecule given in Fig. F.8,1. The HC_1R_1 group in Fig. F.8,1 lies in the plane of the paper and the HC_3R_2 group is perpendicular to that plane.

1.   Which point group does the molecule belong to when
(a) $R_1 = R_2 = H$?
(b) $R_1 = R_2$ and both differ from hydrogen?
(c) R_1, R_2 and H are all different?

Consider now a substituted allene molecule for which R_1, R_2, and H all are different. Assume that both groups R_1 and R_2 are saturated alkyl groups.

2.   How many π electrons does the molecule in Fig. F.8, 1 contain?
We will now perform a Hückel π-electron calculation on the "twisted" allene-type molecule, that is, the molecule where the HC_1R_1 group forms an angle θ with the plane of the paper. In the Hückel calculation it is thus only the $p\pi$ orbital on carbon in the $HC_1 R_1$ group that changes position in space when θ varies. For $\theta = 0$ all β integrals are set equal to a constant β. The β integrals in the twisted configuration, $\beta(\theta)$, are given by the relation

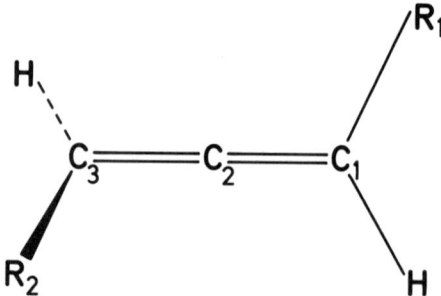

Fig. F.8, 1. The geometry of dialkyl-substituted allene.

$$\frac{\beta(\theta)}{\beta} = \frac{S(\theta)}{S(\theta = 0)} \qquad \text{(F.8, 1)}$$

where $S(\theta)$ is the overlap integral between the neighboring $2p\pi$ orbitals. We assume that overlap integrals can be neglected except when evaluating the β integrals.

3. Determine the Hückel MO energies and plot these for θ in the range $0 \leq \theta \leq \pi$.
4. Which values of θ correspond to the most stable and most unstable conformation of twisted alkyl-substituted allene molecules?
5. How large is the π-rotational barrier (expressed in terms of β) between the molecule in Fig. F.8, 1 and the enantiomeric compound, that is, the molecule that has H and R_1 interchanged in the $HC_1 R_1$ group?
6. Which geometry would the simple π-electron description predict for a twisted allene-type molecule in the first electronically excited state?

F.9 The *cis*- and *trans*-1, 3-butadiene molecules are displayed in Fig. F.9, 1. The cis-trans isomerization of 1, 3-butadiene can be considered as an internal rotation around the $C_2 - C_3$ bond.

1. What is the point group symmetry of *cis*-butadiene, *trans*-butadiene, and the intermediate compound for which $0 < \theta < \pi$?

We will perform a Hückel π-electron calculation in order to estimate the magnitude of the rotational barrier for the cis-trans isomerization of butadiene. We assume that all the Hückel α parameters are equal to a constant α. For *cis*-butadiene all β parameters are set equal to a constant β. For $\theta \neq 0$ the β_{23} parameter is assumed to vary with θ, whereas the β_{12} and β_{34} parameters are independent of θ and equal to β. We assume that the β_{23} parameter is proportional to the overlap between the neighboring $2p\pi$ atomic orbitals [see Eq.(F.8,1)] and that overlap integrals can be neglected otherwise in the calculation.

2. Determine the ground-state Hückel π-electron energy, E_{Π}, as a function of the angle θ and the parameters α and β.
3. Plot E_{Π} versus θ and indicate the location of *cis*- and *trans*-butadiene together with the "transition state" for the isomerization reaction.

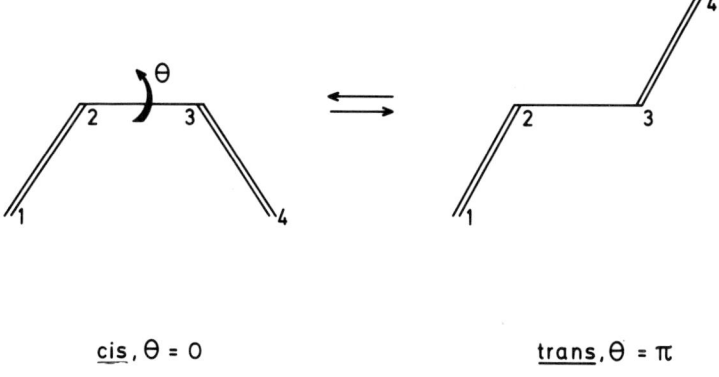

Fig. F.9, 1. The geometry of *cis*- and *trans*-1, 3-butadiene.

4. Determine (in units of β) the π-electron contribution to the height of the energy barrier ΔE (rotational barrier) for the cis-trans isomerization reaction.

5. Determine (in units of β) the π-electron contribution to the torsional force constant of *cis*- and *trans*-butadiene around the $C_2 - C_3$ bond.

The numerical value usually adopted for the β-parameter depends on the property to be calculated. For ground-state properties, such as the rotational barrier, the value is normally chosen within the range -18 to -36 kcal mole^{-1} (see, e.g., Salem 1966).

6. Compute the rotational barrier and torsional force constant for the two limiting values of β.

Experimentally, the height of the rotational barrier relative to the ground state of *trans*-butadiene is determined to be $\Delta E = 5.0$ kcal mole^{-1} (Aston *et al.* 1946). The torsional force constant of *trans*-butadiene is measured to be $f = 7.03$ kcal mole^{-1} rad^{-2} (Herzberg 1966).

7. Compare the theoretical and experimental values for f and ΔE. What is the reason for the large discrepancies between the calculated and experimental values?

The trans form is more stable than the cis form. The difference in energy is 2.3 eV (Aston *et al.* 1946).

8. Argue qualitatively why this is the case and why this difference can never be reproduced by consideration of the π-electron system alone.

F.10 In this problem we will examine the changes in the singlet excitation spectrum of benzene caused by a single substitution in the benzene ring. To describe these changes we use the π-electron approximation. It turns out to be advantageous to use the *complex* Hückel MO's of benzene:

$$\psi_r = \frac{1}{\sqrt{6}} \sum_{\mu=0}^{5} \exp\left(\frac{2\pi i r \mu}{6}\right)\phi_\mu, \qquad r = 0, \pm1, \pm2, 3 \qquad \text{(F.10, 1)}$$

where ϕ_μ indicates the carbon $2p\pi$ atomic orbital on center μ. ψ_r represents linear combinations of the real MO's in Eq. (F.1, 3) for $N = 6$. The corresponding orbital energies are (see Eq. (F.1, 5))

$$\varepsilon_r = \alpha + 2\beta\cos\left(\frac{2\pi r}{6}\right), r = 0, \pm1, \pm2, 3 \qquad \text{(F.10, 2)}$$

1. Show that the MO's transform according to the following irreducible representation of D_{6h}:

$$\psi_0(a_{2u}), \psi_{\pm 1}(e_{1g}), \psi_{\pm 2}(e_{2u}), \text{ and } \psi_3(b_{2g})$$

2. Write the electronic configuration, the molecular term symbol, and a single determinantal wave function for the ground state of benzene.

The excited states of benzene, originating from promotion of one electron from the highest occupied molecular orbital, $\psi_{\pm 1}$, to the lowest unoccupied molecular orbital, $\psi_{\pm 2}$, are degenerate in the Hückel approximations. This degeneracy is lifted by the electron repulsion.

3. Determine the molecular term symbols for the singly excited states originating from promotion of one electron from $\psi_{\pm 1}$ to $\psi_{\pm 2}$.

In the following we will consider the singlet states from question 3. The wave functions for these states and for the ground state can be expressed as linear combinations of the following Slater determinants:

$$V_{0,0} = |\psi_0 \bar\psi_0 \psi_1 \bar\psi_1 \psi_{-1} \bar\psi_{-1}| \qquad \text{(F.10, 3)}$$

$$V_{-1,-2} = \frac{1}{\sqrt{2}}(|\psi_0 \bar\psi_0 \psi_1 \bar\psi_1 \psi_{-1} \bar\psi_{-2}| - |\psi_0 \bar\psi_0 \psi_1 \bar\psi_1 \bar\psi_{-1} \psi_{-2}|) \qquad \text{(F.10, 4)}$$

$$V_{1,2} = \frac{1}{\sqrt{2}}(|\psi_0 \bar\psi_0 \psi_{-1} \bar\psi_{-1} \psi_1 \bar\psi_2| - |\psi_0 \bar\psi_0 \psi_{-1} \bar\psi_{-1} \bar\psi_1 \psi_2|) \qquad \text{(F.10, 5)}$$

$$V_{-1,2} = \frac{1}{\sqrt{2}}(|\psi_0 \bar\psi_0 \psi_1 \bar\psi_1 \psi_{-1} \bar\psi_2| - |\psi_0 \bar\psi_0 \psi_1 \bar\psi_1 \bar\psi_{-1} \psi_2|) \qquad \text{(F.10, 6)}$$

$$V_{1,-2} = \frac{1}{\sqrt{2}}(|\psi_0 \bar\psi_0 \psi_{-1} \bar\psi_{-1} \psi_1 \bar\psi_{-2}| - |\psi_0 \bar\psi_0 \psi_{-1} \bar\psi_{-1} \bar\psi_1 \psi_{-2}|) \qquad \text{(F.10, 7)}$$

The singlet states are given as (Salem 1966, p. 398)

$$|{}^1B_{2u} > (4.72 \text{ eV}) = \frac{1}{\sqrt{2}\,i}(V_{1,-2} - V_{-1,2}) \qquad \text{(F.10, 8)}$$

$$|{}^1B_{1u} > (5.96 \text{ eV}) = \frac{1}{\sqrt{2}}(V_{-1,2} + V_{1,-2}) \qquad \text{(F.10, 9)}$$

$$|^1E_{1u} \rangle (6.76 \text{ eV}) = \begin{cases} -\dfrac{1}{\sqrt{2}}(V_{-1,-2} + V_{1,2}) \\ \dfrac{1}{\sqrt{2}i}(V_{1,2} - V_{-1,-2}) \end{cases}$$ (F.10, 10)

where the numbers in parentheses are the energies of the states relative to the ground state.

Introduction of a substituent in the benzene molecule will change the magnitudes and perhaps the ordering of the lowest excited singlet states. The effect of a single substituent may be treated in a model where the α parameter for the benzene carbon atom nearest to the substituent group changes to $\alpha + \Delta\alpha$. All other α parameters and the bond parameters β are left unchanged. The Hamiltonian for the substituted benzene molecule can thus be written as

$$\hat{H} = \hat{H}_0 + \hat{H}_1$$ (F.10, 11)

with the perturbation being

$$\hat{H}_1 = \sum_{i=1}^{6} \hat{h}_1(i)$$

and where $\hat{H}_0$ is the Hamiltonian for benzene. $\hat{H}_1$ is determined through its matrix element

$$\langle \phi_\mu | \hat{h}_1 | \phi_{\nu} \rangle = \Delta\alpha_0 \delta_{\mu\nu} \delta_{\mu 0}$$ (F.10, 12)

The $\mu = 0$ labels the atomic orbital of the carbon atom at which the single substitution takes place.

4.   Determine the matrix elements of $\hat{H}_1$ within the basis given in Eqs. (F.10, 3–7).

5.   Perform a first-order perturbation calculation in $\hat{H}_1$ of the total energies for the ground state and for the excited states whose wave functions are given in Eqs. (F.10, 8–10). Show that the excitation energies at this level of approximation are the same for a singly-substituted and unsubstituted benzene molecule.

6.   Determine by performing a second-order perturbation calculation the shift in the *lowest* singlet excitation energy ($|^1A_{1g}\rangle \rightarrow |^1B_{2u}\rangle$) for the singly-substituted benzene. The sum over states in the second-order perturbation calculation should include only the singlet states that have explicitly been considered in this problem.

The lowest singlet excitation energy for a number of substituted benzenes are (Weast 1977)

aniline	4.40 eV
phenol	4.45 eV
toluene	4.60 eV
chlorobenzene	4.55 eV.

It is normally argued that the electron-donating ability of the functional groups [see, e.g., Moore and Barton (1978)] decreases from NH_2 to CH_3 such that we have the ordering

$$NH_2 \gtrsim OH > Cl \gtrsim CH_3$$ (F.10, 13)

7.   Is the magnitude of the lowest singlet excitation energy of benzene relative to those of the substituted benzenes in agreement with the prediction in question 6? Are the relative energies of the lowest excitation energy of the four singly substituted benzene compounds as expected from Eq. (F.10, 13)?
The lowest singlet excitation energy of pyridine is 4.96 eV (Kasha 1961).

8.   Why is the ordering of the lowest singlet excitation energy of benzene and of pyridine not in agreement with the prediction of question 6?

F.11 In this problem we analyze the geometry of the cyclopentadienyl radical, C_5H_5. We use the Hückel π-electron approximation and we assume initially that the symmetry of the cyclopentadienyl radical is D_{5h} (see Fig. F.11, 1).

1.   Determine the Hückel MO's and orbital energies of the cyclopentadienyl radical.

2.   Determine the term symbol of the ground state of the cyclopentadienyl radical.

The ground state of the cyclopentadienyl radical will be Jahn–Teller distorted (why?). In the Jahn–Teller distorted radical all bond lengths will change. In this problem, however, we consider a simplified treatment where we assume that the distortion causes equal changes in the lengths of two non-neighboring carbon–carbon bonds and leaves all other carbon–carbon bonds unchanged. We further assume that the change in the carbon–carbon bond length can be described by a change $\delta\beta$ in the β and that only the HOMO's are affected by the change in geometry (the frontier orbital approximation). Notice that the HOMO is doubly degenerate in the present case. The Hamiltonian for the Jahn–Teller distorted molecule may thus be written as

$$\hat{H} = \hat{H}_0 + \hat{H}_1$$

where $\hat{H}_1 = \sum_{i=1}^{5} \hat{h}_1(i)$ can be considered a perturbation operator that is defined through its matrix elements in the atomic orbital representation

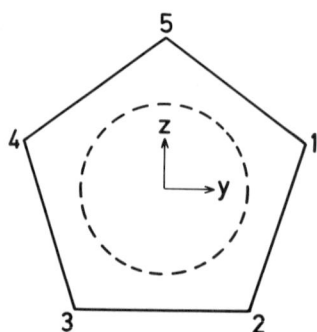

Fig. F.11, 1. The geometry of the cyclopentadienyl radical.

$$\langle \phi_\mu | \hat{h}_1 | \phi_\nu \rangle = \begin{cases} \delta\beta & \text{if } \mu \text{ and } \nu \text{ are neighbors and both are the} \\ & \text{end points of the bond that changes length} \\ 0 & \text{otherwise} \end{cases}$$

3. Determine by carrying out a first-order perturbation calculation in $\hat{h}_1$ which of the HOMO's will be occupied by the unpaired electron if
   (i) the carbon–carbon bond lengths are extended;
   (ii) the carbon–carbon bond lengths are reduced.
   [*Hint*: One of the MO's has a node on atom 5 (the sine solution, ψ_1^S). The mathematically simplest solution is obtained by varying the carbon–carbon bonds 1–2 and 3–4.]

4. Show that only a *reduction* of two opposite bond lengths will lead to a lower *total* π-electron energy in the frontier orbital approximation.

5. What is the point group symmetry of the distorted pentadienyl radical?

6. Determine the irreducible representations spanned by the two HOMO's in the point group of the distorted radical. Which of the two molecular orbitals has the lowest orbital energy if the bond lengths of the distorted cyclopentadienyl radical are reduced [question 3, point (ii)]?

The result of question 6 could equally well have been obtained by using group theoretical arguments which we will show in the last part of the problem.

7. Determine in the point group of the distorted pentadienyl radical the irreducible representations spanned by the MO's determined in question 1.

8. Show that the reduction in point group symmetry will cause a splitting of the two HOMO's.

9. What is the energy ordering of the MO's in the point group of the distorted molecule? Indicate the electronic configuration and ground-state term symbol.

F.12 In this problem we examine the geometry of 1, 3-cyclobutadiene in the ground and lowest excited states. We use the Hückel π-electron approximation and assume initially that cyclobutadiene is a square molecule belonging to the D_{4h} point group.

1. Determine the Hückel MO's and the irreducible representations spanned by these MO's in D_{4h}.

2. Determine the ground-state π-electron configuration and the molecular terms originating from this configuration. Would you expect a Jahn–Teller distortion of cyclobutadiene in any of these states?

3. Determine the term symbols for the states originating from the three singly excited electronic configurations.

Ab initio calculations (Borden *et al.* 1978) and experimental spectral data have established that the two lowest states of cyclobutadiene have $^3A_{2g}$ and $^1B_{2g}$ symmetries. Contrary to the prediction of Hund's rule, the $^1B_{2g}$ state seems to lie lower than $^3A_{2g}$ by approximately 0.8 eV at the square geometry (Borden 1975). The states of 1, 3-cyclobutadiene that originate from the ground-state electronic configuration will not show a normal (first-order) Jahn–Teller distortion (see question 2). However, if there are close-lying states that will "mix" (i.e., form "bonding" and "antibonding" states) when the symmetry of the molecule is lowered, it may still be energetically favorable for the molecule to have lower point group symmetry. This type of energy lowering is referred to as the *pseudo-Jahn–Teller* (Salem 1966) or the *second-order Jahn–Teller "effect"* (Pearson 1976).

As a rule of thumb we may use that if the lowering of point group symmetry allows mixing of states originating from the same electronic configuration, then there will be a second-order Jahn–Teller distortion of the molecule. If the mixing is between states from different electronic configurations, a geometry distortion is less likely.

For cyclobutadiene we have three possible in-plane vibrations which may lead to lower symmetry conformations, namely:

(i) a *rectangular* distortion, which lengthens two opposite bonds and shortens the remaining two;

(ii) a *rhombohedral* distortion, which can be represented by leaving the position of two opposite atoms unchanged and moving the other two atoms simultaneously inward or outward along the diagonal connecting the two atoms;

(iii) a *trapezoidal* distortion, which lengthens one C–C bond, shortens the opposite bond with the same amount, and leaves the remaining two bonds unchanged in length.

4. Determine the point group of the distorted cyclobutadiene in cases (i)–(iii).
5. Which of the states originating from the four electronic configurations in questions 2 and 3 will mix with the $^1B_{2g}$ and $^3A_{2g}$ states if we perform a rectangular, a rhombohedral, or a trapezoidal distortion of square cyclobutadiene, respectively?

Ab initio calculations (Borden *et al.* 1978) show that cyclobutadiene is rectangular in its ground state ($^1B_{2g}$) but square in the excited $^3A_{2g}$ state.

6. Is this in agreement with the second-order Jahn–Teller distortions predicted in question 5?

F.13 In this problem we perform a Pariser–Parr–Pople (PPP) calculation on the allyl cation.

1. Determine the sets of symmetry orbitals and the Hückel MO's and Hückel MO energies for the allyl cation.
2. Determine the Hückel charge–bond order matrix.

We will now determine the set of self-consistent MO's within the PPP approximation and use the Hückel charge–bond order matrix to form an initial guess of the PPP Fock matrix $\mathbf{F}^{(0)}$. The valence ionization potential for carbon is I_C and the effective π-electron charge on the carbon atoms, Z_C, is 1. The resonance integrals, β^c, between two neighboring carbon atoms are -2.32 eV and the β parameters are zero for non-neighbor atoms. The electron repulsion parameters are determined from the Mataga–Nishimoto (1957) approximation to be

$$\gamma_{\mu\nu} = \begin{cases} 10.84 \text{ eV} & \mu = \nu \ (\gamma_{11}) \\ 5.28 \text{ eV} & \mu \text{ and } \nu \text{ neighbors } (\gamma_{12}) \\ 3.84 \text{ eV} & \mu \text{ and } \nu \text{ non-neighbors } (\gamma_{13}) \end{cases}$$

All results in the following questions should be expressed in terms of I_C.

3. Form the zeroth-order PPP Fock matrix, $\mathbf{F}^{(0)}$.
4. Solve the eigenvalue problem for $\mathbf{F}^{(0)}$.

5.   Determine the new charge–bond order matrix.

Continue the iterative process described through questions 3–5 until the difference between the individual charge–bond order matrix elements in two consecutive iterations is less than 0.0005 a.u.

6.   What are the self-consistent PPP MO's and orbital energies?

F.14 In this problem we will use the Pariser–Parr–Pople (PPP) method to calculate the lowest triplet–triplet excitation energy of 1, 3-cyclobutadiene.

1.   Argue that the Hückel π MO's of problem F.12, question 1, are also the self-consistent field (SCF) PPP MO's.

2.   Express the SCF PPP molecular orbital energies in terms of the PPP parameters I_C, β^c, γ_{11}, and γ_{12} (for a definition of these quantities, see the introduction to this section).

Let us now compare the PPP orbital energies and the Hückel orbital energies that were calculated in problem F.12, question 1.

3.   Show that the Hückel α and β parameters may be uniquely identified in terms of PPP parameters by requiring that the Hückel and PPP orbital energies must be equal.

We found in problem F.12 that $^3A_{2g}$ was the lowest state of triplet spin multiplicity and that cyclobutadiene was quadratic in this state.

4.   Determine the *two* lowest excited electronic configurations that yield excited states which can be reached in an electric dipole allowed transition from the $^3A_{2g}$ state. What is the Hückel excitation energy for an excitation from $^3A_{2g}$ to these two excited states?

5.   Determine the $M_S = 1$ component of the single determinantal wave function that represents the $^3A_{2g}$ state and the two "allowed" excited states from question 4.

6.   Determine the PPP excitation energy (i.e., the *total* energy difference for the excitation) for the excitation from $^3A_{2g}$ to the two "allowed" excited states. The result should be expressed in terms of β^c and γ_{13} [Remember that the PPP method uses the "zero differential overlap" (ZDO) approximation.]

7.   Does the inclusion of the electronic repulsion terms increase or decrease the excitation energies?

By comparing the answers to questions 4 and 6, we see that the excitation from $^3A_{2g}$ to the two considered excited states are degenerate both in the Hückel and in the PPP approximation.

8.   Would a configuration interaction (CI) calculation that includes the two "allowed" excited states lift the degeneracy of the two excitation energies?

F.15 The ethylene molecule is known to have a dipole allowed π–π^* absorption transition at 1615Å (Platt *et al.* 1949). The oscillator strength for this transition is $f = 0.3$ (Zelikoff and Watanabe 1953). In this problem we will calculate the excitation energy and the oscillator strength for this transition at various levels of approximation using the π-electron approximation.

Let ϕ_A and ϕ_B be the two *nonorthogonal* $2p\pi$ atomic orbitals of ethylene.

1.   Determine the normalized π MO's and their transformation property in D_{2h}. Determine the MO energies using a Hückel approach where the overlap S between the atomic $2p\pi$ orbitals *is* included. Express the result in terms of the parameters α, β, and the overlap S.

Using the experimental C–C bond length of ethylene (1.366Å) the overlap is calculated to be (Mulliken *et al.* 1949)

$$S = \int \phi_A \phi_B \, d\tau = 0.27$$

when ϕ_A and ϕ_B are approximated by Slater-type orbitals. For the resonance integrals we will use the value $\beta = -3.10$ eV while the parameter α is set equal to -10.3 eV.

2. Use the values of α, β, and S to determine the excitation energy for the π–π^* transition of ethylene within the Hückel approximation.

3. Determine a Slater determinantal wave function for the ground (N) and first excited singlet state (V) of ethylene. Determine further the term symbol for these states. What is the polarization direction of the N-to-V transition (the π–π^* transition)?

4. Determine the transition moment within the dipole length formulation for the N-to-V transition, using the wave functions from the previous question.

5. Determine the dipole length oscillator strength for the π–π^* transition within the Hückel approximation.

Let us now consider the π–π^* transition using the Pariser–Parr–Pople (PPP) method. Recall that $S = 0$ as a result of ZDO approximation.

6. Why are the π MO's the same in the PPP and the Hückel methods? Determine the PPP MO energies expressed in terms of I_C, β^c, γ_{11}, and γ_{12}.

Let $\hat{H}$ be the Hamiltonian for ethylene, $| \text{N} \rangle$ the determinant that represents the ground-state wave function, $|\Psi_1^2\rangle$ a singly excited state wave function in which an electron is excited from the occupied MO ψ_1 into the virtual MO ψ_2, and $| \text{Z} \rangle$ the wave function for the corresponding doubly excited state.

7. Show that

$$\langle \Psi_1^2 | \hat{H} | \Psi_1^2 \rangle = E_N + \varepsilon_2 - \varepsilon_1 - J_{12} + \Lambda K_{12} \qquad \text{(F.15, 1)}$$

where E_N is the energy of the ground state, ε_k the orbital energy of MO k, J_{12} and K_{12} the Coulomb and exchange integrals, respectively, and $\Lambda = 0$ for the triplet excited state $|^3\Psi_1^2\rangle$ and $\Lambda = 2$ for the singlet excited state $|^1\Psi_1^2\rangle$.

A standard choice of the PPP parameters is (Parr 1963)

$$\beta^c = -2.93 \text{ eV}$$

$$\gamma_{11} = 10.53 \text{ eV}$$

$$\gamma_{12} = 7.38 \text{ eV}$$

8. Determine, for the $|\text{N}\rangle \rightarrow | \text{V} \rangle$ (i.e., $|^1\Psi_1^2\rangle$), the excitation energy and the dipole length oscillator strength within the PPP approximation.

The experimental energy difference between the first excited singlet and triplet state of ethylene (the "singlet–triplet" separation) is 3.0 eV (Parr 1963).

9. Determine the singlet–triplet separation in ethylene in (i) the Hückel and (ii) the PPP method.

We will now improve on the PPP calculation by performing a full CI calculation for ethylene within the basis $\{\phi_A, \phi_B\}$.

10. Show that

$$\langle {}^1\Psi_1^2 | \hat{H} | N \rangle = 0 \tag{F.15, 2}$$

$$\langle Z | \hat{H} | Z \rangle = E_N + 2\varepsilon_2 - 2\varepsilon_1 - 4J_{12} + J_{11} + J_{22} + 2K_{12} \tag{F.15, 3}$$

$$\langle Z | \hat{H} | N \rangle = K_{12} \tag{F.15, 4}$$

11. Determine the correlation energy in the full CI approximation for the ground state $|N\rangle$ of ethylene.

12. Determine the ground state $|N\rangle$ and the excited state $|V\rangle$ wave functions within the full CI approximation.

13. Determine the transition moment and oscillator strength for the lowest allowed singlet transition $(\pi-\pi^*)$ in ethylene within the full CI approximation. Compare this result with the Hückel result, the PPP single configurational result of question 8, and the experimental results.

SOLUTIONS

F.1 1. Trial solutions:

$$c_{k\mu} = A \sin(k\mu) \quad \text{or} \quad c_{k\mu} = B \cos(k\mu) \tag{F.1, I}$$

Boundary condition $c_{k\mu} = c_{k(N+\mu)}$ which inserted in Eq. (F.1, I) implies

$$k = \frac{2p\pi}{N}, \quad p = 0, \pm 1, \pm 2, \dots$$

All *different* solutions for $c_{p\mu}$ are obtained with the restrictions on p given in Eq. (F.1, 4). The A and B constants are determined from the normalization conditions.

2. The Hückel secular equation for a cyclic system is

$$c_{p(\mu-1)} + xc_{p\mu} + c_{p(\mu+1)} = 0 \tag{F.1, II}$$

where

$$x = \frac{\alpha - \varepsilon_p}{\beta}$$

and $\mu \pm 1$ labels neighboring atoms to carbon atom μ. Inserting either of the $c_{p\mu}$ solutions of Eq. (F.1, 3) in Eq. (F.1, II) gives

$$x = -2 \cos \frac{2\pi p}{N} \quad \text{Q.E.D.}$$

N even: $p = 0$ and $N/2$ nondegenerate (the sine solution vanishes), all other p doubly degenerate.

N odd: $p = 0$ nondegenerate (the sine solution vanishes), all other p doubly degenerate.

3. The ground-state π-electron energy is

$$E_{\Pi} = 2\alpha + 4\beta + 4 \sum_{p=1}^{n} (\alpha + 2\beta \cos \frac{2\pi p}{4n + 2})$$

$$= (4n + 2)\alpha + 4\beta + 8\beta \frac{\sin \dfrac{n\pi}{4n + 2} \cos \dfrac{(n + 1)\pi}{4n + 2}}{\sin \dfrac{\pi}{4n + 2}} \qquad \text{(F.1, III)}$$

$$= (4n + 2)\alpha + \frac{4\beta}{\sin \dfrac{\pi}{4n + 2}}$$

where we have used (Gradshteyn and Ryzhik 1965, p. 31)

$$\sum_{p=1}^{n} \cos pt = \frac{\sin \dfrac{nt}{2} \cos \dfrac{(n + 1)t}{2}}{\sin \dfrac{t}{2}} \qquad \text{(F.1, IV)}$$

Thus

$$E_D = \frac{E_{\Pi} - (4n + 2)(\alpha + \beta)}{4n + 2} \qquad \text{(F.1, V)}$$

and insertion of Eq.(F.1, III) in Eq.(F.1, V) gives Eq. (F.1, 6).

4.

$$\sin \frac{\pi}{4n + 2} \rightarrow \frac{\pi}{4n + 2} \qquad \text{for } n \rightarrow \infty$$
$$E_D \approx \frac{4\beta}{\pi} - \beta = 0.273\,\beta \qquad \text{for } n \rightarrow \infty \qquad \text{(F.1, VI)}$$

5.

$$\frac{\partial E_D}{\partial N} = \frac{-4\beta}{N^2 \sin^2 \dfrac{\pi}{N}} \left(\sin \frac{\pi}{N} - \frac{\pi}{N} \cos \frac{\pi}{N} \right)$$

For large values of N

$$\sin \frac{\pi}{N} - \frac{\pi}{N} \cos \frac{\pi}{N} \approx (\frac{1}{2} - \frac{1}{6}) \frac{\pi^3}{N^3} > 0$$

Since $\beta < 0$, E_D increases as N becomes larger (i.e., it becomes less negative) and the annulenes become less stable.

6.

$$P_{\mu\nu} = \frac{2}{4n+2} + 2\sum_{p=1}^{n}\left(\cos\frac{2\pi p\mu}{4n+2}\cos\frac{2\pi p\nu}{4n+2}\right.$$

$$\left.+ \sin\frac{2\pi p\nu}{4n+2}\sin\frac{2\pi p\nu}{4n+2}\right)\frac{2}{4n+2}$$

$$= \frac{2}{4n+2}\left(\frac{1}{2} + \sum_{p=1}^{n}\cos\frac{2\pi p(\mu-\nu)}{4n+2}\right) \qquad \text{(F.1, VII)}$$

$$= \frac{4}{4n+2}\left(\frac{1}{2} + \frac{\sin\frac{\pi n(\mu-\nu)}{4n+2}\cos\frac{\pi(n+1)(\mu-\nu)}{4n+2}}{\sin\frac{\pi(\mu-\nu)}{4n+2}}\right)$$

The last equality follows from Eq. (F.1, IV). Equation (F.1, VII) can straightforwardly be reduced to Eq. (F.1, 7).

7.

$$P_{\mu\nu} \rightarrow \frac{\sin\frac{\pi(\mu-\nu)}{2}}{\frac{\pi(\mu-\nu)}{2}}$$

F.2 1. Trial solution: $c_{p\mu} = A\sin(a\mu + b)$.

$$c_{p0} = 0 \quad \text{implies} \quad b = \pi q;\; q = 0, \pm1, \pm2, \ldots$$

Thus, disregarding phase factors,

$$c_{p\mu} = A\sin a\mu$$

$$c_{p(N+1)} = 0 \quad \text{implies} \quad a = \frac{p\pi}{N+1}, \qquad p = 0, \pm1, \pm2, \ldots \text{ (F.2, I)}$$

There are N *different* solutions to the Hückel secular problem which we may obtain by restricting p as indicated in Eq. (F.2, 2). A is determined from the normalization condition.

2. The Hückel secular equation in Eq. (F.1, II) is still valid, since $c_{p0} = c_{p(N+1)} = 0$. Inserting Eq. (F.2, 2) in Eq. (F.1, II) gives Eq. (F.2, 3).

3. Using Eq. (F.2, 3) we find that

$$E_{\Pi} = 2\sum_{p=1}^{N/2}\left(\alpha + 2\beta\cos\frac{p\pi}{N+1}\right)$$

$$= N\alpha + 4\frac{\sin\frac{N\pi}{4(N+1)}\cos\frac{1}{2}\left(\frac{N}{2}+1\right)\frac{\pi}{N+1}}{\sin\frac{\pi}{2(N+1)}}$$

$$= N\alpha + \frac{2\beta}{\sin \dfrac{\pi}{2N+2}} - 2\beta$$

where we have used Eq. (F.1, IV). Thus

$$E_{\mathrm{D}} = \frac{E_{\mathrm{II}} - N(\alpha + \beta)}{N}$$

which gives Eq. (F.2, 4).

4. Since

$$\frac{2}{N \sin \dfrac{\pi}{2N+2}} \to \frac{4}{\pi} \qquad \text{for} \quad N \to \infty$$

we find that

$$E_{\mathrm{D}} \to \frac{4\beta}{\pi} - \beta \qquad \text{for} \quad N \to \infty$$

which is identical to Eq. (F.1, VI).

F.3 1. D_{3h}
 2. C_{3v}
 3. Symmetry orbitals in D_{3h}

$$\chi_1(a_2'') = \frac{1}{\sqrt{3}}(\phi_1 + \phi_3 + \phi_5)$$

$$\chi_2(a_2'') = \frac{1}{\sqrt{3}}(\phi_2 + \phi_4 + \phi_6)$$

$$\chi_3(e'') = \frac{1}{\sqrt{6}}(2\phi_3 - \phi_1 - \phi_5)$$

$$\chi_3'(e'') = \frac{1}{\sqrt{2}}(\phi_1 - \phi_5)$$

$$\chi_4(e'') = \frac{1}{\sqrt{6}}(2\phi_4 - \phi_2 - \phi_6)$$

$$\chi_4'(e'') = \frac{1}{\sqrt{2}}(\phi_2 - \phi_6)$$

The Hückel secular determinant for a_2'' symmetry (χ_1, χ_2) is

$$\begin{vmatrix} \alpha + 2\beta' - E & \beta \\ \beta & \alpha - E \end{vmatrix} = 0$$

The orbital energies become

$$\varepsilon_\pm(a_2'') = \alpha + \beta' \pm (\beta^2 + \beta'^2)^{1/2}$$

In the e'' symmetry class, we obtain two identical 2×2 secular determinants from the sets $\{\chi_3, \chi_4\}$ and $\{\chi_3', \chi_4'\}$, respectively. The orbital energies become

$$\varepsilon_\pm(e'') = \alpha - \tfrac{1}{2}\beta' \pm \tfrac{1}{2}(4\beta^2 + \beta'^2)^{1/2}$$

Since
$(1)\varepsilon_+(a_2'') \leq \varepsilon_+(e'')$
$(2)(\beta'^2 + \beta^2)^{1/2} + \tfrac{1}{2}(\beta'^2 + 4\beta^2)^{1/2} \geq -\tfrac{3}{2}\beta',$ or

$$\varepsilon_+(a_2'') = \alpha + \beta' + (\beta^2 + \beta'^2)^{1/2}$$

$$\geq \alpha + \beta' - \frac{3}{2}\beta' - \tfrac{1}{2}(\beta'^2 + 4\beta^2)^{1/2}$$

$$= \varepsilon_-(e'')$$

we have the energy ordering indicated in Fig. F.3, I.

$$E_{II} = 6\alpha - 2[(\beta^2 + \beta'^2)^{1/2} + (4\beta^2 + \beta'^2)^{1/2}] \qquad (\text{F.3, I})$$

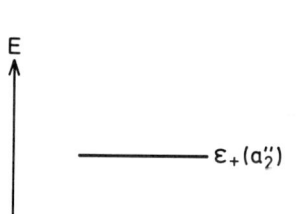

$$\varepsilon_+(e'')$$

E

$$\varepsilon_+(a_2'')$$

$$\varepsilon_-(e'')$$

$$\varepsilon_-(a_2'')$$

Fig. F.3, I. Hückel orbital energy diagram for trimethylene cyclopropane.

4.

$$\frac{\partial E_{\mathrm{II}}}{\partial \beta} = 2p_{12} + 2p_{34} + 2p_{56} = 6p_{12} \qquad \text{(F.3, II)}$$

Using Eqs. (F.3, I–II) we obtain

$$p_{12} = \frac{1}{3}[(1 + k^2)^{-\frac{1}{2}} + 4(4 + k^2)^{-\frac{1}{2}}] \quad \text{where } k = \frac{\beta'}{\beta} \qquad \text{(F.3, III)}$$

p_{13} may be evaluated in a similar way, and we find

$$p_{13} = \frac{1}{6}\frac{\partial E_{\mathrm{II}}}{\partial \beta'} = \frac{1}{3}[(1 + k^{-2})^{-\frac{1}{2}} + (1 + 4k^{-2})^{-\frac{1}{2}}] \qquad \text{(F.3, IV)}$$

For $k = 1$, $p_{12} = 0.862$ and $p_{13} = 0.385$.

5. $E_{\mathrm{D}} = 1.30\ \beta.$

6. The difference in delocalization energy indicates that benzene is more aromatic than trimethylene cyclopropane [see, however, Murrell *et al.* (1970), Chap. 15.5 for the validity of this rule]. Using the standard bond order–bond length relation [see, e.g., Levine (1974), Eq. 15.163]:

$$r_{rs}(\text{Å}) = 1.521 - 0.186\ p_{rs}$$

we obtain $(k = 1)$

$$r_{12} = 1.366\ \text{Å}$$

$$r_{13} = 1.449\ \text{Å}$$

and thus we have typical alternating single and double bond character in trimethylene cyclopropane, that is, the molecule is not aromatic. It is experimentally confirmed that r_{13} is much larger than r_{12} and that trimethylene cyclopropane has characteristic alternating single and double bond character.

7. $|\beta'| < |\beta|$ or $k < 1.$

8. Using Eqs. (F.3, III) and (F.3, IV) we obtain

$$\frac{d}{dk}(p_{12} - p_{13}) = -\frac{1}{3}[k(1 + k^2)^{-\frac{3}{2}} + 4k(4 + k^2)^{-\frac{3}{2}}$$

$$+ k^{-3}(1 - k^{-2})^{-\frac{3}{2}} + 4k^{-3}(1 + 4k^{-2})^{-\frac{3}{2}}] < 0$$

9. For $\beta = \beta'$, $k = 1$ and question 8 thus implies that $p_{12} - p_{13}$ is larger for $k < 1$, that is, the aromaticity decreases when we explicitly consider the effect of bond alteration on β.

F.4 1. The choice of coordinate axes is shown in Fig. F.4, I.
Symmetry orbitals in C_{2v}:

$$\chi_1(b_1) = \phi_1; \quad \chi_2(b_1) = \frac{1}{\sqrt{2}}(\phi_2 + \phi_5); \quad \chi_3(b_1) = \frac{1}{\sqrt{2}}(\phi_3 + \phi_4)$$

$$\chi_4(a_2) = \frac{1}{\sqrt{2}}(\phi_2 - \phi_5); \quad \chi_5(a_2) = \frac{1}{\sqrt{2}}(\phi_3 - \phi_4)$$

2. The Hückel secular determinant of b_1 symmetry (χ_1, χ_2, χ_3) is

$$\begin{vmatrix} \alpha + 2 - \varepsilon & 0.8\sqrt{2}\beta & 0 \\ 0.8\sqrt{2}\beta & \alpha - \varepsilon & \beta \\ 0 & \beta & \alpha + \beta - \varepsilon \end{vmatrix} = 0 \qquad \text{(F.4, I)}$$

Dividing Eq. (F.4, I) by β gives

$$\begin{vmatrix} x + 2 & 0.8\sqrt{2} & 0 \\ 0.8\sqrt{2} & x & 1 \\ 0 & 1 & x + 1 \end{vmatrix} = 0 \qquad \text{(F.4, II)}$$

where $x = (\alpha - E)/\beta$.
Equation (F.4, II) gives the third-order polynomial

$$x^3 + 3x^2 - 0.28\,x - 3.28 = 0$$

the roots of which may be determined by using the procedure described in the Appendix. The orbital energies are given in Fig. F.4, II together with the orbital energies of a_2 symmetry. The total energy is

$$E_{\text{II}} = 6\alpha + 9.130\,\beta$$

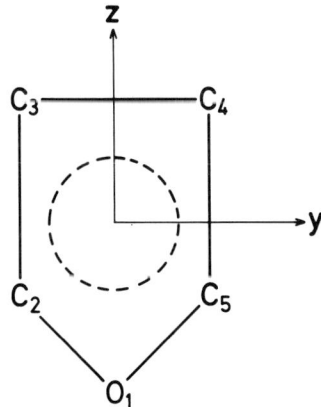

Fig. F.4, I. The geometry and choice of coordinate system of furan.

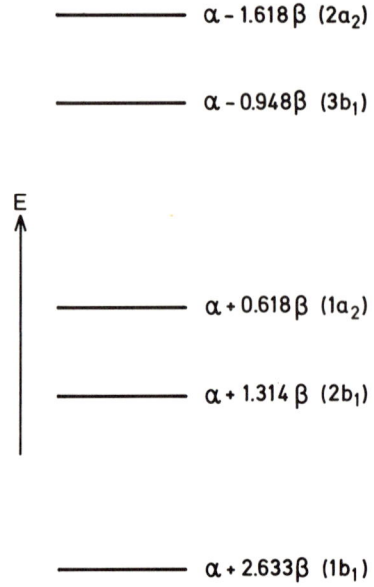

$$\text{———— } \alpha - 1.618\,\beta \quad (2a_2)$$

$$\text{———— } \alpha - 0.948\beta \quad (3b_1)$$

E

$$\text{———— } \alpha + 0.618\,\beta \quad (1a_2)$$

$$\text{———— } \alpha + 1.314\,\beta \quad (2b_1)$$

$$\text{———— } \alpha + 2.633\beta \quad (1b_1)$$

Fig. F.4, II. Hückel orbital energy diagram for furan.

3. Using Koopmans' theorem we can assign the π ionization potentials to ionizations out of the orbitals $1a_2$, $2b_1$, and $1b_1$ (increasing binding energy). That the π orbital energy ε_{1b_1} is lower than four σ orbital energies shows that the σ–π separability is questionable for furan. The Hückel approximation cannot describe the ionization out of the lowest π orbital of b_1 symmetry correctly. If the ionization out of the $1a_2$ and $2b_1$ orbitals was used to determine the parameters α and β, we obtain $\beta = -2.055$ eV and the ionization out of the ψ_{1b_1} orbital then would give an ionization potential of 13.0 eV. We would then, on the basis of the Hückel calculation, incorrectly assign the experimental peak at 13.0 eV to be an ionization out of the $1b_1$ orbital.

4. The orbitals of b_1 symmetry have amplitudes on oxygen. From the secular equations we find

$$\psi_{1b_1} = 0.836\,\phi_1 + 0.331\,(\phi_2 + \phi_5) + 0.202\,(\phi_3 + \phi_4)$$

$$\psi_{2b_1} = -0.443\,\phi_1 + 0.190\,(\phi_2 + \phi_5) + 0.605\,(\phi_3 + \phi_4)$$

The ψ_{1b_1} orbital is predominantly located on oxygen. The breakdown of the σ–π separability is thus caused by the introduction of the heteroatom oxygen.

F.5 1. The point group is D_{4h}. The C_2' axis and σ_v planes contain the atoms $O_4 - C_4$, $C_2 - O_2$ or $O_1 - C_1$, $C_3 - O_3$. The C_2'' axis and σ_d planes are perpendicular to the sides of the square spanned by the carbon atoms.

Symmetry orbitals:

$$\chi_1^C(a_{2u}) = \tfrac{1}{2}\{\phi_1 + \phi_2 + \phi_3 + \phi_4\};$$

$$\chi_2^C(b_{2u}) = \tfrac{1}{2}\{\phi_1 - \phi_2 + \phi_3 - \phi_4\}$$

$$\chi_3^C(e_g) = \frac{1}{\sqrt{2}}\{\phi_1 - \phi_3\}; \qquad \chi_4^C(e_g) = \frac{1}{\sqrt{2}}\{\phi_2 - \phi_4\}$$

and a similar set spanned by the oxygen π orbitals.

2. The Hückel secular determinant of b_{2u} symmetry becomes (χ_1^C, χ_2^O)

$$\begin{vmatrix} \alpha - 2\beta - \varepsilon & \beta \\ \beta & \alpha + \beta - \varepsilon \end{vmatrix} = 0$$

The orbital energies of b_{2u} symmetry together with the ones of a_{2u} and e_g symmetry are given in Fig. F.5, I.

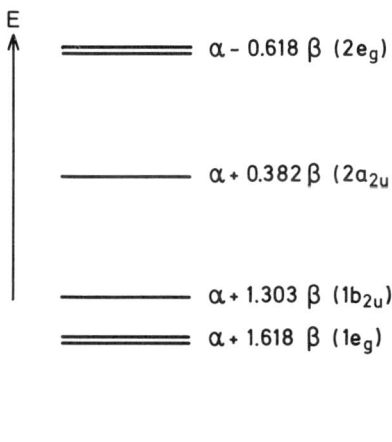

Fig. F.5, I. Hückel orbital energy diagram for squaric acid.

3. Ground state: $1a_{2u}^2\,1e_g^4\,1b_{2u}^2\,2a_{2u}^2$; $^1A_{1g}$
   Lowest excited singlet state:

   $$1a_{2u}^2\,1e_g^4\,1b_{2u}^2\,2a_{2u}^1\,2e_g^1;\ ^1E_u + \text{another state}$$

4. (x, y) transform as E_u. The transition is thus dipole allowed.
5. Jahn–Teller distortion of the molecule in the (degenerate) excited state splits the ^1E_u state into two nondegenerate states giving two close-lying absorption peaks.

F.6 1. The Coulson–Rushbrooke pairing theorem for alternant hydrocarbons gives

$$\psi_5 = -\,0.334\,\phi_1 + 0.308\,\phi_2 + 0.130\,\phi_3 - 0.394\,\phi_4 + 0.130\,\phi_5$$
$$+\, 0.308\,\phi_6 - 0.394\,\phi_7 + 0.595\,\phi_8$$
$$\psi_6 = -\,0.500\,\phi_2 + 0.500\,\phi_3 - 0.500\,\phi_5 + 0.500\,\phi_6$$
$$\psi_7 = -\,0.354\,\phi_1 + 0.354\,\phi_3 - 0.500\,\phi_4 + 0.354\,\phi_5 + 0.500\,\phi_7$$
$$-\,0.354\,\phi_8$$
$$\psi_8 = 0.513\,\phi_1 - 0.394\,\phi_2 + 0.329\,\phi_3 - 0.308\,\phi_4 + 0.329\,\phi_5$$
$$-\,0.394\,\phi_6 - 0.308\,\phi_7 + 0.144\,\phi_8 \qquad\qquad \text{(F.6, I)}$$

$\varepsilon_5 = \alpha - 0.662\,\beta$
$\varepsilon_6 = \alpha - 1.000\,\beta$
$\varepsilon_7 = \alpha - 1.414\,\beta$
$\varepsilon_8 = \alpha - 2.136\,\beta$

2. $\Psi(^1A') = |\psi_1\bar{\psi}_1\psi_2\bar{\psi}_2\psi_3\bar{\psi}_3\psi_4\bar{\psi}_4|$
   $\Psi(^3A', M_S = 1) = |\psi_1\bar{\psi}_1\psi_2\bar{\psi}_2\psi_3\bar{\psi}_3\psi_4\psi_5|$

3. Since styrene is an alternant hydrocarbon, $q_\mu = 1$ for all atoms in the ground state. The ψ_5 orbital contains the same expansion coefficients as the ψ_4 orbital and the atomic charge orders of the first excited triplet state therefore become identical to the ground-state charge orders, that is, all $q_\mu = 1$ for the first excited triplet state.

4. The spin density, ρ_μ, is the contribution from ψ_4 and ψ_5 to the charge density: that is,

   $\rho_1 = 2 \times 0.334^2 = 0.223$
   $\rho_2 = \rho_6 = 0.190$
   $\rho_3 = \rho_5 = 0.034$
   $\rho_4 = \rho_7 = 0.310$
   $\rho_8 = 0.708$

5. Carbon atom 4 has the largest spin density at the nucleus and the experimental assignment is thus as predicted by Eq. (F.6, 1).

F.7 1. We use the coordinate system in Fig. F.7, I.
Symmetry orbitals in C_{2v}:

$$\chi_1(b_1) = \frac{1}{\sqrt{2}}(\phi_{O1} + \phi_{O2}); \qquad \chi^2(b_1) = \phi_N$$

$$\chi_3(a_2) = \frac{1}{\sqrt{2}}(\phi_{O1} - \phi_{O2})$$

Hückel MO's and orbital energies:

$$\psi_{1b_1} = \frac{1}{2}(\phi_{O1} + \sqrt{2}\,\phi_N + \phi_{O2}); \qquad \varepsilon_{1b_1} = -14.0 \text{ eV}$$

$$\psi_{1a_2} = \frac{1}{\sqrt{2}}(\phi_{O1} - \phi_{O2}); \qquad \varepsilon_{1a_2} = -7.0 \text{ eV}$$

$$\psi_{2b_1} = \frac{1}{2}(\phi_{O1} - \sqrt{2}\,\phi_N + \phi_{O2}); \qquad \varepsilon_{2b_1} = 0.0 \text{ eV}$$

2. Electron affinity; EA $= E(NO_2) - E(NO_2^-) = -\varepsilon_{1a_1} = 7.0$ eV
Ionization potential; IP $= E(NO_2^+) - E(NO_2) = -\varepsilon_{1a_1} = 7.0$ eV
Excitation energy; $\Delta E = E(NO_2^*) - E(NO_2) = \varepsilon_{2b_1} - \varepsilon_{1a_1}$
$= 7.0$ eV

3.

Table F.7,I
Charge Orders, q_μ, and Bond Orders, $p_{\mu v}$, for NO_2, NO_2^-,
and NO_2^+ in Their Ground States

System	q_O	q_N	p_{ON}	p_{OO}
NO_2	1	1	$1/\sqrt{2}$	0
NO_2^-	1.5	1	$1/\sqrt{2}$	$-1/2$
NO_2^+	0.5	1	$1/\sqrt{2}$	$1/2$

4. Both oxygen atoms have a negative net charge of -0.5, and the K^+
ion will therefore be bound to NO_2^- on the positive side of the z-
axis (see Fig. F.7, I).

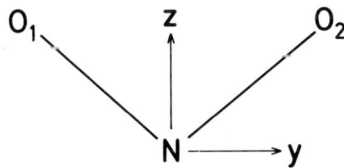

Fig. F.7, I. The choice of coordinate system for NO_2.

5. The unpaired electron from hydrogen is added in the $1a_2$ orbital, which is purely an oxygen MO.

6.

Table F.7,II
Charge Orders, q_μ, and Bond Orders, $p_{\mu\nu}$, for Singly Excited States of NO_2

Excited state	q_O	q_N	p_{ON}	p_{OO}
$\|1\rangle = \|\psi_{1a_1} \to \psi_{2b_2}; {}^2B_1\rangle$	0.75	1.5	$\sqrt{2}/4$	0.75
$\|2\rangle = \|\psi_{1b_1} \to \psi_{1a_2}; {}^2B_1\rangle$	1.25	0.5	$\sqrt{2}/4$	−0.75
$\|3\rangle = \|\psi_{1b_1} \to \psi_{2b_1}; {}^2A_2(2) + {}^4A_2\rangle$	1	1	0	0

Table F.7, II shows that NO_2 is likely to dissociate into $NO + O$ in state $|3\rangle$.

7. p_{ON} are smaller in the excited states and we expect a weaker NO bond strength in both of the excited states than in the ground state.

8. The ground state $|0\rangle$ is of ${}^2A_2$ symmetry. All transitions to the singly excited states are dipole allowed, since $A_2 \otimes B_1 = B_2$ and $A_2 \otimes A_2 = A_1$ and y transforms as B_2 and z as A_1.

9. The excitation from $|0\rangle$ to the ${}^2A_2$ components of $|3\rangle$ are polarized in the z-directions. The polarization direction of the incident light must thus be in the z-direction (see Fig. F.7, I) in order to cause a dissociation of NO_2 into $NO + O$.

F.8 1. (a) D_{2d}; (b) C_2; (c) C_1

2. Allene has two perpendicular π-electron systems, each containing two π electrons, that is, four π electrons in all.

3. In Fig F.8,I we have plotted the $2p\pi$ orbitals that are the basis functions for the Hückel π-electron calculation. The functions ϕ_2 and ϕ_3 form an angle of $\pi/2$ and ϕ_1 and ϕ_2 are in the same plane. When $R_1 C_1 H$ is rotated an angle θ, the new direction of ϕ_4 is as indicated by the dashed $2p\pi$ orbital. From Fig. F.8, I we see that $\beta_{24} = \beta \sin \theta$ and $\beta_{34} = \beta \cos \theta$.

The Hückel secular determinant within the $\{\phi_1, \phi_2, \phi_3, \phi_4\}$ basis is

$$\begin{vmatrix} \alpha - \varepsilon & \beta & 0 & 0 \\ \beta & \alpha - \varepsilon & 0 & \beta \sin \theta \\ 0 & 0 & \alpha - \varepsilon & \beta \cos \theta \\ 0 & \beta \sin \theta & \beta \cos \theta & \alpha - \varepsilon \end{vmatrix} = 0$$

which gives

$$\varepsilon_1(\theta) = \alpha + \beta(1 + \sin \theta)^{1/2}; \qquad \varepsilon_2(\theta) = \alpha + \beta(1 - \sin \theta)^{1/2}$$

$$\varepsilon_3(\theta) = \alpha - \beta(1 - \sin \theta)^{1/2}; \qquad \varepsilon_4(\theta) = \alpha - \beta(1 + \sin \theta)^{1/2}$$

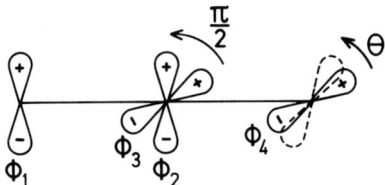

Fig. F.8, I. The $2p\pi$ orbitals that are used as basis functions for a Hückel π-electron calculation of allene.

The orbital energies are plotted as a function of θ in Fig. F.8, II.

4.

$$E_0 = 4\alpha + 2\beta[(1 + \sin\theta)^{1/2} + (1 - \sin\theta)^{1/2}]$$

Maximum total π-electron energy for $\theta = \pi/2$ (least stable conformation).

Minimum total π-electron energy for $\theta = 0, \pi$ (most stable conformation).

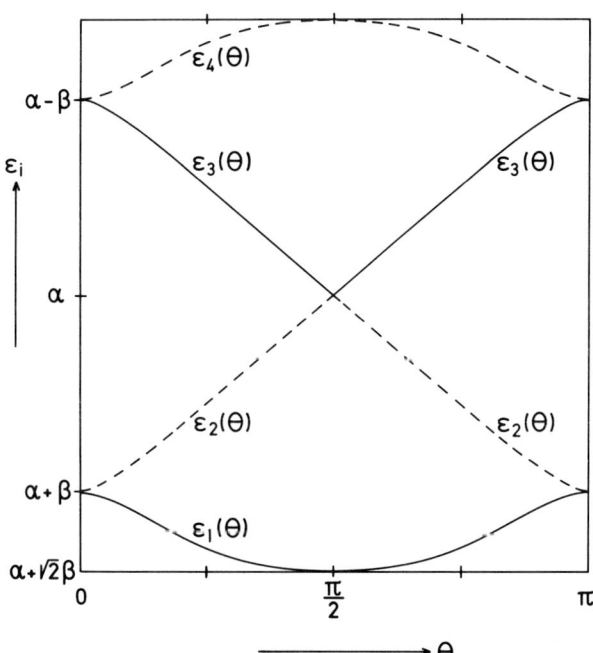

Fig. F.8, II. The Hückel orbital energies of allene as a function of the rotation angle θ.

5.

$$\Delta E = E_0\left(\theta = \frac{\pi}{2}\right) - E_0(\theta = 0) = 2(\sqrt{2} - 2)\beta$$

6. The total energy for the first excited state is

$$E_1 = 4\alpha + 2\beta(1 + \sin\theta)^{1/2}$$

which has its minimum value for $\theta = \pi/2$, that is a planar conformation of allene.

F.9 1. *cis*-butadiene: C_{2v}
     trans-butadiene: C_{2h}
     $0 < \theta < \pi$: C_2

   2. Symmetry orbitals in C_2 point group:

$$\chi_1(a) = \frac{1}{\sqrt{2}}(\phi_1 - \phi_4); \qquad \chi_2(a) = \frac{1}{\sqrt{2}}(\phi_2 - \phi_3)$$

$$\chi_3(b) = \frac{1}{\sqrt{2}}(\phi_1 + \phi_4); \qquad \chi_4(b) = \frac{1}{\sqrt{2}}(\phi_2 + \phi_3)$$

$\beta_{23} = \beta\cos\theta$
The Hückel secular determinant for a symmetry (χ_1, χ_2) is

$$\begin{vmatrix} \alpha - \varepsilon & \beta \\ \beta & \alpha - \beta\cos\theta - \varepsilon \end{vmatrix} = 0$$

the roots of which are

$$\varepsilon_\pm(a) = \alpha + \frac{1}{2}\beta\left(\cos\theta \pm \sqrt{4 + \cos^2\theta}\right)$$

The secular determinant of b symmetry (χ_3, χ_4) is

$$\begin{vmatrix} \alpha - \varepsilon & \beta \\ \beta & \alpha + \beta\cos\theta - \varepsilon \end{vmatrix} = 0$$

yielding

$$\varepsilon_\pm(b) = \alpha + \frac{1}{2}\beta\left(-\cos\theta \pm \sqrt{4 + \cos^2\theta}\right)$$

Since $\sqrt{4 + \cos^2\theta} > |\cos\theta|$ and $\beta < 0$, the two lowest roots are $\varepsilon_+(a)$ and $\varepsilon_+(b)$. Thus $E_{II} = 4\alpha + 2\beta\sqrt{4 + \cos^2\theta}$

   3. See Fig. F.9, I.
   4.

$$\Delta E = E_{II}\left(\theta = \frac{\pi}{2}\right) - E_{II}(\theta = 0) = (4 - 2\sqrt{5})\beta = -0.472\,\beta$$

$$(\text{F.9, I})$$

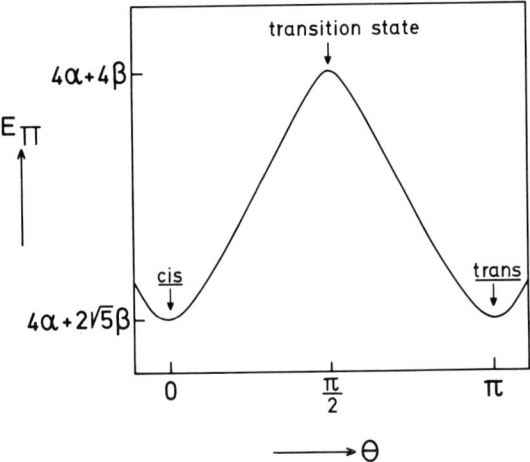

Fig. F.9, I. The rotational barrier for the interconversion of *cis*-1, 3-butadiene to *trans*-1, 3-butadiene.

5. Torsional force constant $f = (\partial^2 E_{\Pi}/\partial^2 \theta)_{\theta=0 \text{ or } \pi}$

$$\frac{\partial^2 E_{\Pi}}{\partial^2 \theta} = -\frac{\beta}{\sqrt{4 + \cos^2 \theta}}[2 \cos 2\theta + \tfrac{1}{2} \sin^2 2\theta (4 + \cos^2 \theta)^{-1}]$$

$$f = -\frac{2\beta}{\sqrt{5}}$$

(F.9, II)

6.

Table F.9,I
Rotational Energy Barrier and Torsional Force Constant of 1,3-Butadiene Calculated from Eqs. (F.9,I) and (F.9,II), Respectively

β (kcal mole^{-1})	f (kcal mole^{-1} rad^{-2})	ΔE (kcal mole^{-1})
−18	16.1	8.5
−36	32.2	17.0

7. The σ contribution to the rotational barrier and the force constant has to be considered explicitly before a quantitative correct description can be made of these quantities.

8. The energy difference is caused by, for example, "steric" effects which can only be described theoretically if the hydrogen atoms are included explicitly in the calculation.

F.10 1.

Table F.10,I
Character of the Reducible Representation Γ Spanned by the Six $2p\pi$ Atomic Orbitals in Benzene[a]

E	$2C_6$	$3C_3$	C_2	$3C_2'$	$3C_2''$	i	$2S_3$	$2S_6$	σ_h	$3\sigma_d$	$3\sigma_v$
6	0	0	0	−2	0	0	0	0	−6	0	2

[a] $\Gamma = a_{2u} + b_{2g} + e_{1g} + e_{2u}$.

ψ_0 transforms as a_{2u} and ψ_3 as b_{2g} (the only other nondegenerate representation). The character for the C_2 operation of D_{6h} is −2 for e_{1g} and +2 for e_{2u}. Applying the C_2 operation to $\psi_{\pm 1}$ gives

$$C_2\psi_{\pm 1} = -\psi_{\pm 1}$$

The set $\{\psi_{+1}, \psi_{-1}\}$ therefore transform as e_{1g} and the set $\{\psi_{+2}, \psi_{-2}\}$ must then transform as e_{2u}.

2. $a_{2u}^2 e_{1g}^4$; $^1A_{1g}$
$$|^1A_{1g}\rangle = |\psi_0 \bar{\psi}_0 \psi_1 \bar{\psi}_1 \psi_{-1} \bar{\psi}_{-1}|$$

3. $e_{1g} \otimes e_{2u} = B_{1u} + B_{2u} + E_{1u}$
The molecular term symbols are

$$^1B_{1u} + {}^1B_{2u} + {}^1E_{1u} + {}^3B_{1u} + {}^3B_{2u} + {}^3E_{1u}$$

4. Consider initially the diagonal matrix element

$$\langle |\psi_0 \bar{\psi}_0 \psi_1 \bar{\psi}_1 \psi_{-1} \bar{\psi}_{-2} || \hat{H}_1 || \psi_0 \bar{\psi}_0 \psi_1 \bar{\psi}_1 \psi_{-1} \bar{\psi}_{-2} |\rangle$$
$$= 2\langle \psi_0 | \hat{h}_1 | \psi_0 \rangle + 2\langle \psi_1 | \hat{h}_1 | \psi_1 \rangle + \langle \psi_{-1} | \hat{h}_1 | \psi_{-1} \rangle + \langle \psi_{-2} | \hat{h}_1 | \psi_{-2} \rangle$$

$$\text{(F.10, I)}$$

where we have used the Slater–Condon rules for the one-electron operator. Since all MO's have a coefficient $1/\sqrt{6}$ on carbon atom 0 [see Eq. (F.10, 1)], the atom where the substitution takes place, we get

$$\langle \psi_i | \hat{h}_1 | \psi_j \rangle = \frac{1}{6}\Delta\alpha_0, \qquad i, j = 0, \pm 1, \pm 2, 3 \qquad \text{(F.10, II)}$$

Thus, the diagonal matrix element in Eq. (F.10, I) becomes $\Delta\alpha_0$. Using Eq. (F.1, II) and the Slater–Condon rules, we find similarly that

$$\langle V_{0,0}|\hat{H}_1|V_{0,0}\rangle = \langle V_{-1,-2}|\hat{H}_1|V_{-1,-2}\rangle = \langle V_{1,2}|\hat{H}_1|V_{1,2}\rangle$$
$$= \langle V_{-1,2}|\hat{H}_1|V_{-1,2}\rangle = \langle V_{1,-2}|\hat{H}_1 V_{1,-2}\rangle = \Delta\alpha_0$$
$$(\text{F.10, III})$$

$$\langle V_{-1,-2}|\hat{H}_1|V_{1,2}\rangle = \langle V_{-1,2}|\hat{H}_1|V_{1,-2}\rangle = 0 \qquad\qquad (\text{F.10, IV})$$

$$\langle V_{-1,-2}|\hat{H}_1|V_{-1,2}\rangle = \langle V_{1,2}|\hat{H}_1|V_{1,-2}\rangle = \frac{1}{6}\Delta\alpha_0 \qquad\qquad (\text{F.10, V})$$

$$\langle V_{-1,-2}|\hat{H}_1|V_{1,-2}\rangle = \langle V_{1,2}|\hat{H}_1|V_{-1,2}\rangle = -\frac{1}{6}\Delta\alpha_0 \qquad\qquad (\text{F.10, VI})$$

$$\langle V_{0,0}|\hat{H}_1|V_{\pm1,\pm2}\rangle = \frac{\sqrt{2}}{6}\Delta\alpha_0 \qquad\qquad (\text{F.10, VII})$$

5. The first-order correction to the total energy is the average value of $\hat{H}_1$ in the zeroth-order state, that is, Eq. (F.10, 3) when the perturbation calculation is carried out on the ground state and Eqs. (F.10, 8–10) when the perturbation calculation is carried out on the excited states. Using Eqs. (F.10, III–VII) we obtain

$$\langle {}^1\!A_{1g}|\hat{H}_1|{}^1\!A_{1g}\rangle = \langle {}^1\!B_{1u}|\hat{H}_1|{}^1\!B_{1u}\rangle = \langle {}^1\!B_{2u}|\hat{H}_1|{}^1\!B_{2u}\rangle$$
$$= \langle {}^1\!E_{1u}^{(1)}|\hat{H}_1|{}^1\!E_{1u}^{(1)}\rangle = \langle {}^1\!E_{1u}^{(2)}|\hat{H}_1|{}^1\!E_{1u}^{(2)}\rangle = \Delta\alpha_0$$

Since both the ground and the excited states are shifted by the $\Delta\alpha_0$, a first-order perturbation calculation does not affect the excitation energies.

6.

$$\langle {}^1\!A_{1g}|\hat{H}_1|{}^1\!B_{1u}\rangle = \langle {}^1\!B_{2u}|\hat{H}_1|{}^1\!E_{1u}^{(2)}\rangle = \frac{1}{3}\Delta\alpha_0$$

$$\langle {}^1\!A_{1g}|\hat{H}_1|{}^1\!E_{1u}^{(1)}\rangle = -\frac{1}{3}\Delta\alpha_0$$

$$\langle {}^1\!A_{1g}|\hat{H}_1|{}^1\!B_{2u}\rangle = \langle {}^1\!A_{1g}|\hat{H}_1|{}^1\!E_{1u}^{(2)}\rangle - \langle {}^1\!B_{2u}|\hat{H}_1|{}^1\!E_{1u}^{(1)}\rangle$$
$$= \langle {}^1\!B_{2u}|\hat{H}_1|{}^1\!B_{1u}\rangle = 0$$

With $\Delta\alpha_0^2$ given in eV we have the following second-order energy correction to the ${}^1\!A_{1g}$ and ${}^1\!B_{2u}$ states:

$$E^{(2)}({}^1\!A_{1g}) = \frac{1}{9}\Delta\alpha_0^2\left(\frac{1}{-5.96} + \frac{1}{-6.76}\right) = -0.0351\,\Delta\alpha_0^2$$

$$E^{(2)}({}^1\!B_{2u}) = -0.0545\,\Delta\alpha_0^2$$

or a (negative) change of $-0.0194\,\Delta\alpha_0^2$ in the lowest ${}^1\!A_{1g} \rightarrow {}^1\!B_{2u}$ excitation energy.

7. The magnitude of $\Delta\alpha_0$ is proportional to the electron-donating strength of the substituent and we observe as predicted a decrease in the lowest singlet excitation energy with an increase in the electron-donating ability.

8. Pyridine cannot be treated within such a simple model, since *both* the $\beta(\beta_{CC} \rightarrow \beta_{CN})$ and α parameters change substantially when a nitrogen atom is introduced into the benzene ring.

F.11 1. Equations (F.1, 3–5) with $N = 5$ give

$$\psi_0 = \frac{1}{\sqrt{5}}(\phi_1 + \phi_2 + \phi_3 + \phi_4 + \phi_5)$$

$$\psi_1^S = \left(\frac{2}{5}\right)^{1/2}(\sin 72°\phi_1 + \sin 144°\phi_2 - \sin 144°\phi_3 - \sin 72°\phi_4)$$

$$(\text{F.11, I})$$

$$\psi_1^C = \left(\frac{2}{5}\right)^{1/2}(\cos 72°\phi_1 + \cos 144°\phi_2 + \cos 144°\phi_3 + \cos 72°\phi_4 + \phi_5)$$

$$(\text{F.11, II})$$

$$\psi_2^S = \left(\frac{2}{5}\right)^{1/2}(\sin 144°\phi_1 - \sin 72°\phi_2 + \sin 72°\phi_3 - \sin 144°\phi_4)$$

$$\psi_2^C = \left(\frac{2}{5}\right)^{1/2}(\cos 144°\phi_1 + \cos 72°\phi_2 + \cos 72°\phi_3 + \cos 144°\phi_4 + \phi_5)$$

and
$$\varepsilon_0 = \alpha + 2\beta$$
$$\varepsilon_1 = \alpha + 2\beta \cos 72°$$
$$\varepsilon_2 = \alpha + 2\beta \cos 144°$$

The superscripts C and S on ψ_1 and ψ_2 label the cos and sin solutions, respectively, in Eq. (F.1, 3).

2. The five $2p\pi$ atomic orbitals span a reducible representation in D_{5h}, Γ, which can be decomposed into

$$\Gamma = a_2'' + e_1'' + e_2''$$

Considering the dimension of the representations, it is obvious that

$$\psi_0 = \psi_0(a_2'')$$

By applying the C_5 operation to ψ_1 and ψ_2 we find that $\{\psi_1^C, \psi_1^S\}$ spans the irreducible representation e_1'' and that $\{\psi_2^C, \psi_2^S\}$ spans e_2''.

Thus the electronic configuration and term symbol are $(1a_2'')^2(1e_1'')^3$ and $^2E_1''$, respectively.

3. The two degenerate HOMO's are ψ_1^S and ψ_1^C. In a first-order (degenerate) perturbation calculation the energy correction to ε_1, $\varepsilon_1^{(1)}$, is obtained from the secular problem:

$$\begin{vmatrix} \langle\psi_1^S|\hat{h}_1|\psi_1^S\rangle - \varepsilon_1^{(1)} & \langle\psi_1^S|\hat{h}_1|\psi_1^C\rangle \\ \langle\psi_1^C|\hat{h}_1|\psi_1^S\rangle & \langle\psi_1^C|\hat{h}_1|\psi_1^C\rangle - \varepsilon_1^{(1)} \end{vmatrix} = 0$$

Using Eqs. (F.11, I–II) and varying the $C_1 - C_2$ and the $C_3 - C_4$ bond lengths in Fig. F.11, 1 we obtain the following matrix elements

$$\langle\psi_1^S|\hat{h}_1|\psi_1^S\rangle = \frac{8}{5}\sin 72°\sin 144°\,\delta\beta = \frac{2}{\sqrt{5}}\delta\beta \quad \text{(F.11, III)}$$

$$\langle\psi_1^C|\hat{h}_1|\psi_1^C\rangle = -\frac{2}{5}\delta\beta \quad\quad\quad\quad\quad\text{(F.11, IV)}$$

$$\langle\psi_1^C|\hat{h}_1|\psi_1^S\rangle = 0 \quad\quad\quad\quad\quad\quad\quad\text{(F.11, V)}$$

Thus

$$\varepsilon_1^{(1)} = \begin{cases} \dfrac{2}{\sqrt{5}}\delta\beta \\[2mm] -\dfrac{2}{5}\delta\beta \end{cases} \quad\quad\quad\text{(F.11, VI)}$$

Extension of bond lengths ($\delta\beta > 0$): electronic configuration $(\psi_1^C)^2(\psi_1^S)^1$
Reduction of bond lengths ($\delta\beta < 0$): electronic configuration $(\psi_1^S)^2(\psi_1^C)^1$
If we had chosen to extend another set of non-neighboring bond lengths, the matrix elements in Eqs. (F.11, III–V) would have been different [Equations (F.11, V), e.g., would be different from zero.] However, the first-order energy corrections would still be those given in Eq. (F.11, VI).

4. Extension ($\delta\beta > 0$):

$$E_{\text{total}} = 2\varepsilon_0 + 3\varepsilon_1 + \frac{2\sqrt{5} - 4}{5}\delta\beta > 2\varepsilon_0 + 3\varepsilon_1$$

Reduction ($\delta\beta < 0$):

$$E_{\text{total}} = 2\varepsilon_0 + 3\varepsilon_1 + \frac{4\sqrt{5} - 2}{5}\delta\beta < 2\varepsilon_0 + 3\varepsilon_1$$

5. C_{2v}

6.

Table F.11,I
Correspondence between Symmetry Operations and IR's of D_{5h} and C_{2v} Using the Coordinate System in Fig. F.11,1

D_{5h}	C_{2v}
E	E
$5C_2$	$C_2(z)$
σ_h	$\sigma_v'(yz)$
$5\sigma_v$	$\sigma_v(xz)$
a_2''	b_1
e_1''	$a_2 + b_1$
e_2''	$a_2 + b_1$

By applying, for example, $C_2(z)$ to ψ_1^C and ψ_1^S of Eqs. (F.11, I–II) we find that ψ_1^S transforms as a_2, and ψ_1^C transforms as b_1. The energy ordering for a reduction in bond length (see question 3) is thus $\varepsilon(a_2) < \varepsilon(b_1)$.

7. See Table F.11, I.
8. See Fig. F.11, I.
9. $1b_1^2\, 1a_2^2\, 2b_1^1$; $^2B_1$
   Notice that in the orbital diagram in Fig. F.11, I we have obtained the same ordering of the HOMO's as in question 6.

F.12 1. The 1, 3-cyclobutadiene molecule is given in Fig. F.12, I. Eqs. (F.1, 3, 5) with $N = 4$ give

$$\psi_0(a_{2u}) = \frac{1}{2}(\phi_1 + \phi_2 + \phi_3 + \phi_4) \qquad \text{(F.12, I)}$$

$$\psi_1^S(e_g) = \frac{1}{\sqrt{2}}(\phi_1 - \phi_3) \qquad \text{(F.12, II)}$$

$$\psi_1^C(e_g) = \frac{1}{\sqrt{2}}(\phi_4 - \phi_2) \qquad \text{(F.12, III)}$$

$$\psi_2(b_{2u}) = \frac{1}{2}(\phi_2 - \phi_1 - \phi_3 + \phi_4) \qquad \text{(F.12, IV)}$$

and

$$\varepsilon_0 = \alpha + 2\beta, \qquad \varepsilon_1 = \alpha, \qquad \varepsilon_2 = \alpha - 2\beta$$

The superscripts C and S on ψ_1 labels the cos and sin solutions, respectively, in Eq. (F.1, 3). The symmetry labeling in D_{4h} corresponds to a choice of symmetry elements where C_2' and σ_v

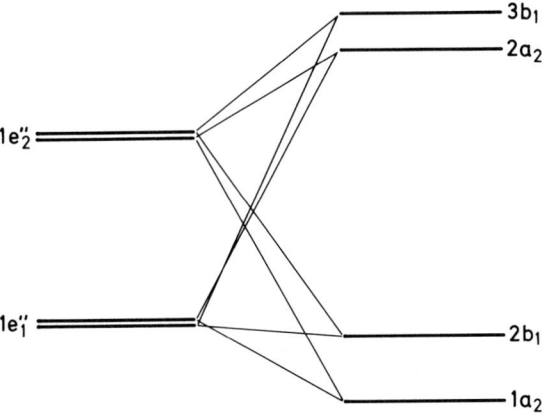

D_{5h} C_{2v}

Fig. F.11, I. The change in the orbital energies of the cyclopentadienyl radical when the symmetry is reduced from D_{5h} to C_{2v}.

contain the carbon atoms C_1 and C_3 or C_2 and C_4 (See the solution to problem F.5, question 1.)

2.

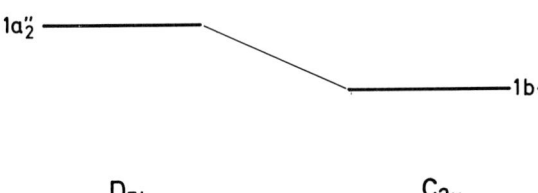

$$a_{2u}^2 e_g^2 ; \; {}^3A_{2g} + {}^1A_{1g} + {}^1B_{1g} + {}^1B_{2g} \qquad (F.12, V)$$

No (first-order) Jahn–Teller distortion of any of these states.

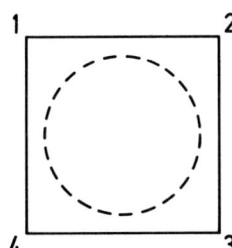

Fig. F.12, I. The geometry of 1, 3-cyclobutadiene.

3.

$$a_{2u}^2 e_g^1 b_{2u}^1;\ {}^1E_u + {}^3E_u \qquad\qquad \text{(F.12, VI)}$$

$$a_{2u}^1 e_g^3;\ {}^1E_u + {}^3E_u \qquad\qquad \text{(F.12, VII)}$$

$$a_{2u}^1 e_g^2 b_{2u}^1;\ {}^5B_{2g} + {}^3B_{2g}(2) + {}^3B_{1g} + {}^3A_{2g} + {}^3A_{1g} + {}^1B_{2g} + {}^1B_{1g} + {}^1A_{2g}$$
$$+ {}^1A_{1g} \qquad\qquad \text{(F.12, VIII)}$$

4. Rectangular: D_{2h} with the C_2'' axis of D_{4h} intact (for the choice of axis, see question 1).

   Rhombohedral: D_{2h} with the C_2' axis of D_{4h} intact.

   Trapezoidal: C_{2v} with one C_2'' axis being the C_2 axis in C_{2v} and σ_d and σ_h being the two σ_v planes in C_{2v}.

5.

Table F.12,I
Transformation Properties of States of Cyclobutadiene

D_{4h} [square]	$\rightarrow$	$D_{2h}(C_2'')$ [rectangle]	or	$D_{2h}(C_2')$ [rhombus]	or	$C_{2v}(\sigma_d)$ [trapezoid]
A_{1g}		A_g		A_g		A_1
A_{2g}		B_{1g}		B_{1g}		A_2
B_{1g}		B_{1g}		A_g		A_2
B_{2g}		A_g		B_{1g}		A_1
E_u		$B_{2u} + B_{3u}$		$B_{2u} + B_{3u}$		$B_1 + B_2$

Rectangular distortion:

$\quad {}^3A_{2g}$ will mix with ${}^3A_{2g} + {}^3B_{1g}$ from the configuration (F.12, VIII).
$\quad {}^1B_{2g}$ will mix with ${}^1A_{1g}$ from the ground-state configuration and ${}^1B_{2g} + {}^1A_{1g}$ from configuration (F.12, VIII).

Rhombohedral distortion:

$\quad {}^3A_{2g}$ will mix with ${}^3A_{2g} + {}^3B_{2g}(2)$ from (F.12, VIII).
$\quad {}^1B_{2g}$ will mix with ${}^1B_{2g} + {}^1A_{2g}$ from (F.12, VIII).

Trapezoidal distortion:

$\quad {}^3A_{2g}$ will mix with ${}^3A_{2g} + {}^3B_{1g}$ from (F.12, VIII).
$\quad {}^1B_{2g}$ will mix with ${}^1A_{1g}$ from the ground-state configuration and ${}^1B_{2g} + {}^1A_{1g}$ from (F.12, VIII).

6. A second-order Jahn–Teller effect would predict the ${}^1B_{2g}$ state to be distorted either to rectangular or to trapezoidal geometry and the ${}^3A_{2g}$ state not to be distorted, that is, to remain a square.

F.13 1. The coordinate axes are chosen as indicated in Fig. F.13, I.
Symmetry orbitals in C_{2v}:

$$\chi_1(b_1) = \frac{1}{\sqrt{2}}(\phi_1 + \phi_3) \qquad (\text{F.13, I})$$

$$\chi_2(b_1) = \phi_2 \qquad (\text{F.13, II})$$

$$\chi_3(a_2) = \frac{1}{\sqrt{2}}(\phi_1 - \phi_3) \qquad (\text{F.13, III})$$

Hückel MO's:

$$\psi_1^{(0)}(b_1) = \frac{1}{2}(\phi_1 + \sqrt{2}\,\phi_2 + \phi_3), \qquad \varepsilon_1 = \alpha + \sqrt{2}\,\beta$$

$$\psi_2^{(0)}(a_2) = \frac{1}{\sqrt{2}}(\phi_1 - \phi_3), \qquad \varepsilon_2 = \alpha$$

$$\psi_3^{(0)}(b_1) = \frac{1}{2}(\phi_1 - \sqrt{2}\,\phi_2 + \phi_3), \qquad \varepsilon_3 = \alpha - \sqrt{2}\,\beta$$

2.

$$q_1^{(0)} = q_3^{(0)} = 0.5, \qquad q_2^{(0)} = 1$$

$$p_{12}^{(0)} = p_{23}^{(0)} = \frac{\sqrt{2}}{2}, \qquad p_{13}^{(0)} = 0.5$$

3. Use Eqs. (F.0, 2–3) to obtain the following matrix elements of the Fock operator in the *atomic orbital* representation (values are in eV).

$$F_{11}^{(0)} = I_C + \frac{1}{2}q_1\gamma_{11} + (q_2 - 1)\gamma_{12} + (q_3 - 1)\gamma_{13} = I_C + 0.79$$

$$F_{33}^{(0)} = F_{11}^{(0)}$$

$$F_{22}^{(0)} = I_C + \frac{1}{2}q_2\gamma_{22} + (q_1 - 1)\gamma_{21} + (q_3 - 1)\gamma_{23} = I_C + 0.14$$

$$F_{12}^{(0)} = \beta_{23} - \frac{1}{2}p_{23}\gamma_{23} = -4.1868$$

$$F_{23}^{(0)} = F_{12}^{(0)}$$

$$F_{13}^{(0)} = -\frac{1}{2}p_{13}\gamma_{13} = -0.96$$

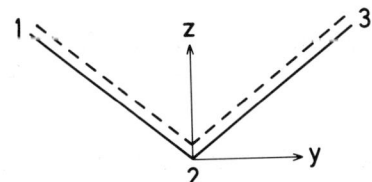

Fig. F.13, I. The choice of coordinate system and atomic labeling for the allyl cation.

In the symmetry orbital representation [Eqs. (F.13, I–III)] $\hat{F}$ is given as

$$\langle \chi_1 | \hat{F} | \chi_1 \rangle = F_{11}^{(0)} + F_{13}^{(0)} = I_C - 0.17 \qquad \text{(F.13, IV)}$$

$$\langle \chi_2 | \hat{F} | \chi_2 \rangle = F_{22}^{(0)} = I_C + 0.14 \qquad \text{(F.13, V)}$$

$$\langle \chi_2 | \hat{F} | \chi_1 \rangle = \sqrt{2}\, F_{23}^{(0)} = -5.9210 \qquad \text{(F.13, VI)}$$

$$\langle \chi_3 | \hat{F} | \chi_3 \rangle = F_{11}^{(0)} - F_{13}^{(0)} = I_C + 1.75 \qquad \text{(F.13, VII)}$$

4. In the a_2 symmetry class $\chi_3(a_2)$ is the self-consistent solution and in the b_1 class we obtain

$$\varepsilon_+^{(1)} = I_C + 5.9060; \qquad \psi_+^{(1)} = 0.4935\,(\phi_1 + \phi_3) - 0.7162\,\phi_2$$
$$\varepsilon_-^{(1)} = I_C - 5.9360; \qquad \psi_-^{(1)} = 0.5065\,(\phi_1 + \phi_3) + 0.6977\,\phi_2$$

5.

$$q_1^{(1)} = q_3^{(1)} = 0.5131, \qquad q_2^{(1)} = 0.9738$$
$$p_{12}^{(1)} = p_{23}^{(1)} = 0.7069, \qquad p_{13}^{(1)} = 0.5131$$

6.

$$\varepsilon_+^{(2)} = I_C + 5.8873; \qquad \psi_+^{(2)} = 0.4925\,(\phi_1 + \phi_3) - 0.7176\,\phi_2$$
$$\text{(F.13, VIII)}$$

$$\varepsilon_-^{(2)} = I_C - 5.9632; \qquad \psi_-^{(2)} = 0.5072\,(\phi_2 + \phi_3) + 0.6968\,\phi_2$$
$$\text{(F.13, IX)}$$

$$q_1^{(2)} = q_3^{(2)} = 0.5145, \qquad q_2^{(2)} = 0.9711$$
$$p_{12}^{(2)} = p_{23}^{(2)} = 0.7068, \qquad p_{13}^{(2)} = 0.5145$$

The SCF cycle has converged and the SCF MO's of b_1 symmetry and their orbital energies are given in Eqs. (F.13,VIII-IX). In the a_2 symmetry class the SCF MO is χ_3 and its orbital energy is $\varepsilon_2 = I_C + 1.7582$.

F.14 1. The Fock operator, $\hat{F}$ transforms as the totally symmetric IR and is thus diagonal in the ψ_i representation in Eqs. (F.12, I–IV), since all four MO's belong either to different IR's or different columns of the degenerate e_g representation.

2.

$$\varepsilon_i = \langle \psi_i | \hat{F} | \psi_i \rangle, \qquad i = 0, 1, 2 \qquad \text{(F.14, I)}$$

Equations (F.0, 2-3) for the electronic configuration $(\psi_0)^2 (\psi_1^S)^1 (\psi_1^C)^1$ with $Z_\mu = 1$, $\mu = 1, 2, 3$ gives

$$q_\mu = 1, \qquad \mu = 1, 2, 3, 4$$

$$p_{12} = 0.5$$

$$p_{13} = p_{24} = 0$$

Thus, in the *atomic* orbital representation the Fock operator is

$$F_{11} = \langle \phi_1 | \hat{F} | \phi_1 \rangle = I_C + \frac{1}{2}\gamma_{11}$$

$$F_{12} = \beta^c - \frac{1}{4}\gamma_{12}$$

$$F_{13} = 0$$

and we obtain, using Eq. (F.14, I), that

$$\varepsilon_0 = F_{11} + 2F_{12} + F_{13} = I_C + 2\beta^c + \frac{1}{2}\gamma_{11} - \frac{1}{2}\gamma_{12}$$

$$\varepsilon_1 = I_C + \frac{1}{2}\gamma_{11}$$

$$\varepsilon_2 = I_C - 2\beta^c + \frac{1}{2}\gamma_{11} + \frac{1}{2}\gamma_{12}$$

3. $\alpha = I_C + \frac{1}{2}\gamma_{11}$, $\beta = \beta^c - \frac{1}{4}\gamma_{12}$

4.

$$a_{2u}^1 e_g^3 ; \; {}^3E_u + \text{another state}$$

$$a_{2u}^2 e_g^1 b_{2u}^1 ; \; {}^3E_u + \text{another state}$$

The Hückel excitation energy for both transitions is -2β.

5.

$$\Psi_0 = |{}^3A_{2g}(a_{2u}^2 e_g^2); M_S = 1\rangle = |\psi_0 \bar{\psi}_0 \psi_1^S \psi_1^C|$$

$$\Psi_1 = |{}^3E_u(a_{2u}^1 e_g^3); M_S = 1\rangle = |\psi_0 \bar{\psi}_1^S \psi_1^S \psi_1^C|$$

$$\Psi_2 = |{}^3E_u(a_{2u}^2 e_g^1 b_{2u}^1); M_S = 1\rangle = |\psi_0 \bar{\psi}_0 \psi_1^S \psi_2|$$

6.

$$\Delta E(0 \to 1) = \langle \Psi_1 | \hat{H} | \Psi_1 \rangle - \langle \Psi_0 | \hat{H} | \Psi_0 \rangle \equiv E_1 - E_0$$
$$\Delta E(0 \to 2) = \langle \Psi_2 | \hat{H} | \Psi_2 \rangle - \langle \Psi_0 | \hat{H} | \Psi_0 \rangle \equiv E_2 - E_0$$

Ψ_1 is obtained from Ψ_0 by promoting one electron from spin orbital $\bar{\psi}_0^S$ to spin orbital $\bar{\psi}_1^S$, while Ψ_2 is obtained by promoting one electron from ψ_1^C to ψ_2. Thus using the Slater–Condon rule for evaluating determinantal matrix elements, we have [see, e.g., Eq. (F.15, VIII)]

$$E_1 = E_0 + \varepsilon_1 - \varepsilon_0 - (\bar{\psi}_1^S \bar{\psi}_1^S \| \bar{\psi}_0 \bar{\psi}_0)$$
$$E_2 = E_0 + \varepsilon_2 - \varepsilon_1 - (\psi_2 \psi_2 \| \psi_1^C \psi_1^C)$$

where ε_i are SCF orbital energies and, for example,

$$(\psi_2 \psi_2 \| \psi_1^C \psi_1^C) = (\psi_2 \psi_2 | \psi_1^C \psi_1^C) - (\psi_2 \psi_1^C | \psi_1^C \psi_2)$$

Using that

$$\varepsilon_1 - \varepsilon_0 = \varepsilon_2 - \varepsilon_1 = -2\beta^c + \frac{1}{2}\gamma_{12}$$
$$(\psi_2 \psi_2 | \psi_1^C \psi_1^C) = (\psi_1^S \psi_1^S | \psi_0 \psi_0) = \frac{1}{4}(\gamma_{11} + 2\gamma_{12} + \gamma_{13})$$
$$(\psi_2 \psi_1^C | \psi_1^C \psi_2) = (\psi_1^S \psi_0 | \psi_0 \psi_1^S) = \frac{1}{4}(\gamma_{11} - \gamma_{13})$$

we obtain

$$\Delta E(0 \to 1) = \Delta E(0 \to 2) = -2\beta^c - \frac{1}{2}\gamma_{13}$$

7. $\gamma_{ij} \geq 0$ and electron repulsion thus reduces the excitation energies.
8. A CI calculation would lift the degeneracy only if $\langle \Psi_2 | \hat{H} | \Psi_1 \rangle$ is different from zero. However,

$$\langle \Psi_2 | \hat{H} | \Psi_1 \rangle = (\psi_0 \psi_1^S | \psi_2 \psi_1^S) = 0$$

in the **ZDO** approximation.

F.15 1. If the coordinate system is chosen as indicated in Fig. F.15, I, that is, with the z-axis perpendicular to the molecular plane, the normalized MO's are

$$\pi: \psi_1(b_{1u}) = (2 + 2S)^{-1/2}(\phi_A + \phi_B) \qquad \text{(F.15, I)}$$

$$\pi^*: \psi_2(b_{2g}) = (2 - 2S)^{-1/2}(\phi_A - \phi_B) \qquad \text{(F.15, II)}$$

and

$$\varepsilon_1 = \frac{\alpha + \beta}{1 + S} \qquad \text{(F.15, III)}$$

$$\varepsilon_2 = \frac{\alpha - \beta}{1 - S} \qquad \text{(F.15, IV)}$$

2.

$$\Delta E = \varepsilon_2 - \varepsilon_1 = \frac{2S\alpha - 2\beta}{1 - S^2} = 12.69 \text{ eV} \qquad (\Delta E^{\text{exp}} = 7.68 \text{ eV})$$

3.

$$|N; {}^1A_{1g}\rangle = |\psi_1 \bar{\psi}_1| \qquad \text{(F.15, V)}$$

$$|V; {}^1B_{3u}\rangle = \frac{1}{\sqrt{2}}\left(|\psi_1 \bar{\psi}_2| - |\bar{\psi}_1 \psi_2|\right) \qquad \text{(F.15, VI)}$$

Since x transforms as B_{3u}, the transition is polarized along the internuclear (x) axis.

4. $\mathbf{M} = (M, 0, 0)$

$$M = \langle N; {}^1A_{1g} | \sum_{i=1}^{2} x(i) | {}^1B_{3u}; V \rangle$$

$$= \sqrt{2}\langle \psi_1 | x | \psi_2 \rangle = (2 - 2S^2)^{-1/2}[\langle \phi_A | x | \phi_A \rangle - \langle \phi_B | x | \phi_B \rangle]$$

$$= R(2 - 2S^2)^{-1/2} = 1.896 \text{ a.u.} \qquad \text{(F.15, VII)}$$

where we have used Eqs. (F.15, I–II) and (F.15, V–VI). R is the internuclear spacing, 1.366 Å.

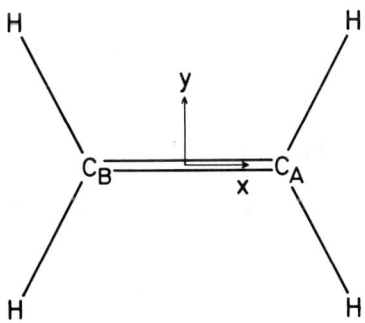

Fig. F.15, I. The choice of coordinate system for ethylene.

5.

$$f_{\mathrm{L}} = \frac{2}{3} M^2 \Delta E = 1.12 \qquad (\text{"too large"})$$

6. Since $\{\psi_1, \psi_2\}$ transform according to different IR's, the Fock operator is diagonal and the $\{\psi_1, \psi_2\}$ are also the PPP MO's.

$$\varepsilon_1 = \langle \psi_1 | \hat{F} | \psi_1 \rangle = I_{\mathrm{C}} + \frac{1}{2}\gamma_{11} + \beta^c - \frac{1}{2}\gamma_{12} \quad (\text{F.15, VIII})$$

$$\varepsilon_2 = \langle \psi_2 | \hat{F} | \psi_2 \rangle = I_{\mathrm{C}} + \frac{1}{2}\gamma_{11} - \beta^c + \frac{1}{2}\gamma_{12} \quad (\text{F.15, IX})$$

7. Replacing a *spin* orbital ψ_i with spin orbital ψ_k in $|N\rangle$ gives

$$\langle \Psi_i^k | \hat{H} | \Psi_i^k \rangle = \langle N | \hat{H} | N \rangle + h_{kk} - h_{ii} - \sum_j (ii\|jj) + \sum_j (kk\|jj)$$

$$- (ii\|kk) \qquad (\text{F.15, X})$$

$$= E_{\mathrm{N}} + \varepsilon_k - \varepsilon_i - (ii\|kk)$$

All sums extend over the spin orbitals which are occupied in $|N\rangle$ and $\hat{h}$ is the one-electron operator.

The triplet state wave function with $M_S = 1$ is obtained by replacing $\bar{\psi}_1$ by ψ_2 in the $|N\rangle$ determinant. Thus from Eq. (F.15, X)

$$\langle {}^3\Psi_1^2 | \hat{H} | {}^3\Psi_1^2 \rangle = E_{\mathrm{N}} + \varepsilon_2 - \varepsilon_1 - (11|22) \qquad \text{Q.E.D.}$$

The singlet wave function is $|V\rangle$ [Eq. (F.15, VI)]. Using Eq. (F.15, X) and the Slater–Condon rules we obtain

$$\langle V | \hat{H} | V \rangle = \frac{1}{2}(\langle |\psi_1 \bar{\psi}_2 \| \hat{H} \| \psi_1 \bar{\psi}_2 |\rangle + \langle |\bar{\psi}_1 \psi_2 \| \hat{H} \| \bar{\psi}_1 \psi_2 |\rangle$$

$$- \langle |\psi_1 \bar{\psi}_2 \| \hat{H} \| \bar{\psi}_1 \psi_2 |\rangle - \langle |\bar{\psi}_1 \psi_2 \| \hat{H} \| \psi_1 \bar{\psi}_2 |\rangle)$$

$$= E_{\mathrm{N}} + \varepsilon_2 - \varepsilon_1 - (11\|22) - (1\bar{1}\|\bar{2}2) \qquad (\text{F.15, XI})$$

$$= E_{\mathrm{N}} + \varepsilon_2 - \varepsilon_1 + 2(12|21) - (11|22) \qquad \text{Q.E.D.}$$

8.

$$\Delta E = E_{\mathrm{V}} - E_{\mathrm{N}}$$

$$= \varepsilon_2 - \varepsilon_1 - (11|22) + 2(12|21)$$

$$= \varepsilon_2 - \varepsilon_1 + \frac{1}{2}\gamma_{11} - \frac{3}{2}\gamma_{12}$$

$$= -2\beta^c + \frac{1}{2}(\gamma_{11} - \gamma_{12})$$

where we have used Eqs. (F.15, VIII–IX) and (F.15, XI).

$$\Delta E = 7.435 \text{ eV} \qquad\qquad \text{(F.15, XII)}$$

The wave functions in the Hückel and the PPP approximation are the same for the N and V states [see Eqs. (F.15, V–VI)] and the transition moment in PPP is thus the one determined in the Hückel approximation in Eq. (F.15, VII) with $S = 0$. The oscillator strength then becomes

$$f_L = \frac{2}{3}\left(\frac{1.366}{0.529\sqrt{2}}\right)^2 \frac{7.435}{27.21} = 0.61$$

f_L is thus a factor of 2 larger than the experimental value.

9.

$$E(^1B_{3u}) - E(^3B_{3u}) = \begin{cases} 0, & \text{Hückel} \\ 2K_{12} = \gamma_{11} - \gamma_{12} = 3.15 \text{ eV}, & \text{PPP} \end{cases}$$

10. Equation (F.15, 2) is valid because the Brillouin theorem is fulfilled. Equation (F.15, 3) may easily be derived by using the relations

$$\langle Z \,|\hat{H}|\, Z\rangle = 2h_{22} + (22|22)$$

$$\langle N|\hat{H}|N\rangle = 2h_{11} + (11|11)$$

$$\varepsilon_1 = h_{11} + (11|11)$$

$$\varepsilon_2 = h_{22} + 2(11|22) - (12|21)$$

Use the Slater–Condon rules to obtain Eq. (F.15, 4).

11. Among the four states that can be constructed from the determinants based on the MO's $\{\psi_1, \psi_2\}(|N\rangle, |Z\rangle, |^1\Psi_1^2\rangle, \text{and}\, |^3\Psi_1^2\rangle)$ only $|N\rangle$ and $|Z\rangle$ will interact in the CI calculation. The matrix elements for the resulting 2×2 CI problem are

$$\langle N\,|\hat{H}|\, N\rangle = E_N$$

$$\langle N\,|\hat{H}|\, Z\rangle = \frac{1}{2}(\gamma_{11} - \gamma_{12})$$

$$\langle Z\,|\hat{H}|\, Z\rangle = E_N - 4\beta^c$$

The CI eigenvalues are

$$E_\pm - E_N = -2\beta^c\left\{1 \pm \left[1 + \frac{1}{16}\left(\frac{\gamma_{11} - \gamma_{12}}{\beta^c}\right)^2\right]^{1/2}\right\} = \begin{cases} 11.928 \text{ eV} \\ -0.208 \text{ eV} \end{cases}$$

Correlation energy for $|N\rangle$: $E_+(\text{CI}) - E_N(\text{SCF}) = -0.208 \text{ eV}$.

12. The full CI wave function for the ground state is the one that corresponds to the eigenvalue -0.208 eV in the previous question; that is

$$|N, CI\rangle = 0.9914| N\rangle - 0.1309 |Z\rangle$$

The full CI wave function for the state $|V\rangle$ is

$$|V, CI\rangle = | V\rangle$$

because $|V\rangle$ is the only state of ${}^1B_{3u}$ symmetry.

13.

$$M = \langle N, CI | \sum_{i=1}^{2} x(i)| V\rangle$$

$$= 0.9914 \langle N | \sum_{i=1}^{2} x(i)| V\rangle - 0.1309 \langle Z | \sum_{i=1}^{2} x(i)| V\rangle$$

(F.15, XIII)

Using Eqs. (F.15, VII) and (F.15, V–VI) we get

$$\langle N | \sum_{i=1}^{2} x(i)| V\rangle = \langle Z | \sum_{i=1}^{2} x(i)| V\rangle = \frac{R}{\sqrt{2}}$$

and Eq. (F.15, XIII) becomes

$$M = 0.8605 \frac{R}{\sqrt{2}} = 1.57 \text{ a.u.}$$

$$\Delta E = E_V - E_N - (E_- - E_N) = E_V - E_-$$

$$= 7.643 \text{ eV} \qquad (\Delta E_{exp} = 7.68 \text{ eV})$$

$$f_L = 0.46 \qquad (f_{exp} = 0.3)$$

G.
CRYSTAL AND LIGAND FIELD THEORY

INTRODUCTION

In the problems in this section we use crystal and ligand field theory to describe the electronic structure of transition metal complexes. *Crystal field theory* is applied in problems G.1, G.2, and G.5 for the "coupling case of the rare earth elements" where the relative magnitude of the atomic electron repulsion, $\sum_{i>j} 1/r_{ij}$, the crystal field from the ligands, $\hat{V}_{cryst}$, and the atomic spin–orbit coupling operator, $\zeta \mathbf{L} \cdot \mathbf{S}$, is

$$\sum_{i>j} \frac{1}{r_{ij}} \gg \zeta \mathbf{L} \cdot \mathbf{S} \gg \hat{V}_{cryst} \tag{G.0, 1}$$

The "weak field"

$$\sum_{i>j} \frac{1}{r_{ij}} \gg \hat{V}_{cryst} \gg \zeta \mathbf{L} \cdot \mathbf{S} \tag{G.0, 2}$$

and the "strong field" limits

$$\hat{V}_{cryst} \gg \sum_{i>j} \frac{1}{r_{ij}} \quad \text{and} \quad \zeta \mathbf{L} \cdot \mathbf{S} \tag{G.0, 3}$$

are considered specifically in problems G.2, G.3, G.5, and in G.3, G.5, respectively, whereas the "intermediate coupling case" where

$$\sum_{i>j} \frac{1}{r_{ij}} \sim \hat{V}_{cryst} \tag{G.0, 4}$$

is treated in problem G.4.

Ligand field theory is used in problems G.5–G.8. In ligand field theory an MO diagram is constructed for the complex using rules A–K of the introduction to section E. The information about the electronic structure of the complex is then determined from the MO diagram. In the present problem set we use ligand field theory to discuss the binding of O_2, N_2, and CO to hemoglobin (G.6) and the conformations and ultraviolet excitation spectrum of XeF_4 (G.7) and XeF_6 (G.8). As in section E, extensive use is made of group theoretical arguments and we mention once more that we have used the group tables of Atkins *et al.* (1970) to prepare the solutions of the problems.

Textbooks that give sufficient background for solving the problems in this section

Atkins (1970). Ballhausen and Gray (1964). DeKock and Gray (1980). Gray (1965). McWeeny (1979). Murrell *et al.* (1970, 1978). Orgel (1960).

Textbooks that give a more elaborate treatment of the topic

Ballhausen (1962, 1979). Griffith (1961). Schläfer and Gliemann (1967). Tinkham (1964).

Problem Description

G.1 Crystal field splitting of Tm^{3+} complexes ("coupling case of the rare earth elements")

G.2 Crystal field splitting of octahedral Ni^{2+} and Pd^{2+} complexes ("weak field" and coupling case of the rare earth elements)

G.3 Crystal field splitting of square planar Ce complexes ("weak" and "strong" fields)

G.4 Crystal field splitting of octahedrally coordinated d^3, d^5, and d^8 complexes ("intermediate coupling"); excitation energies of Ni^{2+} and Cr^{3+} complexes

G.5 Crystal and ligand field treatment of Pu^{2+} complexes for varying strength of the crystal field; ground-state geometry

G.6 The electronic structure of hemoglobin (ligand field theory); affinity of O_2, CO, and N_2 for hemoglobin

G.7 The electronic structure of ground and excited states of XeF_4 (ligand field theory)

G.8 The electronic structure of ground and excited states of XeF_6 (ligand field theory)

G.9 The spin–spin splitting of $[(NH_3)_5CrOHCr(NH_3)_5]^{5+}$

PROBLEMS

G.1 In this problem, we will calculate the crystal field splitting of the ground state of Tm^{3+} complexes. The ground-state electronic configuration of Tm^{3+} is $[Ba]f^{12}$ where [Ba] denotes the closed-shell electronic configuration of the ground state of Ba. We assume that the relative magnitude of the electronic repulsion and spin–orbit coupling is

$$\sum_{i>j} \frac{1}{r_{ij}} \gg \zeta \mathbf{L} \cdot \mathbf{S}$$

1. Determine the degeneracy of the ground state of Tm^{3+} if only the electronic repulsion is considered.

We will now calculate the spin–orbit splitting of the ground state of Tm^{3+}.

2. Determine the energy difference between the highest and the lowest level originating from the ground-state term. The result should be expressed in terms of the spin–orbit parameter ζ.

3. What is the degeneracy of the lowest level?

Tm^{3+} forms a planar, square complex, TmX_4^{3+}, with four ligands X. The relative magnitude of $\hat{V}_{cryst}$ from the four ligands and the spin–orbit coupling is

$$\zeta \mathbf{L} \cdot \mathbf{S} \gg \hat{V}_{cryst}$$

4. Determine the molecular term symbols of the states into which the lowest spin–orbit level of Tm^{3+} will split when we include the effect of $\hat{V}_{cryst}$.

We will now assume that two of the ligands are replaced by ligands Y so that the complex becomes $TmX_2 Y_2^{3+}$. The bond lengths and bond angles of this complex are assumed to be the same as those of TmX_4^{3+}.

5. Determine for both possible point group symmetries of the complex $TmX_2 Y_2^{3+}$ the molecular term symbols of the states into which the lowest spin–orbit level now splits.

G.2 In this problem we will consider the electronic structure of d^8 octahedrally coordinated complexes for various strengths of the crystal field, using crystal field theory. Initially, we consider the Ni^{2+} ion, which has the ground state electronic configuration $[Ar]3d^8$.

1. Determine the ground-state term of Ni^{2+} within the Russell–Saunder's coupling scheme.

The Ni^{2+} ion is now assumed to be located in an octahedral crystal field, $\hat{V}_{cryst}$, of strength

$$\sum_{i>j} \frac{1}{r_{ij}} \gg \hat{V}_{cryst} \gg \zeta \mathbf{L} \cdot \mathbf{S}$$

2. Determine the molecular term symbols for the states into which the ground-state term of Ni^{2+} splits when the crystal field is switched on.

Consider the additional splitting of the crystal field states caused by the atomic spin–orbit coupling.

3. Determine the term symbols for the resulting states.

We will now consider octahedrally coordinated Pd^{2+} complexes for which the relative magnitude of the perturbing operators is

$$\sum_{i>j} \frac{1}{r_{ij}} \gg \zeta \mathbf{L} \cdot \mathbf{S} \gg \hat{V}_{\text{cryst}}$$

The electronic configuration of Pd^{2+} is $[Kr]4d^8$.

4. Indicate the relative energy ordering of the levels originating from the ground-state term.
5. Show that the crystal field operator $\hat{V}_{\text{cryst}}$ will split these levels into states of exactly the same symmetry as those obtained in question 3. Explain why that is the case.

G.3 In this problem we will consider the energy splitting of the ground-state electronic configuration of Ce for various strengths of the crystal field $\hat{V}_{\text{cryst}}$. The electronic configuration of the ground state of the cerium atom is $[Ba]4f^2$.

1. Determine within the Russell–Saunder's coupling scheme the terms that arise from the ground-state electronic configuration. Which term corresponds to the ground state?
2. Indicate the relative energy ordering of the spin–orbit levels originating from the ground-state term.

The cerium atom is perturbed by a crystal field of D_{4h} symmetry, which has the strength

$$\sum_{i>j} \frac{1}{r_{ij}} \gg \hat{V}_{\text{cryst}} \gg \zeta \mathbf{L} \cdot \mathbf{S}$$

3. Determine the molecular term symbols for the states originating from the ground-state term when we include the effect of $\hat{V}_{\text{cryst}}$. Which additional splitting of the ground-state term results if the symmetry of the crystal field is lowered to D_{2h}?

We now assume that Ce is located in a "strong" crystal field of D_{4h} symmetry; that is

$$\hat{V}_{\text{cryst}} \gg \sum_{i>j} \frac{1}{r_{ij}} \quad \text{and} \quad \zeta \mathbf{L} \cdot \mathbf{S}$$

4. Determine the irreducible components of the reducible representation spanned by the f orbitals in the D_{4h} point group.
5. What are the possible term symbols for the ground state of cerium in the "strong field" limit? Which term represents the ground state if Hund's rule is valid?

The f orbitals in real form can be written as

$$f_{z(5z^2-3r^2)} = Y_{30}$$

$$f_{x(5z^2-r^2)} = \frac{1}{\sqrt{2}}[Y_{3-1} + Y_{31}]$$

$$f_{y(5z^2-r^2)} = \frac{i}{\sqrt{2}}[Y_{3-1} - Y_{31}]$$

$$f_{z(x^2-y^2)} = \frac{1}{\sqrt{2}}[Y_{3-2} + Y_{32}]$$

$$f_{xyz} = \frac{i}{\sqrt{2}}[Y_{3-2} - Y_{32}]$$

$$f_{x(x^2-3y^2)} = \frac{1}{\sqrt{2}}[Y_{3-3} + Y_{33}]$$

$$f_{y(y^2-3x^2)} = \frac{i}{\sqrt{2}}[Y_{3-3} - Y_{33}]$$

where Y_{3m} are the complex spherical harmonics with $l = 3$. As shown in question 4, the degeneracy of the f orbitals is lifted by the crystal field.

6. Determine the linear combinations of the f orbitals that transform as the IR's which were determined in question 4.

Assume that the symmetry of the crystal field is reduced to D_{2h}.

7. Which functions from question 6 form bonding and antibonding orbitals as a result of the reduction in symmetry?

G.4 In this problem we examine the crystal field splitting of ground-state electronic configurations of octahedrally coordinated atoms for which the electron repulsion and the crystal field $\hat{V}_{cryst}$ are of comparable magnitude. Initially we consider the Ni^{2+} ion with the ground-state electronic configuration $[Ar]3d^8$.

1. Write the $M_L = 0$ and $M_S = 1$ component of the single configurational wave functions for the ^3F and the ^3P terms originating from the d^8 electronic configuration.

The Ni^{2+} ion is assumed to be surrounded by ligands such that the complex has octahedral point group symmetry.

2. Determine the molecular term symbols for the crystal field states originating from the ^3F and the ^3P terms.

We will now perform a calculation of the crystal field splitting of the two triplet terms, ^3F and ^3P, assuming that the energy splittings caused by the atomic electronic repulsion and by the crystal field is of comparable magnitude. We will describe the energy splitting caused by the electronic repulsion through a parameter X, which is the energy difference between the two triplet terms in the absence of the crystal field. The strength of the crystal field will be described by the crystal field parameter Δ, which is the octahedral crystal field splitting of the d^1 electronic configuration. The matrix representation of $\hat{V}_{cryst}$, in the d^1 configuration, that is, $\langle d_i|\hat{V}_{cryst}|d_j\rangle$ where $i, j = +2, +1, 0, -1, -2$ and d_i denotes the spherical harmonics, is

$$\begin{bmatrix} \frac{1}{10}\Delta & 0 & 0 & 0 & \frac{1}{2}\Delta \\ 0 & -\frac{2}{5}\Delta & 0 & 0 & 0 \\ 0 & 0 & \frac{3}{5}\Delta & 0 & 0 \\ 0 & 0 & 0 & -\frac{2}{5}\Delta & 0 \\ \frac{1}{2}\Delta & 0 & 0 & 0 & \frac{1}{10}\Delta \end{bmatrix} \qquad \text{(G.4, 1)}$$

Table G.4,1
The O_h Crystal Field Splitting of a $^3F(d^2)$ Term

Eigenfunction[a]	Energy	Symmetry (O_h)
$\left(\frac{3}{8}\right)^{1/2} F_{-1} + \left(\frac{5}{8}\right)^{1/2} F_{+3}$		
$\left(\frac{3}{8}\right)^{1/2} F_{+1} + \left(\frac{5}{8}\right)^{1/2} F_{-3}$	$-\frac{3}{5}\Delta$	$^3T_{1g}$
F_0		
$\left(\frac{5}{8}\right)^{1/2} F_{-1} - \left(\frac{3}{8}\right)^{1/2} F_{+3}$		
$\left(\frac{5}{8}\right)^{1/2} F_{+1} - \left(\frac{3}{8}\right)^{1/2} F_{-3}$	$\frac{1}{5}\Delta$	$^3T_{2g}$
$\left(\frac{1}{2}\right)^{1/2}(F_{+2} + F_{-2})$		
$\left(\frac{1}{2}\right)^{1/2}(F_{+2} - F_{-2})$	$\frac{6}{5}\Delta$	$^3A_{2g}$

[a] $F_i = |^3F, M_L = i, M_S = 1\rangle$.

The O_h crystal field splitting of a $^3F(d^2)$ term in the absence of the ^3P term can be determined by carrying out a first-order perturbation calculation in the crystal field. The result of carrying out such a calculation for a d^2 configuration is given in Table G.4, 1 (Murrell *et al.* 1970).

3. Determine as a function of X and Δ the energies of the crystal field states originating from the ^3F and ^3P terms in octahedrally coordinated Ni^{2+} complexes by carrying out a first-order perturbation calculation where the perturbation operator is the sum of the electron repulsion and the crystal field.

4. Plot the energies of the crystal field levels determined in question 3 as a function of Δ. Can the symmetry of the ground state change at any strength of X and Δ within this model?

The three lowest electronic transitions of octahedrally coordinated $Ni(H_2O)_6^{2+}$ and $Ni(NH_3)_6^{2+}$ are 1.05 eV, 1.67 eV, 3.14 eV and 1.33 eV, 2.17 eV, 3.50 eV, respectively (Ballhausen 1962). These transitions have all been assigned to transitions between the ground state and the excited states discussed in the previous question.

5.   Show that Δ and X can be determined such that the excitation energies are reproduced within an accuracy of 0.10 eV for both $Ni(H_2O)_6^{2+}$ and $Ni(NH_3)_6^{2+}$ complexes.

6.   Is the relative magnitude of the two Δ values for $Ni(H_2O)_6^{2+}$ and $Ni(NH_3)_6^{2+}$ as expected from the spectrochemical series? Comment on the difference between the free Ni^{2+} atomic value for X (1.96 eV) and the X value calculated from the spectral date.

The lowest spin allowed transitions in $Cr(H_2O)_6^{3+}$ complexes are 2.16 eV, 3.04 eV, and 4.71 eV (Ballhausen 1962). The ground-state electronic configuration of Cr^{3+} is $[Ar]3d^3$.

7.   Show that these transitions may be described with an accuracy of less than 0.1 eV in the same model as we used for the d^8 configuration.

The crystal field splitting of the terms for an octahedrally coordinated d^5 complex can be calculated by perturbative considerations similar to those used for the d^8 valence configuration. The result of such a calculation is plotted in Fig. G.4, 1.

8.   Use Fig. G.4, 1 to perform an assignment of the ground-state absorption spectrum of $Mn(H_2O)_6^{2+}$ in Fig. G.4, 2 (Murrell et al. 1970) knowing that the ground state of $Mn(H_2O)_6^{2+}$ has A_{1g} symmetry.

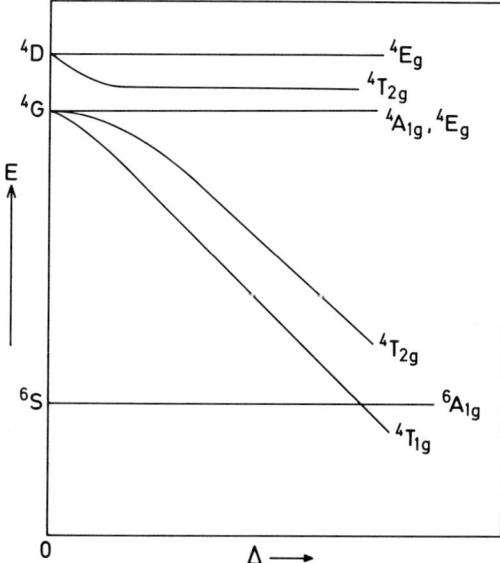

Fig. G.4, 1. Crystal field energy diagram for a d^5 electronic configuration in an octahedral field.

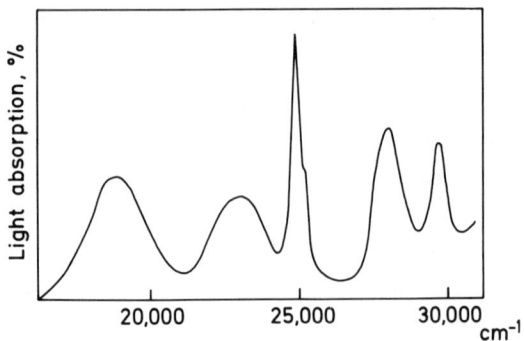

Fig. G.4, 2. The absorption spectrum of $[Mn(H_2O)_6]^{2+}$ (Reprinted from Murrell *et al.* (1970) by permission of John Wiley and Sons, Ltd).

G.5 In this problem we will determine the degeneracy of the ground state of octahedrally coordinated Pu^{2+} complexes for various strengths of the crystal field. We will assume that the ground-state electronic configuration of Pu^{2+} is $[Rn]5f^6$.

1. Determine the ground-state term for Pu^{2+}. What is the degeneracy of the ground state?

2. How would the spin–orbit coupling split the ground-state term, and which of the resulting levels would have the lowest energy?

Very little is known about complexes of the actinides (i.e., elements with a partially filled $5f$ shell). We will examine the complexes of Pu^{2+} for varying strengths of an octahedral crystal field, $\hat{V}_{cryst}$, and we assume initially that

$$\sum_{i>j} \frac{1}{r_{ij}} \gg \zeta L \cdot S \gg \hat{V}_{cryst} \qquad (G.5, 1)$$

3. What would be the molecular term symbol of the ground state for such a complex when $\hat{V}_{cryst}$ is taken into account?

Assume now that the Pu^{2+} central atom is surrounded by ligands which create a stronger crystal field

$$\sum_{i>j} \frac{1}{r_{ij}} \gg \hat{V}_{cryst} \gg \zeta L \cdot S \qquad (G.5, 2)$$

4. Determine the possible term symbols for the ground state of octahedrally coordinated Pu^{2+} complexes. (Do not consider the effect of the spin–orbit coupling.)

Finally, let us consider the strong field limit

$$\hat{V}_{cryst} \gg \sum_{i>j} \frac{1}{r_{ij}} \gg \zeta L \cdot S \qquad (G.5, 3)$$

5. How would the crystal field split the f orbitals?

We will now perform a ligand field treatment of octahedrally coordinated Pu^{2+} complexes. The MO diagram will be constructed from the $5f$ orbitals on Pu^{2+} and the $p\sigma$ and $p\pi$ orbitals on the ligands. Assume that the orbital energies of the ligand p orbitals are less than the $5f$ orbital energy in Pu^{2+}.

6. Construct a MO diagram for octahedrally coordinated Pu^{2+}

Assume that the $p\sigma$ and $p\pi$ orbitals of the ligands were fully occupied before the Pu^{2+} complex was formed.

7. Determine the electronic configuration and the possible term symbols for the ground state of the Pu^{2+} complexes.

Most known octahedrally coordinated actinides do not show Jahn–Teller symmetry distortion of the ground state.

8. Is this in agreement with the results obtained for all three strengths of $\hat{V}_{cryst}$?

G.6 In this problem we will analyze the adhesion of O_2, N_2, and CO to hemoglobin using qualitative MO theory. The electronic structure of hemoglobin will be treated in a model where the iron atom of hemoglobin is surrounded by four ligands such that hemoglobin has D_{4h} symmetry. Each ligand will further be assumed to have unpaired spin, and in this problem each will be described by a hydrogen atom. We assume that it is only necessary to consider the $1s$ orbitals on the hydrogen atoms and the $3d$ and $4s$ orbitals on the iron atoms to describe the bond formation. The lowest ionization potential for the neutral iron atom is 7.90 eV (Moore 1949).

1. Construct a MO diagram for FeH_4, assuming that the orbital energy for $4s_{Fe}$ is slightly lower than the orbital energy for $3d_{Fe}$.

2. What are the possible ground-state electronic configurations and term symbols of FeH_4 ?

An O_2 molecule approaches hemoglobin adiabatically such that (i) the oxygen–oxygen bond is parallel to the plane spanned by FeH_4, (ii) the midpoint of the oxygen–oxygen bond contains the C_4 axis of FeH_4, and (iii) the oxygen–oxygen bond is parallel to a diagonal that contains the atoms H–Fe–H.

3. What is the point group symmetry of the $FeH_4 - O_2$ complex?

In analyzing the bond formation in $FeH_4 - O_2$, we will assume that the perturbation of FeH_4 by O_2 is so small that only changes in the orbital energies of the frontier orbitals of O_2 and FeH_4 need to be considered. The lowest ionization potential for O_2 is 13.1 eV (Siegbahn et al. 1969) corresponding to an ionization out of a π_g orbital.

4. Draw an MO diagram for the $FeH_4 - O_2$ complex that displays the MO's of $FeH_4 - O_2$ that can be formed from the frontier orbitals of O_2 and FeH_4.

5. Does this model predict that O_2 forms a stable complex with hemoglobin?

We will next examine the binding of N_2 to hemoglobin by using the same model as for O_2 binding, that is, only the frontier orbital of N_2 participates in the bond formation. The lowest ionization potential for N_2 is 15.5 eV (Siegbahn et al. 1969), corresponding to an ionization out of an orbital of σ_g symmetry

6. Answer questions 3, 4, and 5 for the $FeH_4 - N_2$ complex.

Consider now the binding of CO to hemoglobin in the same model. The lowest ionization potential for CO is 14.5 eV (Siegbahn *et al.* 1969), corresponding to an ionization out of a σ orbital.

7. Answer questions 3, 4, and 5 for the FeH_4 – CO complex.

If we had chosen to let O_2, N_2, and CO approach FeH_4 such that conditions (i) and (ii) were satisfied but such that (iii) had been replaced by the condition that the oxygen–oxygen bond should be parallel to the C_2'' axis in D_{4h}, the same conclusions about the relative stability of the FeH_4 – O_2, – N_2, and –CO complexes would have been obtained.

G.7 In this problem we will analyze the electronic structure of XeF_4. We will carry out a simple MO description which considers the $2p$ orbitals on fluorine and the $4d$, $5s$, and $5p$ orbitals on xenon. Initially, we assume that XeF_4 has tetrahedral symmetry. An MO diagram based on these orbitals is given in Fig. G.7, 1.

1. Determine the ground-state electronic configuration of XeF_4.

2. Determine the possible ground-state term symbols. Which of these terms has lowest energy if we assume that Hund's rule is valid?

3. Would you expect a Jahn–Teller distortion of XeF_4 in its ground state?

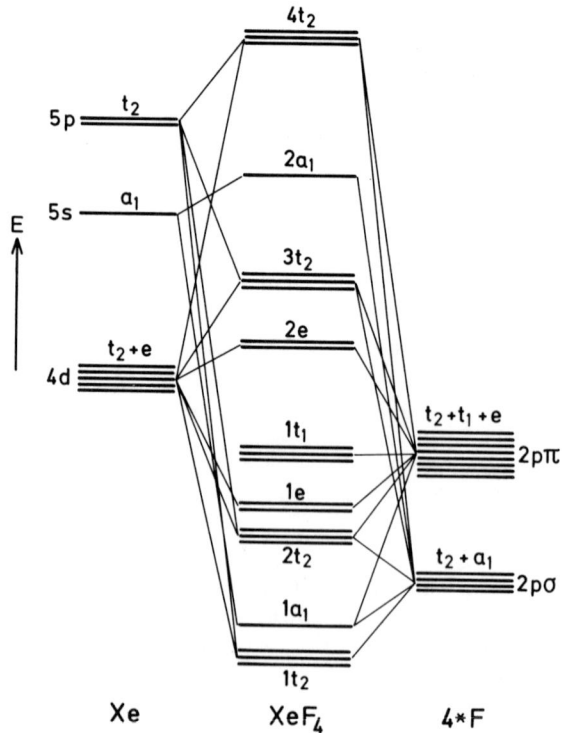

Fig. G.7, 1. MO diagram for a tetrahedrally coordinated XeF_4 molecule.

4.   Would the answers to questions 2 and 3 change if we were to include only the $2p\pi$ orbitals on flourine in the MO description?

5.   Answer question 4 if we include only the $2p\sigma$ orbitals in the MO description.
We will now examine the effect of changing the symmetry of XeF_4 from T_d to D_{4h}. We will describe the changes in the electronic structure in XeF_4 by considering only the changes in energy of the two highest occupied molecular orbitals, $4t_2$ and $2a_1$. Let us first assume that the changes in the crystal field caused by the changes in geometry are so small that it is sufficient to consider changes in the ground state determined in question 2 (weak field limit).

6.   Determine the possible ground-state molecular term symbols.
Assume now that the change of geometry introduces a change in the crystal field that is so large that it is necessary to reconsider the occupation of the $2a_1$ and $4t_2$ MO's in order to determine the ground-state electronic configuration (strong field limit).

7.   Determine then the possible ground-state electronic configurations and the corresponding molecular term symbols.
Experimentally, the ground state of XeF_4 is known to have D_{4h} symmetry and to be of singlet symmetry (Hyman 1963). The lowest ionization potentials of XeF_4 are (Brundle *et al.* 1971) 13.1 eV (a_{1g}) and 13.4 eV (a_{2u}), respectively. The four lowest electronic transitions of singlet and triplet symmetry for XeF_4 are (Hyman 1963; Basch *et al.* 1971)

$$^1A_{1g} \rightarrow {}^1E_u \quad \text{at} \quad 6.8 \text{ eV}$$

$$^1A_{1g} \rightarrow {}^1E_g \quad \text{at} \quad 6.8 \text{ eV}$$

$$^1A_{1g} \rightarrow {}^3E_u \quad \text{at} \quad 5.43 \text{ or } 4.8 \text{ eV}$$

$$^1A_{1g} \rightarrow {}^3E_g \quad \text{at} \approx 6.31 \text{ eV}$$

8.   Are any of the models in questions 6 or 7 able to describe this experimental information? Would a simple theory predict the observation that the triplet excitation energies are smaller than the corresponding singlet excitation energies?

G.8 In this problem we will perform a simple MO description of the XeF_6 molecule. We assume that only the $4d$, $5s$, and $5p$ orbitals on xenon and the $2p\sigma$ orbitals on the ligands participate in the bond formation. We will initially assume that XeF_6 has octahedral symmetry. An MO diagram is given in Fig. G.8, 1.

1.   Argue that the energy ordering of the three highest occupied molecular orbitals will remain unchanged if we include the $2p\pi$ orbitals on the ligands in the calculation.

2.   Determine the term symbol for the ground state of XeF_6.

3.   Determine the term symbol for the first excited singlet state of XeF_6. Is the transition from the ground state to the excited state electric dipole allowed?
We will now examine what happens to the ground and lowest excited singlet state of XeF_6 when we lower the symmetry of the molecule. We assume that it is sufficient to consider the changes in the HOMO and LUMO, that is, the frontier orbitals.

4.   Would you expect a lower ground-state energy of XeF_6 if the symmetry of XeF_6 were D_{3d}?
We will now lower the symmetry to C_{3v}.

5. Would you expect a lower ground-state energy of XeF_6 in C_{3v} than in O_h symmetry?

Experimental investigations of the ultraviolet spectrum of $TeCl_6^{2-}$ (Couch *et al.* 1970), which is electronically isoconfigurational with XeF_6, indicate that the transition discussed in question 3 will split into two close-lying transitions.

6. Show that this splitting can be explained on the basis of the simple model used in the present problem and give the term symbol of the two excited states. Which of these has the lowest energy?

G.9 The magnetic properties of binuclear complexes are often explained on the basis of the spin–spin coupling betweeen the two transition metal ions. The Hamiltonian for the spin–spin interaction is often given in the form

$$\hat{V} = -J\mathbf{S}_1 \cdot \mathbf{S}_2 \tag{G.9, 1}$$

where J is an exchange coupling constant and $\mathbf{S}_i$, $i = 1, 2$, is the electronic spin of ligands 1 and 2, respectively. Since J is very small (10^{-3} eV), close-lying states will result from the spin–spin coupling.

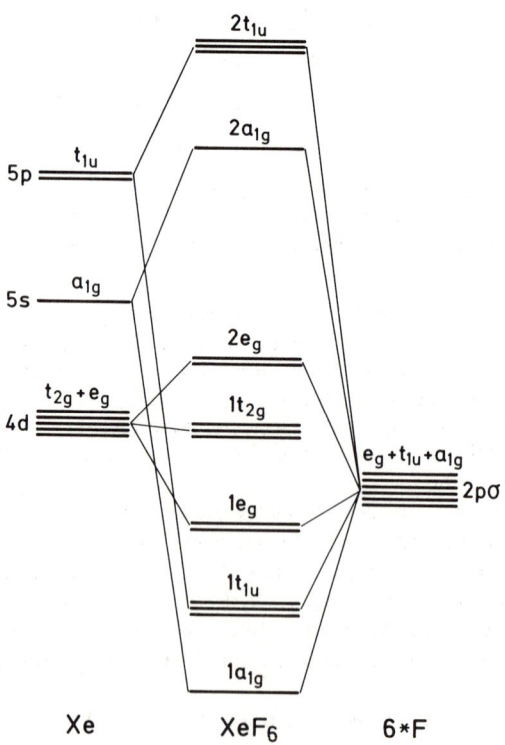

Fig. G.8, 1. MO diagram for an octahedrally coordinated XeF_6 molecule.

Consider the binuclear complex $[(NH_3)_5CrOHCr(NH_3)_5]^{5+}$ where both Cr^{3+} ions are assumed to have a d^3 valence electronic configuration.

1. Determine the ground-state term for Cr^{3+} .

2. Calculate the relative splitting of the ground-state terms of Cr^{3+} caused by $\hat{V}$ by performing a first-order perturbation calculation in $\hat{V}$. Express the result in terms of the constant J.

SOLUTIONS

G.1 1. ^3H; (33)

2. We have inverted multiplet splitting of an f^{12} configuration, that is the lowest level is $^3H_6$ and the highest level is $^3H_4$. Through first order in the spin–orbit coupling the energy splitting is proportional to $\frac{1}{2}\zeta[J(J+1) - L(L+1) - S(S+1)]$. Thus

$$\Delta E(^3H_6 - {}^3H_4) = 11\zeta.$$

3. $^3H_6$; (13)

4. The configuration f^{12} has even parity and the molecular states of TmX_4^{3+} must thus be gerade states in D_{4h}. It is therefore sufficient to consider D_4 point group symmetry. The $^3H_6$ level is spanned by $\{Y_{6M_J}, M_J = -6, -5, \ldots, 6\}$ The character, $\chi^{(6)}$, for the rotations of Y_{6M_J} are obtained from the formula in the Appendix.

D_4	E	$2C_4$	$C_2(=C_4^2)$	$2C_2'$	$2C_2''$
$\chi^{(6)}$	13	-1	1	1	1

and $^3H_6$ thus splits into states that transform as the following irreducible representations in D_{4h}:

$$2A_{1g} + A_{2g} + 2B_{1g} + 2B_{2g} + 3E_g$$

5. Two possible geometries for $TmX_2Y_2^{3+}$:

$$C_{2v}(C_2'', \sigma_d \text{ conserved}): 4A_1 + 3A_2 + 3B_1 + 3B_2$$
$$D_{2h}(C_2', \sigma_v \text{ conserved}): 4A_g + 3B_{1g} + 3B_{2g} + 3B_{3g}$$

G.2 1. ^3F

2. $^3F \rightarrow {}^3A_{2g}, {}^3T_{2g}, {}^3T_{1g}$.

3. The spin operator with $S = 1$ transforms as an "axial" vector, that is, a tensor operator of rank 1 that does not change sign under inversion. In the O_h point group this means that it transforms as T_{1g}.

The total wave function transforms as the direct product ($\otimes$) of the spin and spatial symmetries, and we have that

$$^3A_{2g} \rightarrow T_{1g} \otimes A_{2g} = T_{2g}$$

$$^3T_{2g} \rightarrow T_{1g} \otimes T_{2g} = A_{2g} + E_g + T_{1g} + T_{2g}$$

$$^3T_{1g} \rightarrow T_{1g} \otimes T_{1g} = A_{1g} + E_g + T_{1g} + T_{2g}$$

However, the $^3T_{2g}$ and $^3T_{1g}$ states do not split into four energetically different states as the group theoretical argument indicates, but only into three energetically different states (Ballhausen 1962, p. 89). This can be seen by carrying out a first-order perturbation calculation on the $^3T_{2g}$ or the $^3T_{1g}$ state with $\zeta \mathbf{L} \cdot \mathbf{S}$ as the perturbation operator. We would then find that the E_g state and one of either the T_{1g} or the T_{2g} states becomes degenerate. The $^3T_{2g}$ and the $^3T_{1g}$ states therefore split into three energetically different states with the degeneracy 1, 3, and 5, respectively.

4. $^3F \rightarrow {}^3F_4, {}^3F_3, {}^3F_2$ (decreasing energy ordering, since we have normal multiplet structure).

5. $^3F_4 \rightarrow A_{1g} + E_g + T_{1g} + T_{2g}$ (since the angular momentum quantum number is $J = 4$)

$$^3F_3 \rightarrow A_{2g} + T_{1g} + T_{2g} \qquad (J = 3)$$

$$^3F_2 \rightarrow E_g + T_{2g} \qquad (J = 2)$$

We must obtain the same results as in question 3, since the total number of states and the symmetry of these states do not depend on the order in which the operators $\hat{V}_{\text{cryst}}$ and $\zeta \mathbf{L} \cdot \mathbf{S}$ are applied as long as they both are applied.

G.3 1. $^3H, {}^3F, {}^3P, {}^1I, {}^1G, {}^1D, {}^1S$
The ground-state term is ^3H.

2. $^3H_4(-6\zeta), {}^3H_5(-\zeta), {}^3H_6(5\zeta) \qquad \zeta > 0$ (normal multiplet)

3. $^3H \rightarrow {}^3A_{1g} + 2\,{}^3A_{2g} + {}^3B_{1g} + {}^3B_{2g} + 3\,{}^3E_g$ gerade, since ^3H originates from f^{12} configuration with even parity. $D_{4h} \rightarrow D_{2h}$ removes all spatial degeneracy ($^3E_g \rightarrow {}^3B_{2g} + {}^3B_{3g}$).

4.

$$f(l = 3) \rightarrow a_{2u} + b_{1u} + b_{2u} + 2e_u$$

ungerade, since f orbitals have odd parity.

5. All the configurations $a_{2u}^2, b_{1u}^2, b_{2u}^2, b_{2g}^2$ give a $^1A_{1g}$ ground state. e_u^2 gives $^1A_{1g}, {}^3A_{2g}, {}^1B_{1g}, {}^1B_{2g}$ states of which $^3A_{2g}$ is expected to be the lowest according to Hund's rule.

6. Let Γ_z, Γ_{z^2}, ... denote the IR's of D_{4h} which transform as z, z^2, ...
We then have that the f orbitals transform in the following way:

$$z(5z^2 - r^2) \sim \Gamma_z \otimes \Gamma_{z^2} = a_{2u} \otimes a_{1g} = a_{2u}$$

$$z(x^2 - y^2) \sim \Gamma_z \otimes \Gamma_{x^2-y^2} = a_{2u} \otimes b_{1g} = b_{2u}$$

$$xyz \sim \Gamma_z \otimes \Gamma_{xy} = a_{2u} \otimes b_{2g} = b_{1u}$$

$$\left. \begin{array}{l} x(5z^2 - r^2) \\ y(5z^2 - r^2) \end{array} \right\} \sim \Gamma_{(x,y)} \otimes \Gamma_{z^2} = e_u \otimes a_{1g} = e_u$$

$$\left. \begin{array}{l} x(x^2 - 3y^2) = x(2(x^2 - y^2) - (x^2 + y^2)) \\ y(y^2 - 3x^2) = y(-2(x^2 - y^2) - (y^2 + x^2)) \end{array} \right\}$$

$$\sim \left\{ \begin{array}{l} \Gamma_{(x,y)} \otimes \Gamma_{x^2-y^2} = e_u \otimes b_{1h} \\ \Gamma_{(x,y)} \otimes \Gamma_{x^2+y^2} = e_u \otimes a_{1g} \end{array} \right\} = e_u$$

7.

D_{4h}	$\rightarrow$	$D_{2h}(C_2')$	or	$D_{2h}(C_2'')$
a_{2u}		b_{1u}		b_{1u}
b_{1u}		a_u		b_{1u}
b_{2u}		b_{1u}		a_u
e_u		$b_{2u} + b_{3u}$		$b_{2u} + b_{3u}$

Bonding and antibonding orbitals will thus be formed between the b_{1u} orbitals for distortions to both kinds of D_{2h} symmetry.

G.4 1. The electronic configurations d^8 and d^2 give the same terms; thus from the microstate diagram in Table D.1, I we have

$$|{}^3F(M_L = 2, M_S = 1)\rangle = |d_2 d_0| \quad\quad (G.4, I)$$

By applying L_-^2 to $|d_2 d_0|$ we arrive at the $M_L = 0$ component. The normalized wave function is

$$|{}^3F(M_L = 0, M_S = 1)\rangle = \frac{1}{\sqrt{5}}(2|d_1 d_{-1}| + |d_2 d_{-2}|) \quad (G.4, II)$$

Therefore

$$|{}^3P(M_L = 0, M_S = 1)\rangle = \frac{1}{\sqrt{5}}(|d_1 d_{-1}| - 2|d_2 d_{-2}|) \quad\quad (G.4, III)$$

2. ${}^3F \rightarrow {}^3A_{2g} + {}^3T_{2g} + {}^3T_{1g}$
${}^3P \rightarrow {}^3T_{1g}$

3. Only the $^3T_{1g}$ component of the ^3F term will interact with the ^3P state. According to Table G.4, 1, F_0, [i.e., Eq. (G.4, II)], transforms as T_{1g}. Using Eqs. (G.4, II and III) and (G.4, 1), as well as the Slater–Condon rule for matrix elements over the one-electron operator $\hat{V}_{cryst} = \sum_{i=1}^{2} \hat{V}_{cryst}^{(i)}$, we obtain for a d^2 configuration

$$\langle {}^3F(M_L = 0, M_S = 1) | \hat{V}_{cryst} + \sum_{i>j} \frac{1}{r_{ij}} | {}^3P(M_L = 0, M_S = 1) \rangle$$

$$= \langle {}^3F(M_L = 0, M_S = 1) | \hat{V}_{cryst} | {}^3P(M_L = 0, M_S = 1) \rangle$$

$$= \frac{2}{5} \langle |d_1 d_{-1}| | \hat{V}_{cryst} | |d_1 d_{-1}| \rangle - \frac{2}{5} \langle |d_2 d_{-2}| | \hat{V} | |d_2 d_{-2}| \rangle$$

$$= -\frac{2}{5} \Delta$$

Since the Ni^{2+} configuration is d^8, we must change the sign on Δ in the results obtained for the d^2 configuration in order to obtain the Ni^{2+} results. We therefore have the following first-order secular determinant of T_{1g} symmetry:

$$\begin{vmatrix} \frac{3}{5}\Delta - E^{(1)} & \frac{2}{5}\Delta \\ \frac{2}{5}\Delta & X - E^{(1)} \end{vmatrix} = 0$$

The eigenvalues are

$$E_{\pm}^{(1)}({}^3T_{1g}) = \frac{X}{2} + \frac{3}{10}\Delta \pm \frac{1}{2}\sqrt{X^2 - \frac{6}{5}X\Delta + \Delta^2} \quad (G.4, IV)$$

The total energies of the other states are given in Table G.4, 1:

$$E({}^3T_{2g}) = -\frac{1}{5}\Delta \qquad\qquad (G.4, V)$$

$$E({}^3A_{2g}) = -\frac{6}{5}\Delta \qquad\qquad (G.4, VI)$$

4. The energies are plotted in Fig. G.4, I. It follows from Eq. (G.4, IV) that

$$E_{\pm}^{(1)}({}^3T_{1g}) \approx \begin{cases} \frac{4}{5}\Delta \\ -\frac{1}{5}\Delta \end{cases} \quad \text{for} \quad \Delta \gg X$$

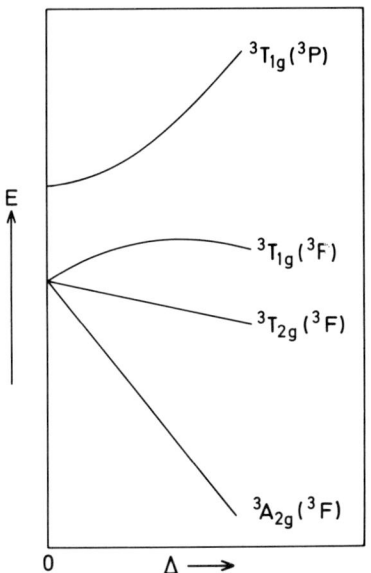

Fig. G.4, I. Crystal field energy diagram for a d^8 electronic configuration in an octahedral field.

The symmetry thus remains the same for any value of X and Δ.

5. Using Eqs. (G.4, V) and (G.4, VI) we obtain

$$E(^3T_{2g}) - E(^3A_{2g}) = \Delta \qquad \text{(G.4, VII)}$$

and from Eqs. (G.4, IV and VI):

$$E_+(^3T_{1g}) + E_-(^3T_{1g}) - 2E(^3A_{2g}) = X + 3\Delta \quad \text{(G.4, VIII)}$$

From Eqs. (G.4, VII and VIII) we obtain $\Delta = 1.05$ eV and $X = 1.66$ eV for $Ni(H_2O)_6^{2+}$ and $\Delta = 1.33$ eV and $X = 1.68$ eV for $Ni(NH_3)_6^{2+}$, which gives the excitation spectrum in Table G.4, I.

Table G.4,I
Excitation Energies (in eV) for $Ni(H_2O)_6^{2+}$ and $Ni(NH_3)_6^{2+}$

Transition	$Ni(H_2O)_6^{2+}$	$Ni(NH_3)_6^{2+}$
$^3A_{2g} \rightarrow {}^3T_{2g}$	1.05	1.33
$^3A_{2g} \rightarrow {}^3T_{1g}(^3F)$	1.74	2.15
$^3A_{2g} \rightarrow {}^3T_{1g}(^3P)$	3.08	3.52

6. $\Delta(NH_3) > \Delta(H_2O)$, as expected from the spectrochemical series.

The "spectral" X values [1.66 eV for $Ni(H_2O)_6^{2+}$ and 1.68 eV for $Ni(NH_3)_6^{2+}$] are much smaller than the free atom value (1.96 eV). In spite of that, the two-parameter model (X, Δ) gives an excellent description of the excitation spectrum. This illustrates what has been learned from this and many other semiempirical methods—that the parameters which enter in semiempirical models should *not* be obtained from separate experiments or calculations, but rather from the application of the semiempirical model to the properties it is supposed to describe.

7. The d^3 valence configuration gives rise to the terms ^4F, ^4P, ^2H, ^2G, ^2F, $^2D(2)$, ^2P. Spin allowed transition thus takes place between the ^4F and ^4P terms. The diagram in Fig. G.4, I for a d^8 configuration would be the same for a d^3 configuration, except that the spin multiplicity of the terms in Fig. G.4, I would be replaced by a spin multiplicity of 4. (Both the d^3 and d^8 diagrams are obtained from the d^2 diagram by interchanging electrons with holes.) Using Eqs. (G.4, IV–VIII) we obtain the excitation spectrum in Table G.4, II.

8. See Table G.4,III.

Table G.4,II
Excitation Energies (in eV) for Cr $(H_2O)_6^{3+}$

Transition	Experiment	Calculation
$^4A_{2g} \to {}^4T_{2g}$	2.16	2.16
$^4A_{2g} \to {}^4T_{1g}(^4F)$	3.04	3.01
$^4A_{2g} \to {}^4T_{1g}(^4P)$	4.71	4.74

Table G.4,III
Assignment of the Excitation Spectrum of $Mn(H_2O)_6^{2+}$ in Fig. G.4,2 (Heidt et al. 1958)

Frequency (cm^{-1})	Assignment
18,870	$^6A_{1g} \to {}^4T_{1g}(^4G)$
23,120	$^6A_{1g} \to {}^4T_{2g}(^4G)$
24,960	$^6A_{1g} \to {}^4A_{1g}(^4G)^a$
25,275	$^6A_{1g} \to {}^4E_g(^4G)^a$
27,980	$^6A_{1g} \to {}^4T_{2g}(^4D)$
29,750	$^6A_{1g} \to {}^4E_g(^4D)$

^{a}This assignment may be reversed.

G.5 1. ^7F (49)

2. ^7F_J, $J = 0, 1, \ldots, 6$; $J = 0$ is the ground state.

3. $^7F_0 \rightarrow A_{1g}$ since $J = 0$.   (G.5, I)

4. $^7F \rightarrow {}^7A_{2g} + {}^7T_{1g} + {}^7T_{2g}$ since $\cdot L = 3$.   (G.5, II)

5. $f(l = 3) \rightarrow a_{2u} + t_{1u} + t_{2u}$ since f orbitals have odd parity.

6. The MO diagram is displayed in Fig. G.5, I.

7. $\ldots 2t_{1u}^6\, 1a_{2u}^2\, 2t_{2u}^4$; $^1A_{1g}$, ^1E_g, $^3T_{1g}$, $^1T_{2g}$ (G.5, III)

8. The ground state must be spatially nondegenerate in O_h. Otherwise, we would expect a Jahn–Teller distortion to a lower point group symmetry. When the magnitude of $\hat{V}_{\text{cryst}}$ is given by Eq. (G.5, 1) [$^7A_{1g}$ ground state, Eq. (G.5, I)] and by Eq. (G.5, 2) [$^7A_{2g}$ ground

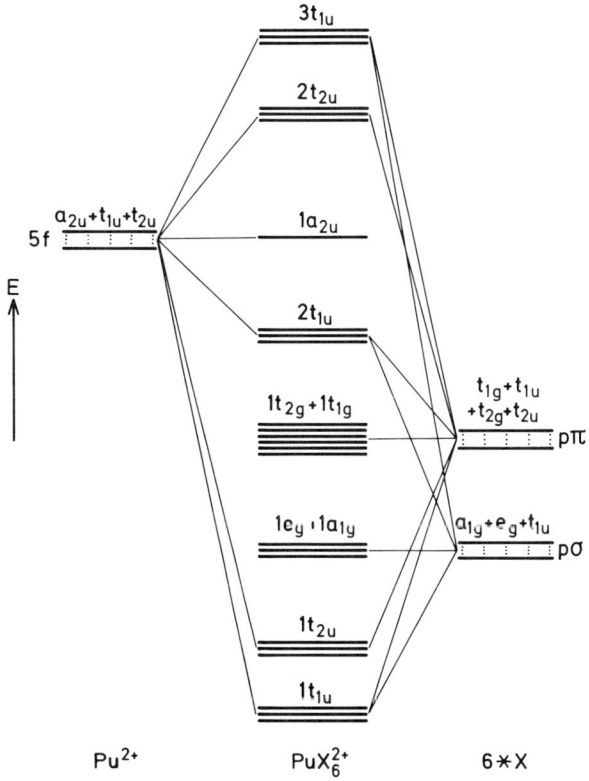

Fig. G.5, I. MO diagram for an octahedrally coordinated PuX_6^{2+} complex.

state, Eq. (G.5, II)], a Jahn–Teller distortion would thus not occur. The strong field limit, Eq. (G.5, 3), is less likely to give a spatially nondegenerate ground state in O_h, since Hund's rule predicts a spatially degenerate $^3T_{1g}$ ground state [see Eq. (G.5, III)]. A Jahn–Teller distortion would thus be expected in the strong field limit.

G.6 1. Following Atkins *et al.* (1970) we use the convention that the $\sigma_{v'}$, C_2' symmetry elements of D_{4h} contain atoms and that C_2'' and σ_d do not contain atoms. This yields the MO diagram in Fig. G.6, I.

2. The electronic configuration of neutral iron is [Ar] $3d^6 4s^2$, which implies that the possible electronic configurations of FeH_4 are

$$\cdots 2a_{1g}^2 e_g^1 b_{2g}^1; \,^3E_g + \text{another state} \,(^1E_g)$$

$$\cdots 2a_{1g}^2 e_g^2; \,^3A_{2g} + \text{other states} \,(^1A_{1g}, \,^1B_{1g}, \,^1B_{2g})$$

$$\cdots 2a_{1g}^2 b_{2g}^2; \,^1A_{1g}$$

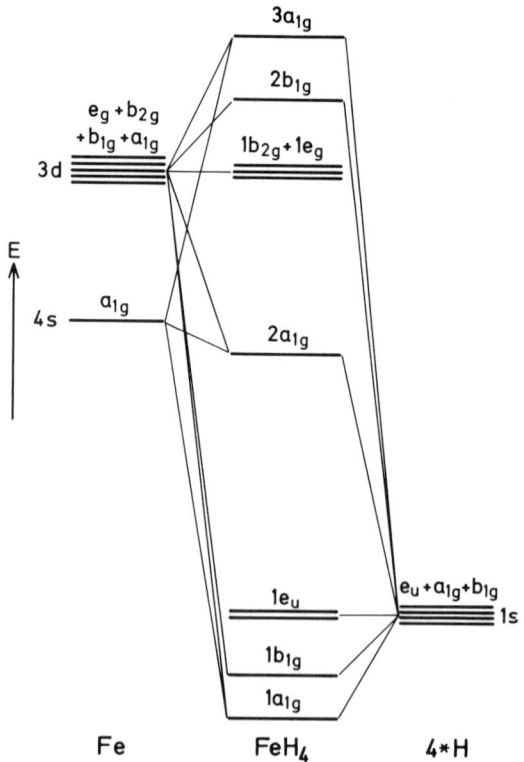

Fig. G.6, I. MO diagram for a square planar coordinated FeH_4 molecule.

3. C_{2v}
4. We choose the coordinate system such that the z-axis coincides with the C_4 axis of D_{4h} and the y-axis is in the direction of the oxygen–oxygen bond. The IR's spanned by the frontier orbitals before and after bonding are given in Table G.6, I and the MO diagram is displayed in Fig. G.6, II.
5. The electronic configuration is $\cdots 1a_2^2\, 1b_2^2$. Thus both the electron pairs from O_2 (π_g^2) and from hemoglobin are moved into orbitals of

Table G.6,I
Transformation Properties of the Frontier Orbitals of FeH$_4$ and O_2 in C_{2v}

$D_{\infty h}$	$\rightarrow$	C_{2v}	D_{4h}	$\rightarrow$	$C_{2v}(C_2', \sigma_v)$
π_g		$a_2 + b_2$	e_g		$b_1 + b_2$
σ_g		a_1	b_{2g}		a_2

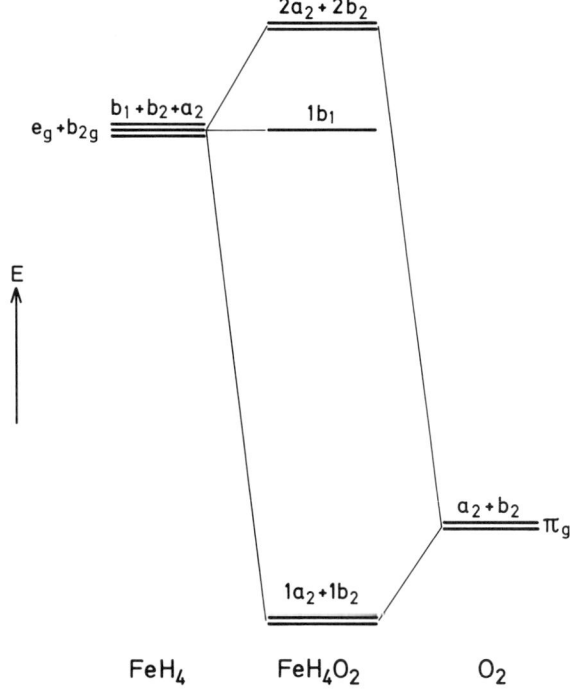

Fig. G.6, II. The interaction between the HOMO's of FeH$_4$ and O_2 in an FeH$_4$ – O_2 complex.

the $FeH_4 - O_2$ complex with lower energies, and we expect a strong adhesion of O_2 to hemoglobin [rule K.3 in the introduction to section E].

6. C_{2v} and the MO diagram in Fig. G.6, III.
   The electronic configuration is $\cdots 1a_1^2 \, 1b_1^2$ or other equivalent configurations with $1b_1$ interchanged with $1b_2$ or/and $1a_2$. Following the arguments in question 5, we have a nonbonding situation.

7. C_s
   When the σ plane is chosen as the y–z plane in the C_{2v} point group we have the transformation properties in Table G.6, II.
   From the MO diagram in Fig. G.6, IV we see that the electronic configuration is $\cdots 1a'^2 \, 1a''^2$, that is, CO bonds to hemoglobin.

Fig. G.6, III. The interaction between the HOMO's of FeH_4 and N_2 in an $FeH_4 - N_2$ complex.

Table G.6,II
Relations between IR's
of C_{2v} and C_s

C_{2v}	$\rightarrow$	$C_s(yz)$
a_1		a'
a_2		a''
b_1		a''
b_2		a'

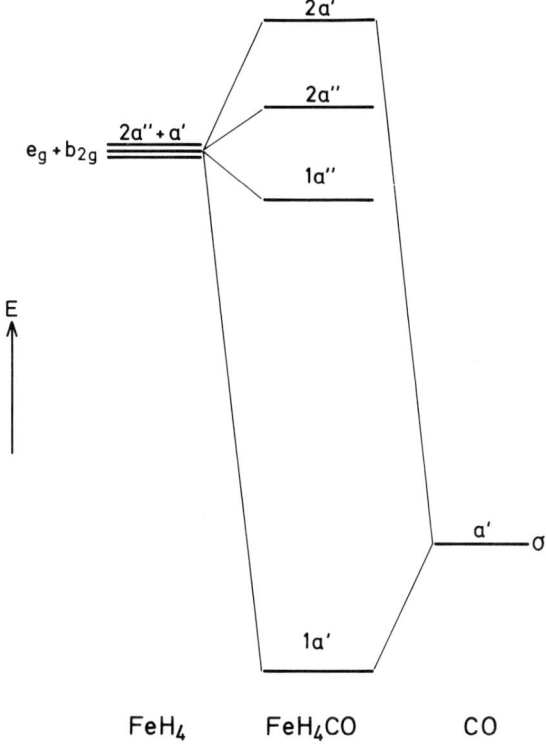

Fig. G.6, IV. The interaction between the HOMO's of FeH_4 and CO in an $FeH_4 - CO$ complex.

G.7 1. $\cdots 3t_2^6\, 2a_1^2\, 4t_2^2$

2. $^1A_1 + {}^1E + {}^1T_2 + {}^3T_1$; ground state: $^3T_1$

3. The spatially degenerate $^3T_2$ state is Jahn–Teller distorted.

4. The HOMO would still be of t_2 symmetry and no changes would therefore be observed in the results of questions 2 and 3.

5. The same answer as in question 4.

6. $^3T_{1g} \rightarrow {}^3A_{2g} + {}^3E_g$; gerade, since the molecular terms originate from a t_{2g}^2 or a t_{2u}^2 configuration.

7. $t_2 \rightarrow a_{2u} + e_u$; ungerade, since t_2 is mainly the $5p$ orbital on Xe. $a_1 \rightarrow a_{1g}$; gerade, since a_1 is mainly the $5s$ orbital on Xe. The possible ground-state electronic configurations and term symbols are given in Fig. G.7, I.

8. The models in questions 6 and 7.I [see Fig. G.7,I] predict the incorrect spin symmetry of the ground state. Model 7.IV predicts incorrect ionization potentials. Models 7.II and III predict correct

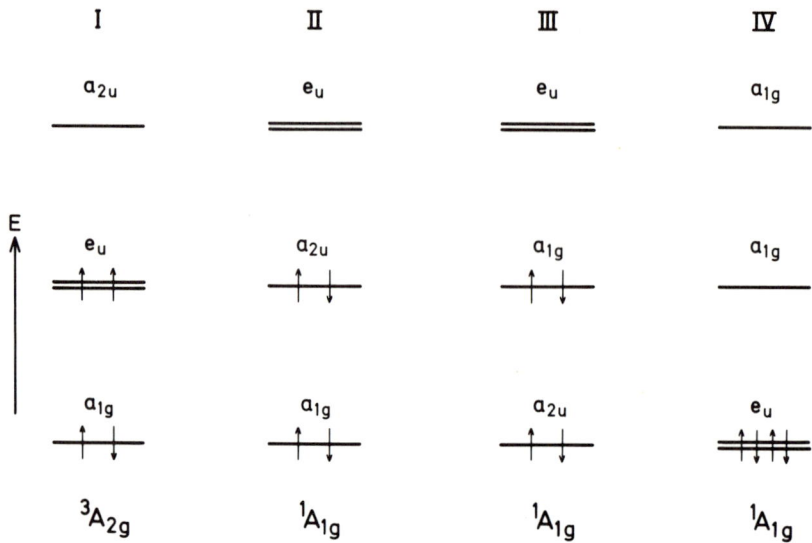

Fig. G.7, I. Possible ground-state configurations and molecular terms for a D_{4h} coordinated XeF_4 molecule.

spin symmetry and ionization potentials. (Within the considered model it has no meaning to pay any attention to the very small difference in ionization potential.) The spatial symmetry of the two lowest excited states in both 7.II and III would be $e_u \otimes a_{1g} = E_u$ and $e_u \otimes a_{2u} = E_g$, in agreement with experiment.

According to Eq. (F.15, 1)

$$\Delta\,^3E(i \rightarrow n) = \Delta^1 E(i \rightarrow n) - 2(ni|in)$$

where $\Delta\,^3E$ and $\Delta\,^1E$ label the triplet and singlet excitation energy in the Hartree–Fock model when an electron is promoted from orbital i to orbital n. Since $(ni|in)$ is positive, $\Delta\,^3E(i \rightarrow n) < \Delta\,^1E(i \rightarrow n)$.

G.8 1. The $2p\pi$ ligand orbitals transform as $t_{1g} + t_{1u} + t_{2g} + t_{2u}$. The $2p\pi(t_{1u})$ orbital interacts with the $2p\sigma$ and $5p(t_{1u})$ orbitals. The $2p\pi(t_{2g})$ orbital interacts with the $4d(t_{2g})$ orbital. Because these interactions are of π type, it is highly unlikely that the energy ordering of the three highest orbitals will change. This energy ordering has its origin in σ-type interactions. The t_{1g} and t_{2u} will form nonbonding MO's.

2. $\cdots 2e_g^4 2a_{1g}^2 ; \,^1A_g$

3. $\cdots 2e_g^4 2a_{1g}^1 2t_{1u}^1 ; \,^1T_{1u}(+\,^3T_{1u})$

$\,^1A_{1g} \rightarrow \,^1T_{1u}$ is dipole allowed since (x, y, z) transforms as T_{1u}.

4.

O_h	$\to$	D_{3d}
a_{1g}		a_{1g}
t_{1u}		$a_{2u} + e_u$

Thus, no additional interaction occurs among the frontier orbitals when the symmetry is reduced from O_h to D_{3d} and the ground-state total energy is not expected to change.

5.

D_{3d}	$\to$	C_{3v}
a_{1g}		a_1
a_{2u}		a_1
e_u		e

The HOMO and LUMO interact in C_{3v}, giving rise to the electronic configuration $\cdots 1a_1^2$, and we therefore expect lower ground-state total energy in C_{3v} (see Fig. G.8, I).

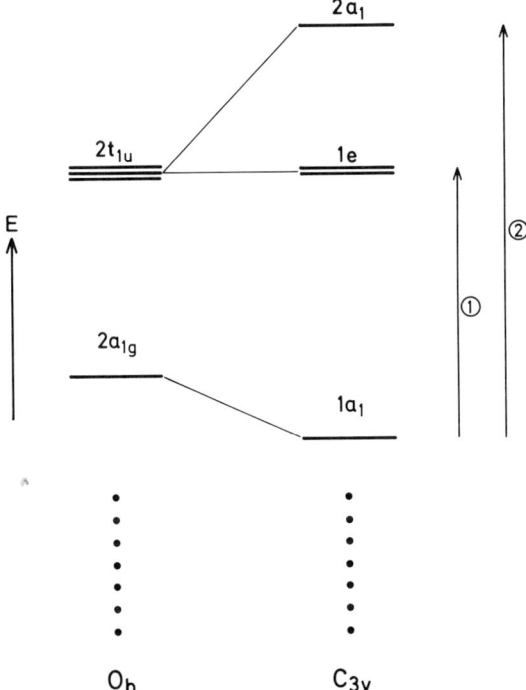

Fig. G.8, I. The interaction between the highest occupied and lowest unoccupied molecular orbital in XeF_6 when an octahedrally coordinated molecule is distorted to C_{3v} symmetry.

6. Excitation (1) in Fig. G.8, I is of symmetry $^1E_1(^3E_1)$ and is lower in energy than the $^1A_1$ excitation, (2).

G.9 1. ^4F

2. $S_1 \cdot S_2 = \frac{1}{2}(S^2 - S_1^2 - S_2^2)$ where S is the total electronic spin. Using that $S_1 = \frac{3}{2}$ and $S_2 = \frac{3}{2}$ we find that the total spin quantum numbers are $S = 3, 2, 1, 0$. The operator in Eq. (G.9, 1) is diagonal in the representation $|S_1 S_2 S\rangle$. Choosing these states as zeroth-order states, we have that the first-order correction is

$$\langle S_1 S_2 S | \hat{V} | S_1 S_2 S \rangle = \frac{1}{2}J[S(S + 1) - \frac{3}{2}(\frac{3}{2} + 1)] \quad (G.9, I)$$

Inserting the four different values for S in Eq. (G.9, I) we obtain the energy splitting in Fig. G.9, I.

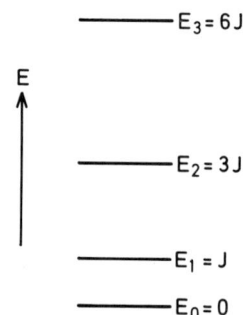

Fig. G.9, I. The energy splitting that results from the spin–spin interaction in the ground state of $[(NH_3)_5CrOHCr(NH_3)_5]^{5+}$.

APPLICATION OF SYMMETRY IN CHEMICAL REACTIONS

INTRODUCTION

Quantitative MO calculations can be used to decide whether a chemical reaction has a large activation energy (is forbidden) or a small activation energy (is allowed). However, when symmetry elements are conserved during a reaction, *qualitative* MO considerations may be used to estimate the magnitude of the activation energy. The concept of "conservation of orbital symmetry" (Woodward and Hoffmann 1970) may then be used to decide whether a reaction is thermally (H.1-4, H.7) or photochemically (H.1, H.5, H.6) allowed. Conservation of orbital symmetry implies an allowed reaction. To determine if a reaction is symmetry allowed it suffices to analyze the symmetry of the orbitals that are directly involved in the bond formation and/or the bond cleavage. These orbitals are called *active* orbitals. For instance, in a reaction where a σ bond is formed it suffices to consider how the σ bonding and σ^* antibonding orbitals (representing the σ bond formation) change and thus to include in the symmetry analysis only the orbitals required to form the σ and σ^* orbitals. If π orbitals are active, all π orbitals of the π skeleton must be included in the symmetry analysis since all the π orbitals then change structure during the reaction. In problem H.2 we illustrate how the orbitals that need to be included in a symmetry analysis may be selected.

When an appropriate selection of orbitals is performed an orbital correlation diagram may be constructed and used to construct a state correlation diagram. If an avoided crossing occurs in the state correlation diagram when correlating the dominant ground-state configuration of the reactants with those of the products, the thermal reaction is expected to have a high activation energy and therefore to be thermally forbidden. If the dominant configurations of the ground state of the reactants and the products correlate, a very small activation energy probably exists and the reaction is expected to be thermally allowed (H.1-4, H.7). In deciding whether a reaction is photochemically allowed, we examine if the dominant configuration of the lowest excited state of a given symmetry of the reactants correlates with the dominant configuration of the lowest excited state of the same symmetry of the products. If they correlate, the reaction is expected to have a small activation energy and thus to be photochemically allowed. If an avoided crossing is observed, the reaction has a high activation energy and is then expected to be photochemically forbidden (H.5, H.6). Very detailed experimental data may in some cases be available about the photophysics and photochemistry of a given reaction. The state correlation diagrams of problems H.5, H.6 are used to interpret such detailed experimental data.

The theoretical background needed to solve the problems in this set is included in most newer textbooks in quantum chemistry. Some elementary textbooks (e.g., Bellamy 1974) consider solely reactions in which either the symmetry element C_2 or σ is conserved. However, the treatment given in these texts is easily generalized so that they also provide sufficient background to solve the problems of the present set even when conservation of other symmetry elements is considered. In the cases where we feel that additional explanation is required to solve the problems we provide it in the problem itself.

Textbooks that give sufficient background for solving the problems in this section

Bellamy (1974). Chandra (1974). Dewar (1969). Levine (1974). Lowe (1978). McWeeny (1979). Murrell, *et al.* (1970, 1978).

Textbooks that give a more elaborate treatment of the topic

Pearson (1976). Simons (1983) Woodward and Hoffmann (1970).

Problem description

H.1 The electrocyclic ring closure: pentadienyl cation→ cyclopentenyl cation

H.2 The thermal addition of $H_2 + N_2 \rightarrow$ *cis-(trans-)* N_2H_2

H.3 The addition of $H_2 + I_2 \rightarrow 2HI$

H.4 The thermal decomposition of $H_2CO \rightarrow H_2 + CO$

H.5 The photochemistry of benzene and benzene isomers: benzene $\rightarrow$ Dewar benzene; benzene $\rightarrow$ prismane; Dewar benzene $\rightarrow$ prismane

H.6 The photochemistry of the ring opening in 1,4-Dewar naphthalene $\rightarrow$ naphthalene

H.7 The catalytic effect of iron in the electrocyclic ring opening of *syn-*tricyclooctadiene iron tricarbonyl $\rightarrow$ bicyclooctatriene iron tricarbonyl

PROBLEMS

H.1 In this problem we will analyze the electrocyclic ring closure of the pentadienyl cation. The active MO's are the five π MO's of the reactant (pentadienyl cation) and the three π MO's and the terminal σ and σ^* MO's of the product, the cyclopentenyl cation. The terminal σ and σ^* MO's are the σ bonding and antibonding MO's that are formed between the two end atoms of the pentadienyl cation during the reaction.

1. Determine the Hückel π MO's of the pentadienyl and the cyclopentenyl cations. Sketch a contour plot of these orbitals and of the terminal σ and σ^* MO's orbitals of the cyclopentenyl cation.

We will now examine the possible reaction paths (conrotatory and disrotatory) for a thermal and a photochemical electrocyclic ring closure of the pentadienyl cation.

2. Classify the symmetry of the orbitals of question 1 with respect to the C_2 symmetry axis which is conserved for a conrotatory reaction path and with respect to the mirror plane which is conserved for a disrotatory reaction path.

3. Draw an orbital correlation diagram for both the conrotatory and the disrotatory reaction path.

We will now construct a state correlation diagram for the conrotatory and disrotatory reaction paths. The state correlation diagram must be constructed such that it shows whether the configurations of the ground and first excited states of the reactant and product correlate with each other or with other excited state configurations.

4. Draw a state correlation diagram for both the conrotatory and the disrotatory reaction path. Include only excited states of singlet symmetry in the state correlation diagram.

5. Which reaction path would a thermal and which would a photochemical reaction follow according to the orbital and according to the state correlation diagram?

We will now consider two examples of electrocyclic reactions that involve the pentadienyl cation. Consider initially the reaction in Fig. H.1,1 (Shoppee and Cooke 1972). When I is treated with acid, III is formed with II as an intermediate.

6. Will the phenyl groups ($\phi = C_6 H_5$) in III end up in a cis or trans configuration?

Consider next the reaction in Fig. H.1,2 (Childs *et al.* 1974, Cabell-Whiting and Hogeveen 1973).

7. Is the reaction in Fig. H.1,2 thermally or photochemically allowed?

H.2 In this problem we will analyze the thermal addition of a hydrogen molecule to a nitrogen molecule to obtain the diimide molecule $N_2 H_2$. The reaction mechanism may be either a planar suprafacial addition reaction, as indicated in Fig. H.2,1, or an antarafacial addition reaction, as shown in Fig. H.2,2 $(0 < \theta < \pi)$ (Woodward and Hoffmann 1970). C_{2v} symmetry is preserved in the

Fig. H.1,1. Electrocyclic reaction involving the pentadienyl cation.

CH₃ CH₃

CH₃—[⊕ ring structure]—

CH₃ CH₃

→

CH₃ [bicyclic structure] —CH₃
CH₃ (⊕) CH₃
CH₃

Fig. H.1,2. Electrocyclic reaction involving the pentadienyl cation.

suprafacial reaction which gives *cis*-diimide. In the antarafacial reaction C_2 symmetry is conserved and the reaction product in the planar configuration is *trans*-diimide, which has the point group symmetry C_{2h}. To describe the change in the molecular binding of the reactants, the intermediate, and the product of the addition reaction, it suffices to consider the MO's that can be formed from the $2p$ atomic orbitals on the nitrogen atoms and the $1s$ atomic orbitals on the hydrogen atoms.

Consider initially the suprafacial addition reaction.

1. Draw MO diagrams for the reactants (N_2 and H_2) and the product (*cis*-diimide).

2. Draw an orbital correlation diagram for the reaction in Fig. H.2,1 indicating the correlation between the MO's of question 1.

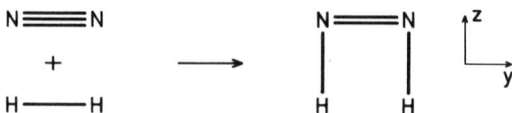

Fig. H.2,1. Reaction path for the suprafacial addition of $N_2 + H_2$ yielding *cis*- $N_2 H_2$.

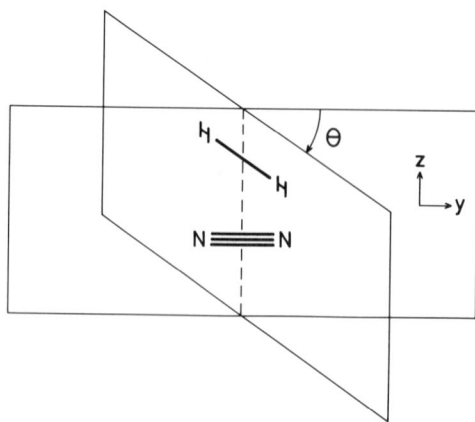

Fig. H.2,2. Reaction path for the antarafacial addition of $N_2 + H_2$ yielding *trans*-$N_2 H_2$.

3. Determine from a state correlation diagram whether the reaction is thermally allowed.

Consider now the antarafacial addition reaction in Fig. H.2,2.

4. Draw an MO diagram for the product of the reaction, *trans*-diimide.

5. Draw an orbital correlation diagram for the reaction in Fig. H.2,2.

6. Determine from a state correlation diagram whether the reaction is thermally allowed.

We will now show that the results of questions 3 and 6 could equally well have been obtained by considering only the correlations between the orbitals that were directly involved in the bond cleavage and bond formation (active orbitals).

Consider initially the suprafacial addition reaction. Two of the three unoccupied and two of the three occupied MO's of N_2 are almost unaffected by the bond cleavages and bond openings in the reaction in Fig. H.2,1 and correlate with orbitals of *cis*- $N_2 H_2$, which has the same bonding character (i.e., σ, σ^*, π, or π^* bonds between the same atoms).

7. Indicate the four MO's of N_2 that have the same bonding character in both N_2 and *cis*- $N_2 H_2$. Indicate also the dominant bonding (or antibonding) character of these orbitals.

8. Determine the dominant bonding character of the orbitals of N_2 and H_2 which completely change bonding character during the suprafacial reaction. Indicate the bonding characters of the orbitals of *cis*- $N_2 H_2$ which correlate with these orbitals.

9. Show that the result of question 3 can be obtained by considering only the active orbitals for the suprafacial addition reaction.

10. Show that the result of question 6 could also have been obtained by considering only the active orbitals for the antarafacial addition reaction.

Even though this problem shows that a thermal reaction of N_2 + H_2 gives *trans*-diimide, the rotational barrier between the cis and trans form is so small that interconversion between the two forms readily occurs and both conformers should thus be relatively easy to make (Ahlrichs and Staemmler 1976).

H.3 At low temperature the reaction

$$H_2 + I_2 \rightarrow 2HI$$

proceeds according to the mechanism (Sullivan 1967)

$$I_2 \rightleftarrows 2I$$

$$I + H_2 \rightleftarrows IH_2$$

$$IH_2 + I \rightleftarrows 2HI$$

Until 1967 it was believed that the reaction proceeded as the concerted reaction with conservation of C_{2v} point group symmetry (suprafacial addition reaction) displayed in Fig. H.3,1. We will show that the reaction in Fig. H.3,1 is thermally forbidden (Hoffmann 1968).

1. Draw an orbital correlation diagram for the C_{2v} concerted reaction path. Include only the active orbitals.

2. Argue—based on the orbital correlation diagram—that the concerted C_{2v} reaction path is thermally forbidden.

The antarafacial addition reaction may similarly be shown to be thermally forbidden (Pearson 1976).

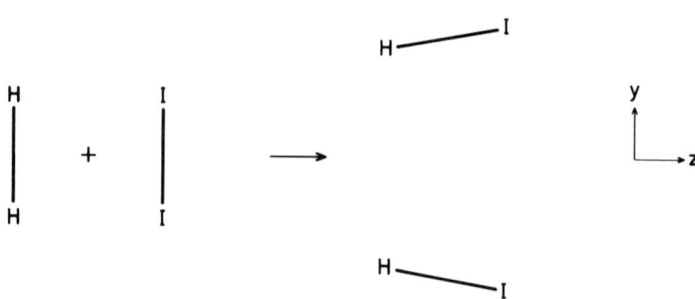

Fig. H.3,1. Reaction path of the suprafacial addition of $I_2 + H_2$ yielding 2HI.

H.4 In this problem we analyze the thermal decomposition of formaldehyde to carbon monoxide and a hydrogen molecule:

$$H_2CO \rightarrow H_2 + CO \qquad (H.4,1)$$

Minimum basis MO calculations give the MO energies in Table H.4,1.
We will initially assume that the reaction proceeds as indicated in Fig. H.4,1.

1. Determine the point group symmetry that is preserved during the decomposition.
2. Determine the IR's spanned by the MO's of formaldehyde, of the hydrogen molecule, and of carbon monoxide in the point group symmetry that is preserved during the reaction path.
3. Draw an orbital correlation diagram for the reaction in Fig. H.4,1.
4. Determine from a state correlation diagram whether the thermal decomposition of formaldehyde is symmetry allowed.

Assume now that the decomposition takes place such that only the reflection symmetry in the x-z plane is preserved during the whole reaction. This would be the case, for example, if CO were tilted out of the plane of the H_2CO molecule (the y-z plane) but were still lying in the x-y plane.

5. Determine from a state correlation diagram if the thermal decomposition of formaldehyde is allowed for this reaction path.

H.5 In this problem we will analyze the (thermal and photochemical) disrotatory electrocyclic ring closures in Fig. H.5,1. According to the principle of least motion the reaction paths for all the reactions must have C_{2v} symmetry

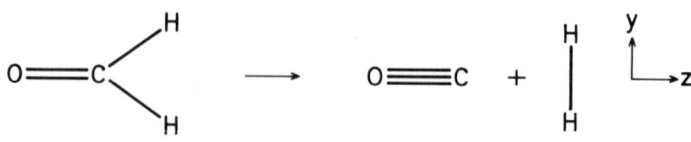

Fig. H.4,1. Reaction path for the thermal decomposition of formaldehyde.

Table H.4,1
Orbital Energies in eV

Formaldehyde[a] (C_{2v})		Carbon monoxide[b] ($C_{\infty v}$)		Hydrogen[c] ($D_{\infty h}$)	
Symmetry	ε_i	Symmetry	ε_i	Symmetry	ε_i
$1a_1$	−561.17	1σ	−565.94	$1\sigma_g$	−16.84
$2a_1$	−310.26	2σ	−311.19	$1\sigma_u$	10.91
$3a_1$	−38.03	3σ	−41.76		
$4a_1$	−22.62	4σ	−20.66		
$1b_2$	−18.39	1π	−16.61		
$5a_1$	−16.14	5σ	−13.74		
$1b_1$	−13.52	2π	6.12		
$2b_2$	−10.76	6σ	23.47		
$2b_1$	6.12				
$6a_1$	10.75				
$3b_2$	14.79				
$7a_1$	21.89				

[a]Newton and Palke (1966).
[b]Brion and Moser (1960).
[c]Problem B.9.

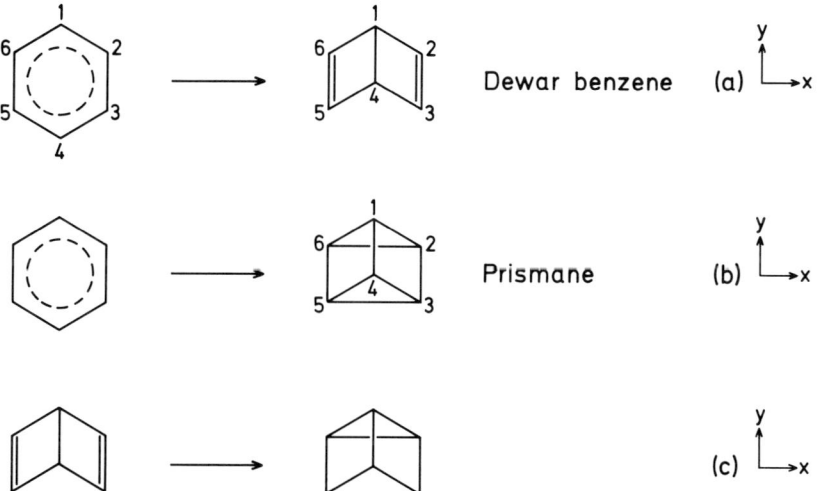

Fig. H.5,1. Reaction paths for the photochemical reactions: (a) benzene → Dewar benzene, (b) benzene → prismane, and (c) Dewar benzene → prismane.

Table H.5,1
Hückel MO's, ψ_i, and $(\varepsilon_i - \alpha)/\beta = X_i$ for Benzene[a]

	ψ_1	ψ_2	ψ_3	ψ_4	ψ_5	ψ_6
$\phi_i \backslash X_i$	2.000000	1.000000	1.000000	−1.000000	−1.000000	−2.000000
1	0.408248	0	0.577350	−0.577350	0	−0.408248
2	0.408248	0.500000	0.288675	0.288675	0.500000	0.408248
3	0.408248	0.500000	−0.288675	0.288675	−0.500000	−0.408248
4	0.408248	0	−0.577350	−0.577350	0	0.408248
5	0.408248	−0.500000	−0.288675	0.288675	0.500000	−0.408248
6	0.408248	−0.500000	0.288675	0.288675	−0.500000	0.408248

[a]From Heilbronner and Straub (1966).

preserved. We consider in our analysis only the active orbitals, that is, the π orbitals of benzene and Dewar benzene and a σ bonding and a σ^* antibonding orbital for each σ bond formed from the π (and π^*) MO's of benzene. The π MO's of benzene are given in Table H.5,1.

1. Determine the irreducible representations spanned by the active orbitals of Fig. H.5,1.

2. Draw an orbital and a state correlation diagram for the C_{2v} reaction path in Fig. H.5,1(a).

3. Draw an orbital and a state correlation diagram for the C_{2v} reaction path in Fig. H.5,1(b).

4. Draw an orbital and a state correlation diagram for the C_{2v} reaction path in Fig. H.5,1(c).

Dewar benzene is 60 kcal mole^{-1} higher in energy than benzene, while prismane is about 90 kcal mole^{-1} higher than benzene (Pearson 1976). Both Dewar benzene and prismane are stable at room temperature.

5. Explain from the state correlation diagram in questions 2-4 why Dewar benzene and prismane are stable molecules.

The lowest excitation energies of benzene having singlet symmetry are $^1B_{2u}$(4.72 eV), $^1B_{1u}$(5.96 eV), and $^1E_{1u}$(6.76 eV) where the numbers in parentheses are the energies of the excited states relative to the ground state (Salem 1966). The excitation energy at 5.96 eV is not uniquely assigned and may be of $^1E_{2g}$ symmetry. The reaction in Fig. H.5,1 (a) takes place if benzene is irradiated by light of wavelength 2000 Å but not by light of wavelength 2600 Å (Haller 1967).

6. Use the state correlation diagram of question 2 to explain that
   (a) light of wavelength 2000 Å is required to make the photochemical reaction take place;
   (b) the second lowest excited state is of $^1B_{1u}$ symmetry and not of $^1E_{2g}$ symmetry.

7. At which wavelength of the exciting light would you expect the reaction in Fig. H.5,1(b) to take place?

8. Is the reaction in Fig. H.5,1(c) photochemically allowed?

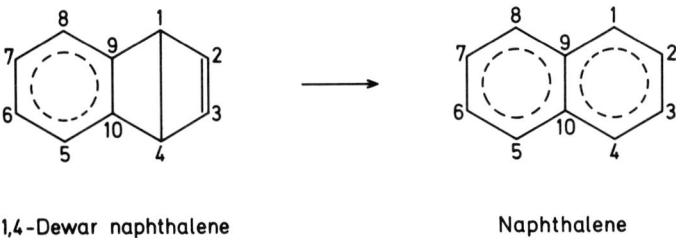

Fig. H.6,1. The labeling of the atomic sites in 1,4-Dewar naphthalene and naphthalene.

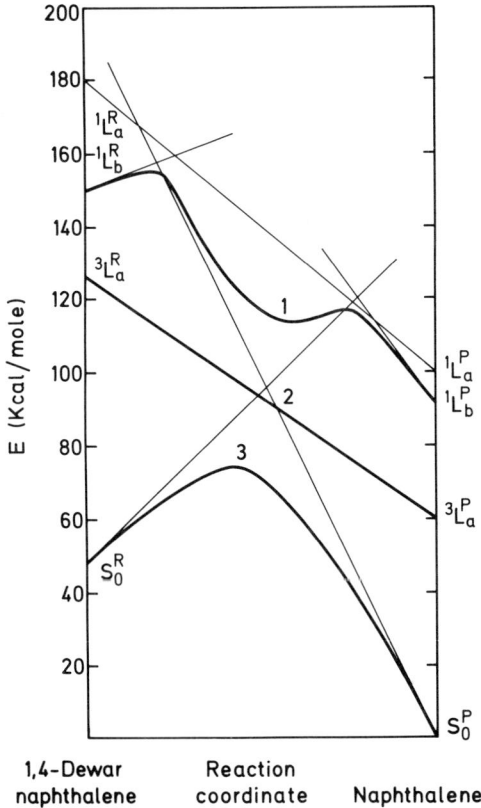

Fig. H.6,2. State correlation diagram for the reaction 1,4-Dewar naphthalene → naphthalene.

Table H.6,1
Hückel MO's, ψ_i^N and $(\varepsilon_i - \alpha)/\beta = X_i$ for Naphtalene Using the Numbering of the AO's, ϕ_i, in Fig. H.6,1[a]

	ψ_1^N	ψ_2^N	ψ_3^N	ψ_4^N	ψ_5^N
$\phi_i \backslash X_i$:	2.302776	1.618034	1.302776	1.000000	0.618034
1	0.300552	0.262866	0.399586	0	0.425325
2	0.230701	0.425325	0.173524	0.408248	0.262866
3	0.230701	0.425325	−0.173524	0.408248	−0.262866
4	0.300552	0.262866	−0.399586	0	−0.425325
5	0.300552	−0.262866	−0.399586	0	0.425325
6	0.230701	−0.425325	−0.173524	0.408248	0.262866
7	0.230701	−0.425325	0.173524	0.408248	−0.262866
8	0.300552	−0.262866	0.399556	0	−0.425325
9	0.461402	0	0.347047	−0.408248	0
10	0.461402	0	−0.347047	−0.408248	0

	ψ_6^N	ψ_7^N	ψ_8^N	ψ_9^N	ψ_{10}^N
$\phi_i \backslash X_i$:	−0.618034	−1.000000	−1.302776	−1.618034	−2.302776
1	−0.425325	0	−0.399586	−0.262866	−0.300552
2	0.262866	0.408248	0.173524	0.425325	0.230701
3	0.262866	−0.408248	0.173524	−0.425325	−0.230701
4	−0.425325	0	−0.399586	0.262566	0.300552
5	0.425325	0	−0.399586	−0.262566	0.300552
6	−0.262866	−0.408248	0.173524	0.425325	−0.230701
7	−0.262866	0.408248	0.173524	−0.425325	0.230701
8	0.425325	0	−0.399586	0.262866	0.300552
9	0	−0.408248	0.347047	0	0.461402
10	0	0.408248	0.347047	0	−0.461402

[a]From Heilbronner and Straub (1966).

H.6 In this problem we will consider the disrotatory electrocyclic ring opening of 1,4-Dewar naphthalene in Fig. H.6,1. Hückel π-electron calculations on naphthalene and benzene give the MO energies in Tables H,6.1 and H,5.1, respectively. The active orbitals in the ring opening reaction are the π orbitals of naphthalene, the π orbitals of the benzene skeleton of 1,4-Dewar naphthalene, the π orbitals of the $C_2 - C_3$ ethylene fragment, and the σ and σ^* orbitals corresponding to the $C_1 - C_4$ σ bond in 1,4-Dewar naphthalene.

1.   Determine within the point group symmetry that is preserved during the disrotatory reaction path the IR's spanned by the active orbitals.

2.   Draw an orbital correlation diagram for the disrotatory ring opening of 1,4-Dewar naphthalene. Use in the construction of the MO's of 1,4-Dewar naphthalene that the π MO's of the benzene and of the ethylene fragments with approximate equal orbital energies and with the same molecular symmetry interact weakly, forming bonding and antibonding orbitals.

A state correlation diagram for the disrotatory electrocyclic ring opening of 1,4-Dewar naphthalene (Carr *et al.* 1976) is displayed in Fig. H.6,2. Curve 1 indicates the reaction path for the lowest excited singlet state; 2 the reaction path for the lowest excited triplet state, and 3 the reaction path for the ground state (schematic). The thin lines connect correlating electronic configurations. The dominant electronic configuration of the state of lowest triplet symmetry, $^3L_a^R$ or $^3L_a^P$, is also the major component in the third state of singlet symmetry for both 1,4-Dewar naphthalene, $^1L_a^R$, and naphthalene, $^1L_a^P$. The lowest excited state of singlet symmetry for 1,4-Dewar naphthalene and naphthalene is denoted by $^1L_b^R$, and $^1L_b^P$, respectively.

3. Use the orbital correlation diagram to determine the electronic configurations corresponding to the thin lines in Fig. H.6,2 in the 1,4-Dewar naphthalene and the naphthalene limits. Indicate the molecular term symbols for the states denoted by 1, 2, and 3 in Fig. H.6,2.

4. Argue that the reaction in Fig. H.6,1 is photochemically but not thermally allowed.

The photochemical ring opening of 1,4-Dewar naphthalene induces fluorescence of *both* 1,4-Dewar naphthalene and naphthalene. The relative intensity of the fluorescence of 1,4-Dewar naphthalene becomes smaller when the wavelength of the exciting light gets smaller. For instance, the ratio between the fluorescence of naphthalene and 1,4-Dewar naphthalene increases by a factor of 1.8 as λ_{ex} decreases from 285 to 255 nm. At λ_{ex} = 214 nm 1,4-Dewar naphthalene undergoes a quantitative ring opening. Only a fraction of the number of naphthalene molecules formed during the photochemical ring opening fluoresces. The ratio between the number of naphthalene molecules that fluoresce and the number that radiationless decay to the ground state of naphthalene is independent of the wavelength used to excite 1,4-Dewar naphthalene (Michl 1977).

5. How would you from Fig. H.6,2 explain that:
    (a) fluorescence is observed for 1,4-Dewar naphthalene?
    (b) the relative intensity of the fluorescence of 1,4-Dewar naphthalene decreases with increasing energy of the irradiating light?
    (c) the ratio between the number of naphthalene molecules which go directly to the ground state and which fluoresce is independent of the energy of the irradiating light?

When 1,4-Dewar naphthalene is irradiated by light, phosphorescence of naphthalene but not of 1,4-Dewar naphthalene is observed. Heating converts 1,4-Dewar naphthalene quantitatively to naphthalene in its ground state.

6. How would you explain that:
    (a) phosphorescence of naphthalene but not of 1,4-Dewar naphthalene is observed?
    (b) the reaction, despite the answer to question 4, proceeds thermally?

H.7 In this problem we will analyze the catalytic effect of iron in the electrocyclic ring opening *syn*-tricyclooctadiene iron tricarbonyl → bicyclooctatriene iron tricarbonyl. The reaction is given in Fig. H.7,1 and is known to take place thermally (Pearson 1976). The reaction may be described as an electrocyclic ring opening of a cyclobutene ring to a butadiene system. The reaction path must be disrotatory because of geometric constraints.

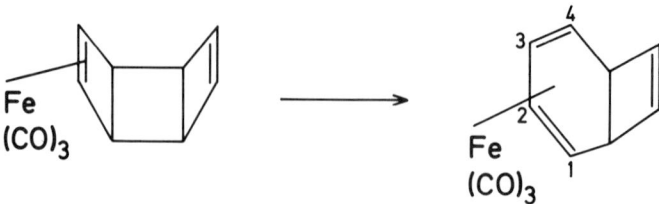

Fig. H.7,1. The geometry of the *syn*-tricyclooctadiene iron tricarbonyl and bicyclooc-tatriene iron tricarbonyl molecules.

First we treat the ring opening in a model where we consider only the cyclobutene → butadiene system and thus disregard the $Fe(CO)_3$ complex.

1. Draw an orbital and a state correlation diagram for this reaction. Assume that C_s symmetry is conserved during the reaction path (i.e., a disrotatory reaction path). Does this model predict that the reaction in Fig. H.7,1 is thermally allowed?

We will now include the d orbitals on the iron atom in the description of the ring opening in Fig. H.7,1. The iron atom is assumed to have the valence electron configuration d^8 (i.e., in the oxidation state zero) and it is located such that both the reactant and the product have a reflection plan as a symmetry element. According to the *principle of least motion*, the reaction takes place such that a reflection plane is conserved during the reaction. Figure H.7,2 illustrates the geometry of the model in which we now consider the reaction in Fig. H.7,1. The d orbital energy of an isolated iron atom is a little higher than the orbital energies of the highest occupied π orbital of cyclobutene and butadiene.

2. Determine from simple MO considerations, based on the magnitude of the overlap, how the d orbitals on iron will split as a result of the interaction with the highest occupied π orbitals of cyclobutene and butadiene. It is convenient to choose the coordinate axes as shown in Fig. H.7,2.

3. Draw an orbital correlation diagram which includes the d orbitals on the iron atom. Draw the corresponding state correlation diagram for the thermal reaction.

4. Is the ring opening thermally allowed in the model which takes into account the d orbitals of Fe?

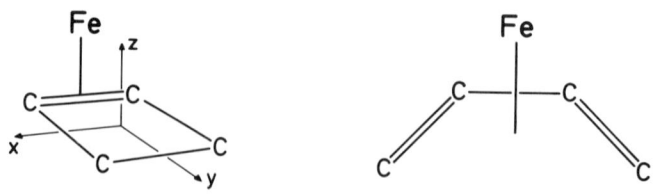

Fig. H.7,2. The model systems that are used to describe the reactant and product in Fig. H.7,1.

SOLUTIONS

H.1 1. The Hückel MO's and orbital energies for the pentadienyl cation
        are obtained from Eqs. (F.2,2-3):

$$\psi_1 = \frac{1}{2\sqrt{3}}(\phi_1 + \sqrt{3}\,\phi_2 + 2\phi_3 + \sqrt{3}\,\phi_4 + \phi_5), \qquad \varepsilon_1 = \alpha + \sqrt{3}\,\beta$$

$$\psi_2 = \frac{1}{2}(\phi_1 + \phi_2 - \phi_4 - \phi_5), \qquad \varepsilon_2 = \alpha + \beta$$

$$\psi_3 = \frac{1}{\sqrt{3}}(\phi_1 - \phi_3 + \phi_5), \qquad \varepsilon_3 = \alpha$$

$$\psi_4 = \frac{1}{2}(\phi_1 - \phi_2 + \phi_4 - \phi_5), \qquad \varepsilon_4 = \alpha - \beta$$

$$\psi_5 = \frac{1}{2\sqrt{3}}(\phi_1 - \sqrt{3}\,\phi_2 + 2\phi_3 - \sqrt{3}\,\phi_4 + \phi_5), \qquad \varepsilon_5 = \alpha - \sqrt{3}\,\beta$$

The three π MO's, π_i, $i = 1, 2, 3$, of the cyclopentenyl cation are the allyl π MO's spanned by the three $p\pi$ orbitals on the nonterminal atoms. These MO's are given in the solution to question 1 of problem F.13. In Fig. H.1,I we have displayed contour plots of all the MO's in question.

2. C_2 point group symmetry is conserved in a conrotatory (CON) reaction whereas C_s point symmetry is conserved in a disrotatory (DIS) reaction. From Fig. H.1,I we obtain the transformation properties of the MO's given in Table H.1,I.

3. See Figs. H.1,II and H.1,III.

4. See Figs. H.1,IV and H.1,V.

5. Both the orbital and the state correlation diagram predict that the thermal electrocyclic reaction is conrotatory. By photochemically exciting one electron of pentadienyl from ψ_2 into ψ_3 we will obtain the cyclopentenyl cation in its first excited state via a disrotatory reaction path. DIS is photochemically allowed according to both Figs. H.1,III and H.1,V.

6. The conrotatory reaction path is thermally allowed and the phenyl groups therefore end up trans to each other.

7. The reaction path must be disrotatory because of geometric constraints. The reaction therefore proceeds photochemically.

H.2 1. The MO diagrams are given in Figs. H.2,I-III. In the construction of the MO diagram for cis- $N_2 H_2$ in Fig. H.2,III we have used that the symmetry orbitals of the 2 * N and the 2 * H fragments have the transformation properties given in Table H.2,I. Furthermore, we have also used that the only interaction which basically occurs

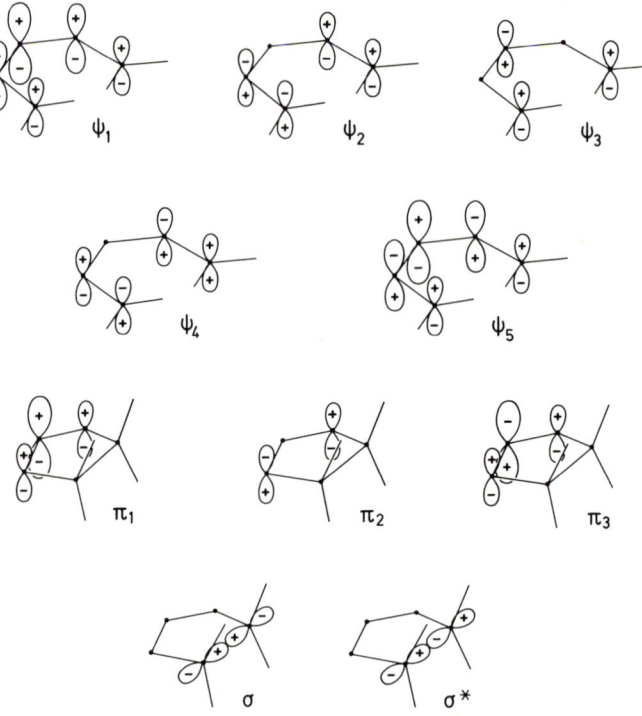

Fig. H.1,I. Contour maps of the active MO's in the electrocyclic ring closure of the pentadienyl cation.

among the orbitals of a_1 symmetry is between the $1\pi_u^{\ddot{}}(a_1)$ and the $1\sigma_g(a_1)$ orbitals [$\sigma(NN)$ bonds are formed], since the $3\sigma_g(a_1)$ orbital, which represents a $\sigma(NN)$ bond, has its amplitude far from the region where molecular binding takes place. Similarly, the $1\sigma_u(b_2)$ and the $1\pi_g^{\ddot{}}(b_2)$ interact strongly, forming $\sigma^*(NH)$ bonds, while the $3\sigma_u(b_2)$ orbital, the $\sigma^*(NN)$ bond, is virtually unaffected.

Table H.1,I
Transformation Properties of the MO's of the Pentadienyl and Cyclopentenyl Cation during the Electrocyclic Ring Closure of the Pentadienyl Cation

Reaction path	Pentadienyl					Cyclopentenyl				
	ψ_1	ψ_2	ψ_3	ψ_4	ψ_5	σ	π_1	π_2	π_3	σ^*
CON (C_2)	b	a	b	a	b	a	b	a	b	b
DIS (C_s)	a'	a''	a'	a''	a'	a'	a'	a''	a'	a''

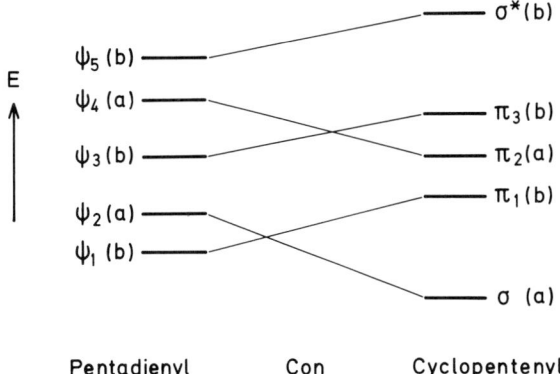

Fig. H.1,II. Orbital correlation diagram for the conrotatory electrocyclic ring closure of the pentadienyl cation.

2. See Fig.H.2,IV.
3. The state correlation diagram in Fig. H.2,V shows that the suprafacial addition reaction is thermally forbidden. This result is independent of the relative energies of the orbital energies of the bonding (and antibonding) MO's of the N_2 and H_2 molecules in Fig. H.2,IV.
4. In constructing the MO diagram for *trans*- $N_2 H_2$ in Fig. H.2,VI we have used that the symmetry orbitals of the orbitals of the $2 * N$ and the $2 * H$ fragments have the transformation properties in Table H.2,II.

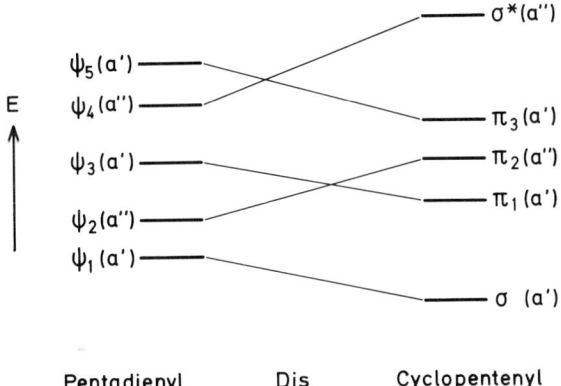

Fig. H.1,III. Orbital correlation diagram for the disrotatory elctrocyclic ring closure of the pentadienyl cation.

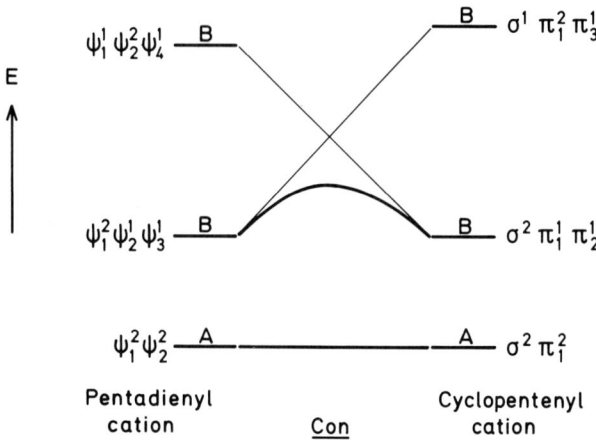

Fig. H.1,IV. State correlation diagram for the conrotatory electrocyclic ring closure of the pentadienyl cation.

We have also used that the only interaction that basically occurs between the orbitals of b_u symmetry is between the $1\sigma_u(b_u)$ and the $1\pi_u^x(b_u)$ orbitals [$\sigma(NH)$ bonds are formed]. Similarly, the $1\sigma_g(a_g)$ and the $1\pi_g^x(a_g)$ orbitals interact strongly, forming $\sigma^*(NH)$ bonds.

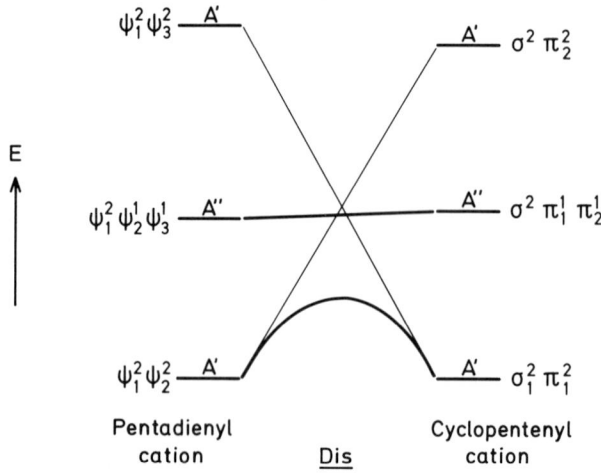

Fig. H.1,V. State correlation diagram for the disrotatory electrocyclic ring closure of the pentadienyl cation.

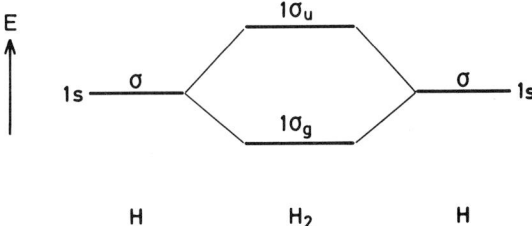

Fig. H.2,I. MO diagram for the H_2 molecule.

5. a_1, a_2 of C_{2v} and a_g, a_u of C_{2h} reduce to a in C_2.
   b_1, b_2 of C_{2v} and b_g, b_u of C_{2h} reduce to b in C_2.
   We therefore have the orbital correlation diagram in Fig. H.2,VII.
6. The state correlation diagram in Fig. H.2,VIII shows that the antarafacial reaction is thermally allowed.
7. The $3\sigma_g$ and $3\sigma_u$ MO's of N_2 are mainly $\sigma(NN)$ and $\sigma^*(NN)$ orbitals, respectively. According to Fig. H.2,IV they correlate with the $1a_1$ and the $3b_2$ orbitals, which also have dominant $\sigma(NN)$ and $\sigma^*(NN)$ character. The $1\pi_u^x (1\pi_g^x)$ orbital correlates with $1b_1 (1a_2)$. For both reactants and products these orbitals have $\pi(NN)$ $[\pi^*(NN)]$ bonding (antibonding) character.

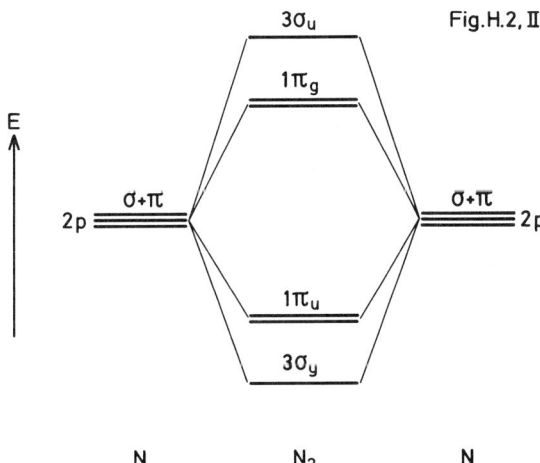

Fig. H.2,II. MO diagram for the N_2 molecule.

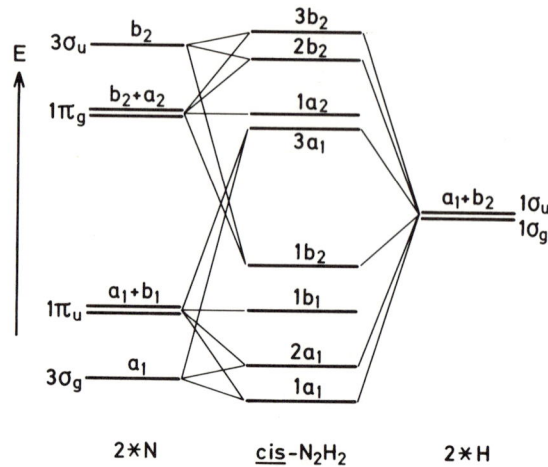

Fig. H.2,III. MO diagram for the *cis*-$N_2 H_2$ molecule.

8. From Figs. H.2,II and H.2,IV we see that the $1\pi_u^z$ and $1\pi_g^z$ orbitals are of the $\pi(NN)$ and $\pi^*(NN)$ type. However, they correlate with the $2a_1$ and $2b_2$ orbitals which are predominantly $\sigma(NH)$ and $\sigma^*(NH)$ orbitals. The $1\sigma_g$ and $1\sigma_u$ orbitals of H_2 are $\sigma(HH)$ and $\sigma^*(HH)$ orbitals, while the orbitals with which they correlate are basically $\sigma(NH)$ and $\sigma^*(NH)$ orbitals; that is, the latter orbitals have a different bonding structure.

Table H.2,I
Tranformation Properties of the Orbitals of the 2∗N and 2∗H Fragments in the C_{2v} Point Group

Orbital	E	$C_2(z)$	$\sigma_v(xz)$	$\sigma_v'(yz)$	IR
$1\sigma_g(2*H)$	1	1	1	1	a_1
$1\sigma_u(2*H)$	1	-1	-1	1	b_2
$3\sigma_g(2*N)$	1	1	1	1	a_1
$1\pi_u^z(2*N)$	1	1	1	1	a_1
$1\pi_u^x(2*N)$	1	-1	1	-1	b_1
$1\pi_g^z(2*N)$	1	-1	-1	1	b_2
$1\pi_g^x(2*N)$	1	1	-1	-1	a_2
$3\sigma_u(2*N)$	1	-1	-1	1	b_2

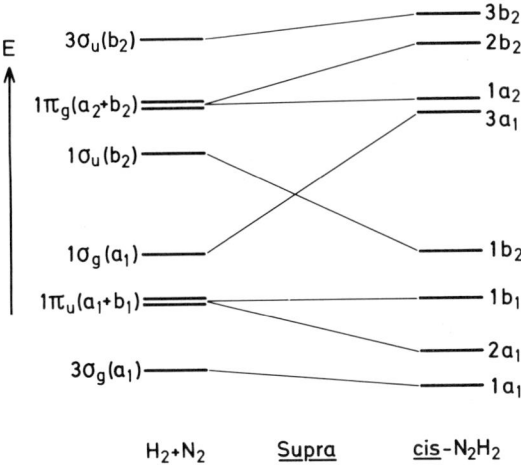

Fig. H.2,IV. Orbital correlation diagram for the suprafacial addition of $N_2 + H_2$ yielding *cis*-$N_2 H_2$.

9. If the $3\sigma_g$, $3\sigma_u$, $1\pi_u^*$, and $1\pi_g^*$ orbitals and the corresponding orbitals in the *cis*- $N_2 H_2$ limit had not been included in the orbital correlation diagram in Fig. H.2,IV we would still have obtained the state correlation diagram in Fig. H.2,V, and the conclusion of question 3 would therefore remain unchanged.

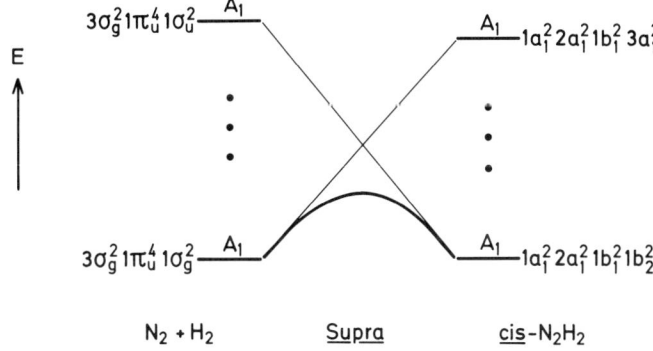

Fig. H.2,V. State correlation diagram for the suprafacial addition of $N_2 + H_2$ yielding *cis*-$N_2 H_2$.

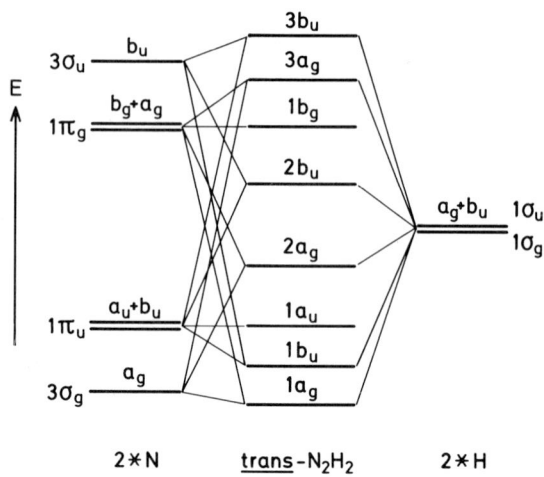

Fig. H.2,VI. MO diagram for *trans*-N_2H_2.

10. Using the same arguments as in question 7, we see that the orbitals in Table H.2,III do not change bonding character during the antarafacial reaction. The neglect of these orbitals in the orbital and state correlation diagrams in Figs. H.2,VII and H.2,VIII, respectively, leads to the same correlation as those given in Fig. H.2,VIII, thus still predicting the reaction in Fig. H.2,2 to be thermally allowed.

   The results of questions 9 and 10 are examples of how it suffices to consider only the active orbitals in order to determine if a reaction is thermally (or photochemically) symmetry allowed. In the suprafacial reaction it suffices, for example, to analyze how the π_u^{-} bonding

Table H.2,II
Transformation Properties of the Orbitals of the 2∗N and 2∗H Fragments in the C_{2h} Point Group

Orbital	E	C_2	i	σ_h	IR
$1\sigma_g(2{*}H)$	1	1	1	1	a_g
$1\sigma_u(2{*}H)$	1	-1	-1	1	b_u
$1\pi_u^z(2{*}N)$	1	1	-1	-1	a_u
$1\pi_u^x(2{*}N)$	1	-1	-1	1	b_u
$1\pi_g^z(2{*}N)$	1	-1	1	-1	b_g
$1\pi_g^x(2{*}N)$	1	1	1	1	a_g

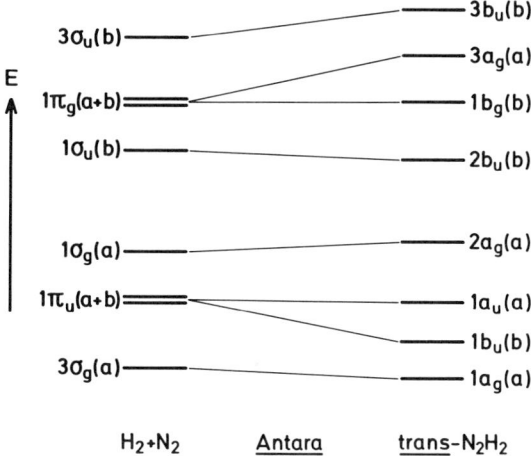

Fig. H.2,VII. Orbital correlation diagram for the antarafacial addition of $N_2 + H_2$ yielding *trans*-$N_2 H_2$

and π_g^z antibonding orbital on N_2 and the $1\sigma_g$ bonding and $1\sigma_u$ antibonding orbital on H_2 correlate with the two $\sigma(NH)$ and the two $\sigma^*(NH)$ MO's of $N_2 H_2$.

H.3 1. We use the coordinate system in Fig. H.3,1. The MO's that can be constructed from $5p$ orbitals on I and the $1s$ orbital on H are the active orbitals. These MO's and their transformation properties are given in Table H.3,I. For each HI molecule the $5p$ orbital on I and

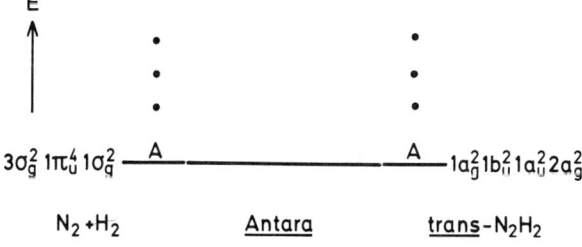

Fig. H.2,VIII. State correlation diagram for the antarafacial addition of $N_2 + H_2$ yielding *trans*-$N_2 H_2$.

Table H.2,III
MO's of N_2 and *trans*-N_2H_2 that Have the Same
Bonding Character during the Reaction in Fig. H.2,2

Orbitals of the reactant	Orbitals of the product	Bonding character
$3\sigma_u$	$3b_g$	$\sigma^*(NN)$
$1\pi_g^z$	$1b_g$	$\pi^*(NN)$
$1\pi_u^z$	$1a_u$	$\pi(NN)$
$3\sigma_g$	$1a_g$	$\sigma(NN)$

the $1s$ orbital on H form one bonding σ MO, one antibonding σ^* MO, and one nonbonding π orbital. The symmetry orbitals of $(HI)_1 + (HI)_2$ which can be constructed from σ, σ^*, and π of the two HI molecules are also given in Table H.3,I. The orbital correlation diagram is displayed in Fig. H.3,I.

2. The ground-state electronic configuration of $H_2 + I_2$ is $H_2\,(\sigma_g^2) + I_2\,(\sigma_g^2\pi_u^4\pi_g^4)$ and it correlates according to Fig. H.3,I with the doubly *excited* configuration of two HI molecules

$$(\sigma_1 + \sigma_2)^2(\sigma_1 - \sigma_2)^2(\pi_1^x + \pi_2^x)^2(\pi_1^x - \pi_2^x)^2(\pi_1^y - \pi_2^y)^2(\sigma_1^* + \sigma_2^*)^2$$

The reaction is thus thermally forbidden.

H.4 1. C_{2v}
2. hydrogen molecule: $1\sigma_g(a_1)$, $1\sigma_u(b_2)$
   carbon monoxide: $\sigma(a_1)$, $\pi(b_1 + b_2)$
3. The orbital correlation diagram is displayed in Fig. H.4,I.

Table H.3,I
MO's of H_2, I_2, and HI + HI, and Their Transformation Properties in the C_{2v} Symmetry Point Group That Is Preserved in the Reaction in Fig. H.2,1

H_2		I_2		$(HI)_1 + (HI)_2$	
Orbital	Symmetry	Orbital	Symmetry	Orbital	Symmetry
$\sigma_g(H_2)$	a_1	$\sigma_g(I_2)$	a_1	$\sigma_1 + \sigma_2$	a_1
$\sigma_u(H_2)$	b_2	$\pi_u(I_2)$	$a_1 + b_2$	$\sigma_1 - \sigma_2$	b_2
		$\pi_g(I_2)$	$a_2 + b_1$	$\pi_1^x + \pi_2^x$	b_1
		$\sigma_u(I_2)$	b_2	$\pi_1^x - \pi_2^x$	a_2
				$\pi_1^y + \pi_2^y$	b_2
				$\pi_1^y - \pi_2^y$	a_1
				$\sigma_1^* + \sigma_2^*$	a_1
				$\sigma_1^* - \sigma_2^*$	b_2

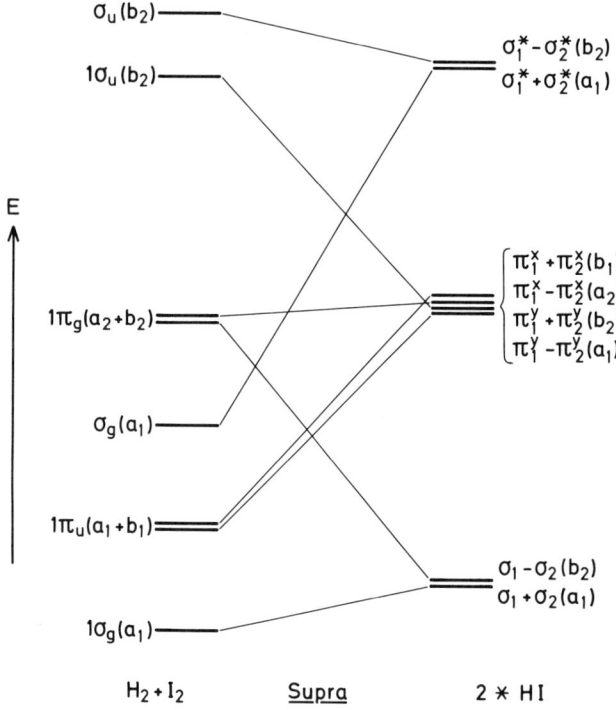

Fig. H.3,I. Orbital correlation diagram for the suprafacial addition of $I_2 + H_2$ yielding 2HI.

4. The $1a_1$ to $4a_1$ orbitals of formaldehyde and the 1σ to 4σ orbitals of carbon monoxide are considered to be core orbitals. H_2CO and $CO + H_2$ have 16 electrons. The state correlation diagram is displayed in Fig. H.4,II, which shows that the decomposition is thermally forbidden.

5. C_s point group symmetry is conserved and the relation between the IR's of C_{2v} and C_s is

C_{2v}	$\rightarrow$	C_s
a_1, b_1	$\rightarrow$	a'
b_2	$\rightarrow$	a''

Thus, the ground-state configuration of H_2CO would still correlate with a highly excited configuration of $CO + H_2$ and the reaction is thermally forbidden.

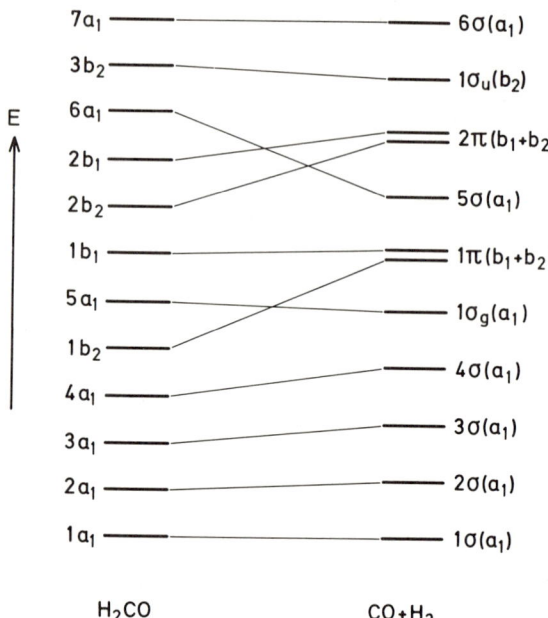

Fig. H.4,I. Orbital correlation diagram for the thermal decomposition of formaldehyde.

H.5 1. In the following π_{ij}^* denotes the antibonding linear combination of $p\pi_i$ and $p\pi_j$ (π AO's on center i and j) chosen such that $p\pi_i$ has positive amplitude in the positive z-direction (see Fig. H.5,1). σ_{ij}^* is an antibonding linear combination of σ_i and σ_j chosen such that σ_i has positive amplitude in the bond direction. In C_{2v} we have the following transformation properties of the MO's which change bonding character during the reaction.

Benzene:

$$\psi_1(a_1),\ \psi_2(b_1),\ \psi_3(b_2),\ \psi_4(a_1),\ \psi_5(a_2),\ \psi_6(b_2)$$

Dewar benzene:

$$\sigma_{14}(a_1),\ \sigma_{14}^*(b_2),\ \pi_{23}+\pi_{65}(a_1),\ \pi_{23}-\pi_{65}(b_1),\ \pi_{23}^*+\pi_{65}^*(b_2),$$
$$\pi_{23}^*-\pi_{65}^*(a_2)$$

Prismane:

$$\sigma_{14}(a_1),\ \sigma_{14}^*(b_2),\ \sigma_{26}+\sigma_{35}(a_1),\ \sigma_{26}-\sigma_{35}(b_2),\ \sigma_{26}^*+\sigma_{35}^*(b_1),$$
$$\sigma_{26}^*-\sigma_{35}^*(a_2)$$

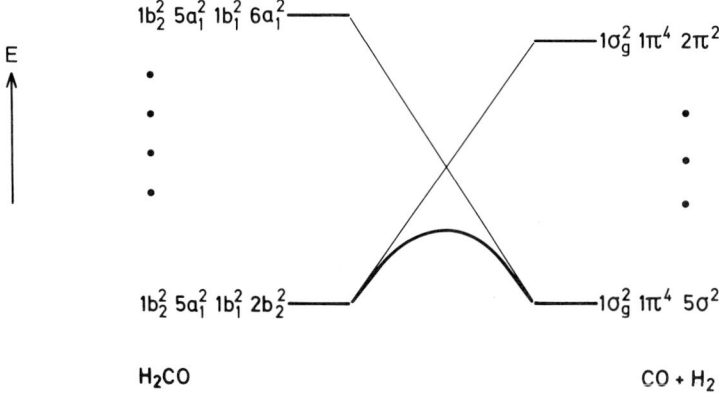

Fig. H.4,II. State correlation diagram for the thermal decomposition of formaldehyde.

2. See Figs. H.5,I and H.5,II.
3. See Figs. H.5,III and H.5,IV.
4. See Figs. H.5,V and H.5,VI.

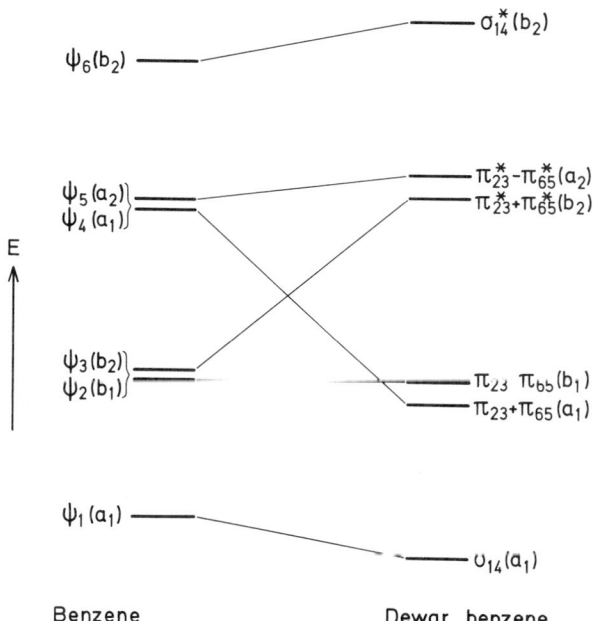

Fig. H.5,I. Orbital correlation diagram for the disrotatory ring closure in the reaction benzene → Dewar benzene.

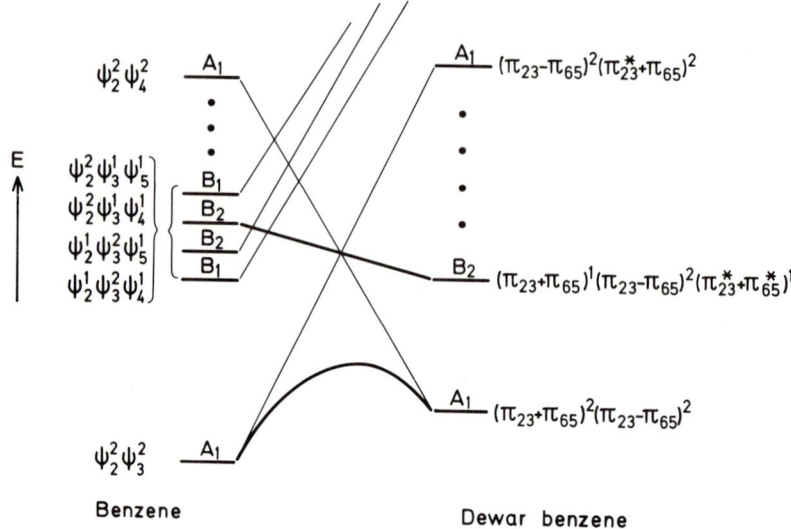

Fig. H.5,II. State correlation diagram for the disrotatory ring closure in the reaction benzene → Dewar benzene.

5. The reactions Dewar benzene → benzene are thermally forbidden. The magnitude of the activation energies prevents a thermal decomposition from taking place even though the reactions are highly exothermic. A similar argument holds for the thermal decomposition of prismane.

6. (a)The following symmetry elements are common to the D_{6h} and the C_{2v} point groups:
    (1)the C_2 axis perpendicular to the molecular plane of benzene;
    (2)the σ_v plane in D_{6h} which passes through carbon atoms 1 and 4 [it becomes $\sigma_v'(xy)$ in C_{2v}]; and
    (3)one of σ_d of D_{6h}, namely, the one which becomes the $\sigma_v(xz)$ plane in C_{2v}.
    This results in the relation between the irreducible representations of D_{6h} and C_{2v} given in Table H.5,I. By photochemically exciting benzene into its second excited state $^1B_{1u}(\sim\ ^1B_2$ in C_{2v}, according to Table H.5,I) the reaction in Fig. H.5,I(a) becomes allowed according to Fig. H.5,II. This requires an energy of 5.96 eV, that is, more than $2600\text{Å} = 4.77$ eV but less than $2000\text{Å} = 6.20$ eV .
    (b)The state correlation diagram in Fig. H.5,II tells that the reaction occurs via a state that has $^1B_2$ symmetry. The ^1E_g state of benzene reduces in C_{2v} to the states $^1A_1 + {}^1A_2$ (see Table H.5,I), none of which gives a photochemically allowed reaction according to Fig. H.5,II.

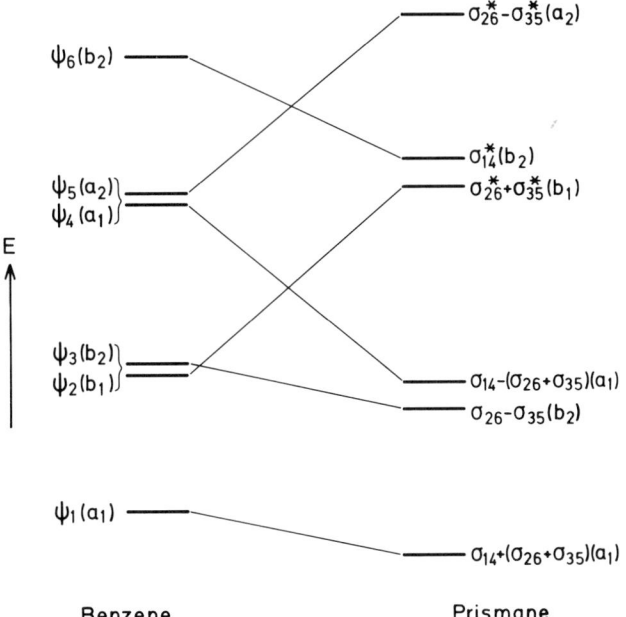

Fig. H.5,III. Orbital correlation diagram for the disrotatory ring closure in the reaction benzene → prismane.

7. Figure H.5,IV shows that benzene during the reaction must be excited into the lowest $^1B_1$ state, which according to Table H.5,I is the $^1B_{2u}$ state of isolated benzene. That requires an energy of 4.72 eV $\sim$ 2600Å. However, it turns out that the activation energy is so high that the reaction does not occur even though it is symmetry allowed. Instead, prismane is formed by the two-step reaction: Fig. H.5,1(a) followed by Fig. H.5,1(c) (Lemal and Lokensgard 1966).

8. Yes, by exciting Dewar benzene into the first excited $^1A_2$ state.

H.6 1. Point group symmetry C_s is conserved and the transformation properties of the orbitals which we consider are as follows
Naphthalene:

$$\psi_1^N(a'),\ \psi_2^N(a'),\ \psi_3^N(a''),\ \psi_4^N(a'),\ \psi_5^N(a''),\ \psi_6^N(a'),\ \psi_7^N(a''),\ \psi_8^N(a'),$$

$$\psi_9^N(a''),\ \psi_{10}^N(a'')$$

Benzene skeleton:

$$\psi_1^B(a'),\ \psi_2^B(a'),\ \psi_3^B(a''),\ \psi_4^B(a'),\ \psi_5^B(a''),\ \psi_6^B(a''),$$

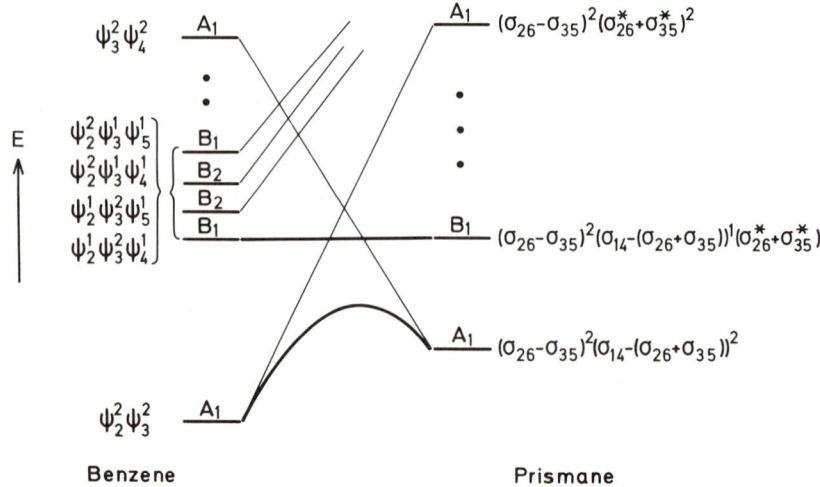

Fig. H.5,IV. State correlation diagram for the disrotatory ring closure in the reaction benzene → prismane.

Ethylene fragment:

$$\pi_{23}(a'), \ \pi_{23}^{*}(a'')$$

"Bridge" orbitals:

$$\sigma_{14}(a'), \ \sigma_{14}^{*}(a'')$$

2. ψ_2^{B} and π_{23} will interact, forming $\psi_2^{B} \pm \pi_{23}$ MO's for 1,4-Dewar naphthalene. The same holds for the antibonding ψ_4^{B} and π_{23}^{*} MO's.

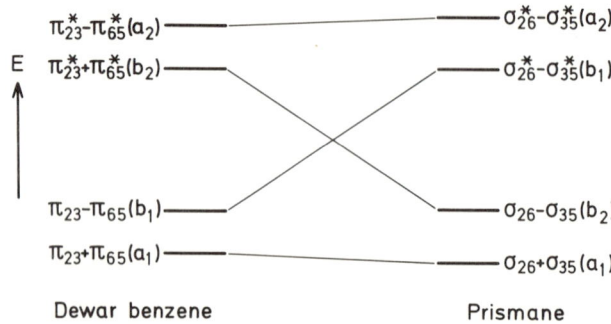

Fig. H.5,V. Orbital correlation diagram for the reaction Dewar benzene → prismane.

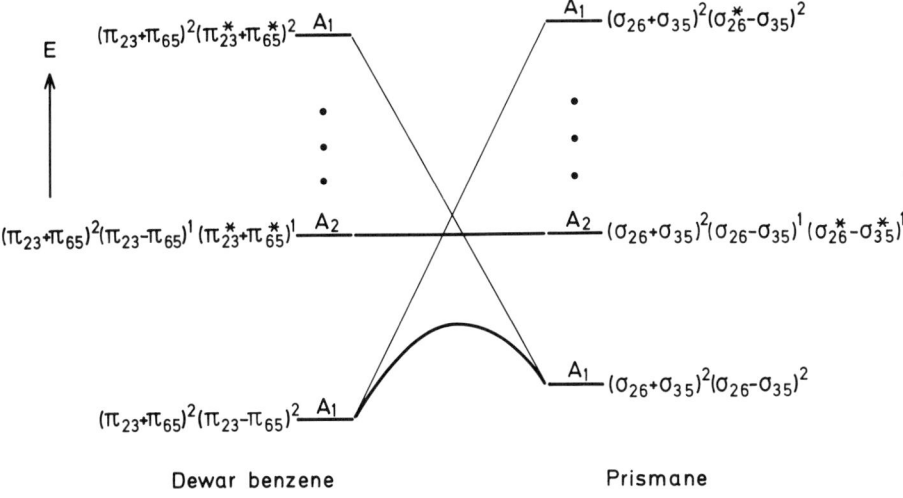

E

$(\pi_{23}+\pi_{65})^2(\pi_{23}^*+\pi_{65}^*)^2$ — A_1

A_1 — $(\sigma_{26}+\sigma_{35})^2(\sigma_{26}^*-\sigma_{35})^2$

$(\pi_{23}+\pi_{65})^2(\pi_{23}-\pi_{65})^1(\pi_{23}^*+\pi_{65}^*)^1$ — A_2

A_2 — $(\sigma_{26}+\sigma_{35})^2(\sigma_{26}-\sigma_{35})^1(\sigma_{26}^*-\sigma_{35}^*)^1$

A_1 — $(\sigma_{26}+\sigma_{35})^2(\sigma_{26}-\sigma_{35})^2$

$(\pi_{23}+\pi_{65})^2(\pi_{23}-\pi_{65})^2$ — A_1

Dewar benzene   Prismane

Fig. H.5,VI. State correlation diagram for the reaction Dewar benzene → Prismane

However, the energy splitting is rather small. The relative energy ordering is given in the correlation diagram in Fig. H.6,I.

3. $\sigma(a')$, $\psi_1^B(a')$, $\psi_1^N(a')$, and $\psi_2^N(a')$ are core orbitals. The occupancy of the other orbitals is given in Table H.6,I. The term symbols are; for state 1: $^1A'$; for state 2: $^3A''$; for state 3: $^1A'$.

4. The barrier in state 3 shows that the reaction is thermally forbidden. However, by irradiating 1,4-Dewar naphthalene and thus exciting it into either of the states $^3L_a^R$ and $^1L_b^R$ we have a symmetry allowed ring opening.

5. (a)The existence of a small local minimum of state 1 at the geometry of 1,4-Dewar naphthalene explains the fluorescence of 1,4-Dewar naphthalene.

(b)From Fig. H.6,2 we see that 285 nm = 100.36 kcal mole^{-1} corresponds to the energy that is required to excite a 1,4-Dewar naphtha-

Table H.5,I
Relation between IR's of D_{6h} and C_{2v}

D_{6h}	E	C_2	$\sigma_d = \sigma_v(xz)$	$\sigma_v = \sigma'_v(yz)$	IR in C_{2v}
E_{1u}	2	-2	0	0	$B_1 + B_2$
E_{2g}	2	2	0	0	$A_1 + A_2$
B_{2u}	1	-1	1	-1	B_1
B_{1u}	1	-1	-1	1	B_2

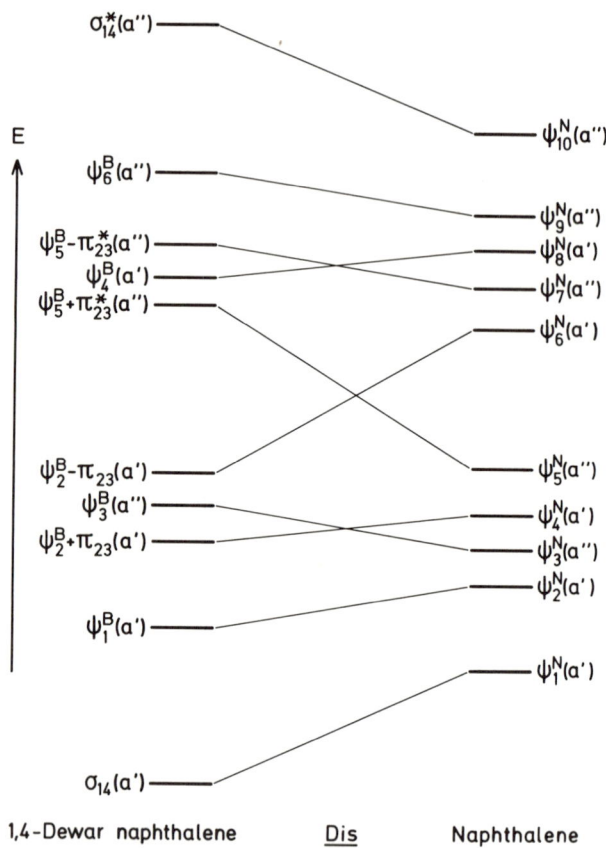

Fig. H.6,I. Orbital correlation diagram for the reaction 1,4-Dewar naphthalene →
naphthalene.

lene molecule from its ground state (S_0^R) to the excited state ($^1L_b^R$).
When the energy of the exciting light becomes larger, for example,
255 nm = 112.16 kcal mole^{-1}, it becomes easier to pass the small
energy barrier and a larger portion of the irradiated 1,4-Dewar
naphthalene molecules therefore get photochemically converted to
naphthalene. At 214 nm = 133.65 kcal mole^{-1} the energy of the
exciting light is so large that 1,4-Dewar naphthalene undergoes a
quantitative ring opening.
(c)The constant ratio is caused by the fact that the molecules that
react photochemically are captured temporarily in the small local
minimum at the intermediate geometry of state 1. A certain fraction

Table H.6,I
Electronic Configurations Giving Rise to the States of Reactant and Product Indicated in Fig. H.6,2

State	Electronic configuration in the 1,4-Dewar napthalene limit	Electronic configuration in the napthalene limit
S_0^R	$(\psi_2^B + \pi_{23})^2 (\psi_3^B)^2 (\psi_2^B - \pi_{23})^2$	$(\psi_3^N)^2 (\psi_4^N)^2 (\psi_6^N)^2$
$\left.\begin{array}{l}{}^3L_a^R, {}^3L_a^P \\ {}^1L_a^R, {}^1L_a^P\end{array}\right\}$	$(\psi_2^B + \pi_{23})^2 (\psi_3^B)^2 (\psi_2^B - \pi_{23})^1 (\psi_5^B + \pi_{23}^*)^1$	$(\psi_3^N)^2 (\psi_4^N)^2 (\psi_5^N)^1 (\psi_6^N)^1$
${}^1L_b^R$	$\begin{cases}(\psi_2^B + \pi_{23})^2 (\psi_3^B)^1 (\psi_2^B - \pi_{23})^2 (\psi_5^B + \pi_{23}^*)^1 \\ (\psi_2^B + \pi_{23})^2 (\psi_3^B)^2 (\psi_2^B - \pi_{23})^1 (\psi_4^B)^1\end{cases}$	$\begin{array}{l}(\psi_3^N)^1 (\psi_4^N)^2 (\psi_5^N)^1 (\psi_6^N)^2 \\ (\psi_3^N)^2 (\psi_4^N)^2 (\psi_6^N)^1 (\psi_8^N)^1\end{array}$
S_0^P	$(\psi_2^B + \pi_{23})^2 (\psi_3^B)^2 (\psi_2^B + \pi_{23}^*)^2$	$(\psi_3^N)^2 (\psi_4^N)^2 (\psi_5^N)^2$
${}^1L_b^P$	$\begin{cases}(\psi_2^B + \pi_{23})^1 (\psi_3^B)^2 (\psi_2^B - \pi_{23})^1 (\psi_5^B + \pi_{23}^*)^2 \\ (\psi_2^B + \pi_{23})^2 (\psi_3^B)^2 (\psi_5^B + \pi_{23}^*)^1 (\psi_5^B - \pi_{23}^*)^1\end{cases}$	$\begin{array}{l}(\psi_3^N)^2 (\psi_4^N)^1 (\psi_5^N)^2 (\psi_6^N)^1 \\ (\psi_3^N)^2 (\psi_4^N)^2 (\psi_5^N)^1 (\psi_7^N)^1\end{array}$

of the reacting molecules then decay nonradiatively to the ground state 3 while the rest passes the small energy barrier and gives naphthalene molecules that fluoresce.

6. (a) The phosphorescence of naphthalene, but not of 1,4-Dewar naphthalene, indicate that there are no energy barriers in state 2.
(b) The reaction is thermally forbidden, but the highly exothermic reaction causes the state correlation diagram to be tilted in such a way that the ground-state energy barrier becomes small.

H.7 1. Using Eqs. (F.2,2) and (F.2,3) we find that the Hückel MO's of butadiene are

$$\psi_1 = 0.3717\,(\phi_1 + \phi_4) + 0.6015\,(\phi_2 + \phi_3), \qquad \varepsilon_1 = \alpha + 1.618\,\beta$$

$$\psi_2 = 0.3717\,(\phi_2 - \phi_3) + 0.6015\,(\phi_1 - \phi_4), \qquad \varepsilon_2 = \alpha + 0.618\,\beta$$

$$\psi_3 = -0.3717\,(\phi_2 + \phi_3) + 0.6015\,(\phi_1 + \phi_4), \qquad \varepsilon_3 = \alpha - 0.618\,\beta$$

$$\psi_4 = 0.3717\,(\phi_1 - \phi_4) - 0.6015\,(\phi_2 - \phi_3), \qquad \varepsilon_4 = \alpha - 1.618\,\beta$$

The transformation properties of ψ_1 to ψ_4 and of the orbitals of cyclobutene are, for butadiene,

$$\psi_1(a'),\ \psi_2(a''),\ \psi_3(a'),\ \psi_4(a'')$$

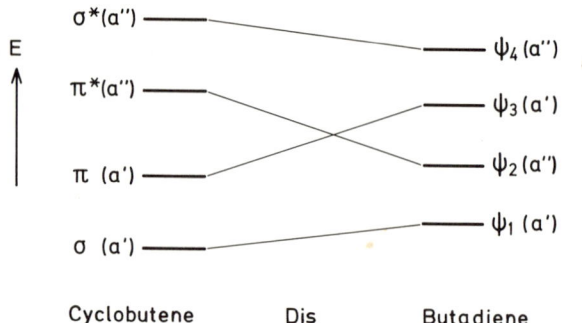

Fig. H.7,I. Orbital correlation diagram for the disrotatory ring opening of cyclobutene.

and for cyclobutene,

$$\sigma(a'), \sigma^*(a''), \pi(a'), \pi^*(a'')$$

The orbital correlation diagram is given in Fig. H.7,I and the state correlation diagram in Fig. H.7,II. Both Figs. H.7,I and II show that the ring opening is thermally forbidden if we disregard the presence of the iron atom.

2. Since the y-z plane is the symmetry plane of the C_s point group, the d orbitals transform as

$$d_{xz}(a''), d_{x^2-y^2}(a'), d_{xy}(a''), d_{zy}(a'), d_{z^2}(a')$$

The highest occupied orbital in cyclobutene is the $\pi(a')$, which interacts with $d_{z^2}(a'), d_{x^2-y^2}(a')$, and $d_{zy}(a')$. d_{z^2} points toward the

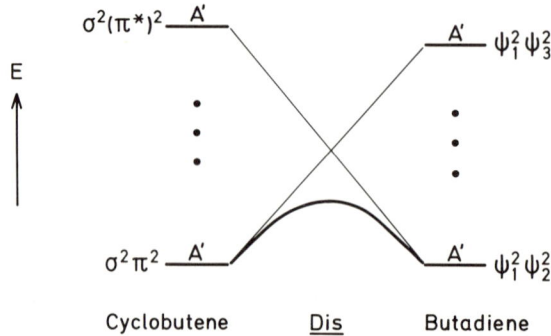

Fig. H.7,II. State correlation diagram for the disrotatory ring opening of cyclobutene.

$\pi(a')$ orbital and thus has a large σ-type overlap with $\pi(a')$. The $d_{x^2-y^2}$ orbital is positive on the x-axis and thus has a π-type overlap with $\pi(a')$, while d_{zy}, with its amplitudes in the diagonals of the y-z plane, has a rather small overlap with $\pi(a')$. The relative splitting of the three d orbitals is thus $\varepsilon(d_{z^2}) > \varepsilon(d_{x^2-y^2}) > \varepsilon(d_{zy})[> \varepsilon(\pi(a'))]$. The orbital energies of the $d_{xz}(a'')$ and $d_{xy}(a'')$ orbitals do not change within this model.

The highest occupied π orbital in butadiene is $\psi_2(a'')$, which interacts with $d_{xy}(a'')$ and $d_{xz}(a'')$ in increasing order (use the same type of arguments as above). The orbital energies of the $d_{x^2-y^2}(a')$, $d_{zy}(a')$, and the $d_{z^2}(a')$ orbitals do not change within this model.

3. The orbital correlation diagram is given in Fig. H.7,III. The system now contains 12 electrons and the state correlation is displayed in Fig.H.7,IV.

4. The inclusion of the d orbitals makes the thermal reaction symmetry allowed, illustrating the catalytic effect of the transition metal (Slegeir *et al.* 1974).

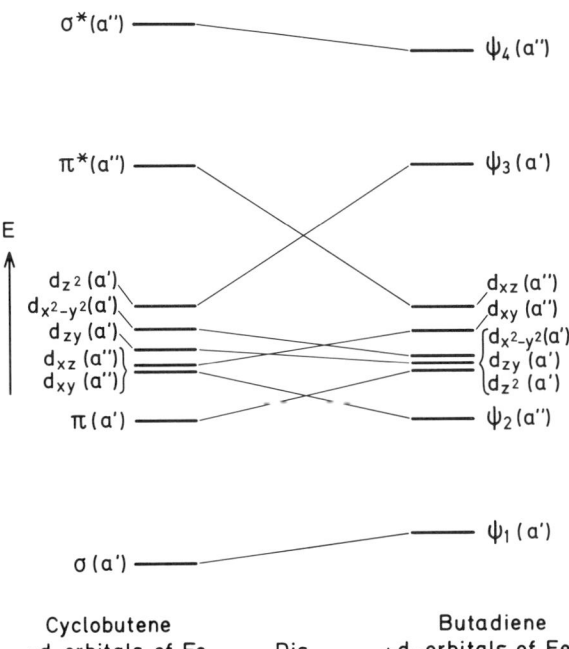

Fig. H.7,III. Orbital correlation diagram for the disrotatory ring opening cyclobutene + iron → butadiene + iron.

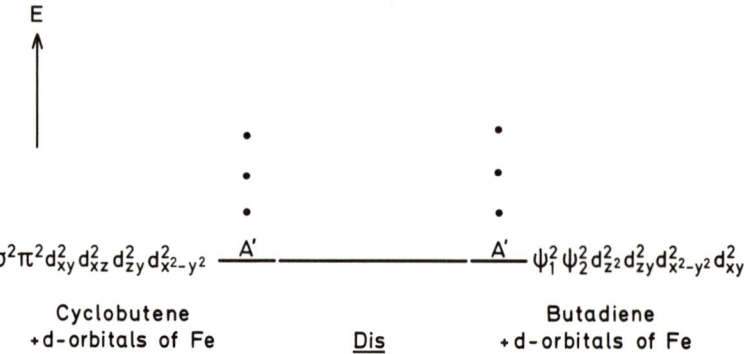

Fig. H.7,IV. State correlation diagram for the disrotatory ring opening cyclobutene + iron → butadiene + iron.

APPENDIX

LIST OF CONCEPTS

Ab initio calculation

A calculation that does not invoke empirical parameters. In particular, all integrals are evaluated without approximations.

Alternant hydrocarbon

Alternant hydrocarbons are those in which no odd-membered rings are present, so that it is always possible to divide the unsaturated carbon atoms into two sets, "starred" atoms and "unstarred" atoms, in such a way that no two atoms of a set are joined by a bond. Alternant hydrocarbon fulfills the Coulson–Rushbrooks pairing theorem (see later).

Avoided crossing

See under (state) correlation diagrams.

Confocal elliptical coordinates (ξ, η, ϕ)

Often used to evaluate integrals for diatomic molecules. Denoting the two focuses (the two nuclear sites) a and b and the distances from the point in question to the two focuses, r_a and r_b, respectively, the three elliptical coordinates are defined as

$$\xi = \frac{r_a + r_b}{R}, \qquad 1 \leq \xi \leq \infty$$

$$\eta = \frac{r_a - r_b}{R}, \qquad -1 \leq \eta \leq 1$$

and ϕ, the azimuthal angle around the internuclear axis ($0 \leq \phi \leq 2\pi$). R is the distance between the two focuses and the range of the variables is indicated above. The volume element in elliptical coordinates is

$$d\tau = \pm \frac{R^3}{8}(\xi^2 - \eta^2)\, d\xi\, d\eta\, d\phi$$

Correlation diagrams

An *orbital* correlation diagram correlates the orbitals of the reactants with the orbitals of the products in a chemical reaction. The chemical reaction may be of a fictitious nature (e.g., the united atom limit → molecular formation → separated atoms limit of a given molecule—problem E.4). On the left-hand side of the orbital correlation diagram the orbital energies of the reactants are displayed in order of increasing energies, and on the right-hand side are displayed the orbital energies of the products. The lowest "reactant" orbital of a given symmetry must correlate with the *lowest* orbital of the same symmetry of the product, and so on. The correlation between the orbitals is shown by drawing lines connecting the orbital energies. Each (nondegenerate) reactant orbital can only correlate with one (nondegenerate) product orbital. The "matching" of the orbital energies is carried out for all orbits in order of increasing orbital energy. Crossing must not occur between lines of the same symmetry.

In a *configuration* correlation diagram the electronic configurations of the reactants are correlated with those of the products.

In a *state* correlation diagram the electronic states of the reactants and products are correlated. States of the same symmetry cannot cross. If a crossing of lines occurs in the configuration correlation diagram, this crossing will not occur in the state correlation diagram if the configurations result in states of the same symmetries. In such a case *an avoided crossing* will be observed in the state correlation diagram. This means that the dominant configuration of a state will change at the internuclear separation at which the avoided crossing occurs.

Correlation energy

The correlation energy for state $|\Psi_i\rangle$ is defined as

$$E_i(\text{corr}) = E_i - E_i(\text{SCF})$$

where $E_i(\text{SCF})$ is the SCF total energy for state $|\Psi_i\rangle$ and E_i is the total energy obtained in the correlation calculation. The complete correlation energy is obtained when E_i is the exact energy of state $|\Psi_i\rangle$ and $E_i(\text{SCF})$ is the Hartree–Fock limit value.

Coulomb integral

$(ii|jj)$ [in the Mulliken notation]

Coulson–Rushbrooke pairing theorem

For an alternant hydrocarbon (see definition earlier) the π MO's of nonzero binding energies ($\varepsilon \neq \alpha$) occur in pairs with opposite energies, that is, with orbital energies in the form $\varepsilon = \alpha \pm x\beta$. The atomic orbital coefficients for the two MO's with $\varepsilon = \alpha \pm x\beta$ are the same for the one set of atoms, the starred atoms, say, and have opposite signs for the unstarred atoms (Coulson and Rushbrooke, 1940). The ground-state charge orders, q_μ, are 1 for all atoms in an alternant hydrocarbon. Furthermore, bond order between atoms of the same set (i.e., atoms separated by an even number of bonds) vanishes (for a proof, see, e.g., Salem 1966, p. 36–43).

Dipole allowed transitions

Transitions for which $\langle \Psi_i | \mathbf{r} | \Psi_f \rangle \neq 0$ where Ψ_i and Ψ_f are the initial and final state wave functions, respectively. $\mathbf{r}$ is the position operator. The polarization direction of a given transition is the component of $\mathbf{r}$ that gives a nonvanishing matrix element. Specific rules for dipole allowed transition in atomic spectra is given in the introduction to section D.

Dirac notation for two electron integrals

$$\langle ij|kl \rangle = \int u_i^*(1)u_j^*(2)\frac{1}{r_{12}}u_k(1)u_l(2)\, d\tau_1\, d\tau_2$$

Double zeta basis

A basis set that uses two basis functions to describe each orbital that is occupied in the atoms that constitute the molecule. For H_2O, for example, a double-zeta basis set consists of two $1s$ functions on each of the hydrogen atoms and two $1s$, $2s$, $2p_x$, $2p_y$, and $2p_z$ functions on oxygen (14 basis functions total).

Electron affinity (EA)

The energy gained when species S captures an electron, that is, the energy required to perform the process

$$S^- \rightarrow S + e^-$$

Thus

$$\text{EA}(S) = E(S) - E(S^-)$$

where $E(S)$ indicates the total energy of species S.

Excitation energy

For an excitation from state Ψ_i to Ψ_f the excitation energy is

$$\Delta E(i \rightarrow f) = E_f - E_i$$

where E_i and E_f are the total energies of states Ψ_i and Ψ_f, respectively.

Exchange integral

$(ij|ij)$ [in the Mulliken notation]

Fluorescence

The emission of radiation caused by an electronic transition from an excited singlet state to a singlet state with lower energy.

Frontier orbitals

The frontier orbitals are the highest occupied MO (HOMO) and the lowest unoccupied MO (LUMO) (Fukui *et al.* 1952). The frontier orbitals may be degenerate. For a molecule with an open shell ground state there is only one (possibly degenerate) frontier orbital.

f-sum rule

The sum of all oscillator strengths, f_{0n}, from a state $|0\rangle$ to all other states $|n\rangle$ obeys the sum rule

$$\sum_{n \neq 0} f_{0n} = N$$

where N is the total number of electrons in the system (also called the Thomas–Reiche–Kuhn sum rule).

Hund's rules

The relative energy of *atomic LS* terms is determined from these rules:
1. The lowest term is that of highest spin multiplicity, that is, largest value of $2S + 1$.
2. When two terms have the same spin multiplicity the lowest term is the one with the largest orbital angular quantum number, L.

These rules are generally only applicable when determining the term with the lowest energy, that is, the ground-state term. The relative energies of excited state terms may deviate from these rules. For *molecules* rule 1 is applicable for the ground state, whereas there is no equivalent of rule 2.

Inverted multiplet

The level with the highest value of J (atoms) has the lowest energy (valid for atoms for which the valence shell is more that half filled).

Ionization potential (IP)

The energy required to ionize species S, that is, to perform the process

$$S \rightarrow S^+ + e^-$$

Thus

$$IP(S) = E(S^+) - E(S)$$

where E (S) indicates the total energy of species S.

Jahn–Teller distortion

All nonlinear molecules in an *orbitally* degenerate electronic state will be unstable with respect to distortion in the nuclear framework. There exists a "Jahn–Teller-distorted" conformation of the molecule with both lower symmetry and lower energy.

Landé interval rule

This rule states that the energy separation of two adjacent multiplet levels is proportional to the larger of the J values involved (valid only in the LS coupling scheme).

Level

The designation of an atomic state with definite values of the quantum numbers

$$LSJ \qquad \text{in the } LS \text{ coupling scheme}$$

and

$$j_1 j_2 J \qquad \text{in the } jj \text{ coupling scheme}$$

Linear molecules and ± symmetry

Consider a linear molecule. Let $\hat{\sigma}_v$ denote the reflection operator corresponding to a reflection in a plane containing the internuclear axis. A reflection in this plan determines the ± symmetry of a Σ state. The operator

$$\hat{O}_+ = \frac{1}{2}(\hat{1} + \hat{\sigma}_v)$$

is a projection operator that projects out the component of a state that has Σ^+ symmetry, whereas the

$$\hat{O}_- = \frac{1}{2}(\hat{1} - \hat{\sigma}_v)$$

projects out the component that has Σ^- symmetry. A projection operator is an operator that, when applied to a given state, projects out a particular symmetry component of the state. Examples of the use of the projection operators $\hat{O}_\pm$ are given in problem E.13.

Matrix elements of Slater determinants

See Slater–Condon rules.

Microstate diagram

A microstate diagram is a table displaying all the microstates that can be constructed from a given electronic configuration, taking into account the Pauli exclusion principle. A microstate is a determinantal wave function that is constructed from the orbitals of the electronic configuration and that has a definite value of $M_L = \sum m_{li}$ and $M_S = \sum m_{si}$. An example is given in Table D.1, I.

Mulliken notation for two-electron integrals

$$(ij|kl) = \int u_i^*(1)u_j(1)\frac{1}{r_{12}}u_k^*(2)u_l(2)\,d\tau_1\,d\tau_2$$

Multiplets

The levels originating from one LS term.

Normal multiplet

The level with the lowest value of J (atoms) has the lowest energy (valid for atoms for which the valence shell is less than half filled).

Oscillator strength

The dipole length form is

$$f_{if}^{L} = \frac{2}{3}|\langle \Psi_i|\mathbf{r}|\Psi_f\rangle|^2(E_i - E_f)$$

where Ψ_i and Ψ_f are the wave functions of the initial and final states, respectively, and E_i and E_f the corresponding total energies. $\mathbf{r}$ is the position vector. The dipole velocity form is

$$f_{if}^{V} = \frac{\frac{2}{3}|\langle \Psi_i|\mathbf{p}|\Psi_f\rangle|^2}{E_f - E_i}$$

where $\mathbf{p}$ is the momentum operator (see problem A.6 for a more extensive description). The oscillator strength fulfills the f-sum rule. For the exact eigenfunctions $f_{if}^{L} = f_{if}^{V}$. This relation is in general not true for approximate eigenfunctions.

Pauli exclusion principle

"No two electrons can have all quantum numbers the same." This means that a spin orbital cannot contain more than one electron and a spatially nondegenerate orbital cannot contain more than two electrons.

Phosphorescence

The emission of radiation caused by an electronic transition from an excited triplet state to a singlet state with lower energy.

$\pi - \pi^*$ transition

An electronic transition from an occupied bonding π MO to an antibonding π MO, called π^*.

Polar coordinates

The polar coordinates (r, θ, ϕ) are defined in the Fig. L.C. 1.
Range of the variables:

$$0 \leq r \leq \infty$$
$$0 \leq \theta \leq \pi$$
$$0 \leq \phi \leq 2\pi$$

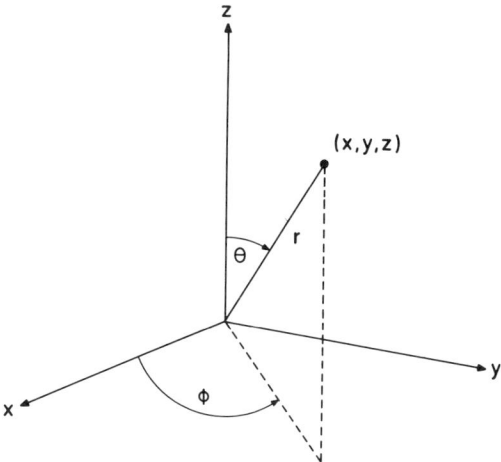

Fig. L.C. 1 Relationships between Cartesian coordinates (x, y, z) and spherical polar coordinates (r, θ, ϕ).

Volume element: $dx\, dy\, dz = r^2 \sin\theta\, d\theta\, d\phi\, dr$

Principle of least motion

This empirical principle states that the lowest activation energy for a reaction is the one that requires the least motion of the nuclei and the least disruption of the original electronic distribution.

Resolution of the identity

For any complete set of states $\{|i\rangle\}$ it holds that

$$\sum_i |i\rangle\langle i| = 1$$

where the summation extends over all states.

Rotational barrier

A molecule in a stable conformation may, through a rotation θ around a single bond, be brought into another stable conformation. The total energy $E(\theta)$ associated with each angle of rotation will exhibit a maximum for an angle (θ_M) somewhere between the start (θ_i) and end (θ_f) conformation.
The difference in energy

$$\Delta E = E(\theta_M) - E(\theta_i)$$

is called the rotational barrier.

Rotation of spherical harmonics

A set of sperical harmonic functions $\{Y_{LM_l}, M_L = -L, -L + 1, \ldots, L\}$ when rotated an angle α around an axis passing through the origin of the coordinate system (normally the atom) spans a representation, the characters of which are

$$\chi^{(L)}(\alpha) = \begin{cases} \dfrac{\sin(L + \frac{1}{2})\alpha}{\sin\dfrac{\alpha}{2}}, & \alpha \neq 0 \\[4mm] 2L + 1, & \alpha = 0 \end{cases}$$

Rydberg states

In a Rydberg state the outermost electron (or electrons) of the electronic configuration occupies a *diffuse* orbital often called a Rydberg orbital. In the Rydberg orbital the electron is at *large average distance* from the nuclei and other electrons, and it thus experiences almost a Coulomb-type potential from an ion core of $(n - 1)$ electrons and n protons, that is, a positive charge of 1. The Rydberg energy levels are therefore *hydrogenlike* and are labeled as hydrogen orbitals. Both atoms and molecules have Rydberg states. Rydberg levels are often (but not always—see, e.g., problem E.4) higher-lying excited states. Other typical characteristics of Rydberg states of diatomic molecules are (1) they have shorter equilibrium internuclear separations than valence states, (2) the vibrational constants are larger (i.e., the potential curve is steeper), and (3) spectroscopic constants vary little through a Rydberg series.

Single configurational wave function

A wave function that is constructed from a single electronic configuration. In many cases this becomes a single determinantal wave function, but in cases where, for example, a particular spin component like a singlet state is wanted, a linear combination of determinants is required to construct the single configuration wave function with the proper symmetry.

Single determinantal wave function

A many-electron wave function that is a *single* Slater determinant.

Single-zeta basis

Basis set that uses *one* atomic orbital [STO or (contracted) Gaussian] to describe each orbital that is occupied in the atoms that constitute the molecule.

Slater–Condon rules

Let an N-electron Slater determinant Ψ_K be defined as

$$\Psi_K = u_{K_1}(1)u_{K_2}(2)\cdots u_{K_1}(N)$$

where the *orthogonal* orbitals (AO or MO) are denoted by $\{u_{K_j}\}$ with the index K labeling the ordering of the orbitals in Ψ_K.

(a) Consider the *one-electron* operator $\hat{Q}_1 = \sum_{i=1}^{N} \hat{q}_1(i)$.

The matrix element of a one-electron operator between two Slater determinants becomes

$$\langle \Psi_K | \hat{Q}_1 \Psi_L \rangle =$$

$$\begin{cases} 0 \\ \qquad \text{if more than one orbital in } \Psi_K \text{ is different from the} \\ \qquad \text{orbitals in } \Psi_L \\ \\ (-1)^{p(a)+p(b)} \langle u_{K_a} | \hat{q}_1 | u_{L_b} \rangle \\ \qquad \text{if all orbitals of } \Psi_L \text{ and } \Psi_K \text{ are the same except for } u_{K_a} \text{ of} \\ \qquad \Psi_K \text{ and } u_{L_b} \text{ of } \Psi_L, \text{ respectively} \\ \\ \sum_{k=1}^{N} \langle u_{K_k} | \hat{q}_1 | u_{L_k} \rangle \\ \qquad \text{if } \Psi_L = \Psi_K \end{cases}$$

In this equation $p(a)$ and $p(b)$ represent the position of orbital u_{K_a} in Ψ_K and orbital u_{L_b} in Ψ_L, respectively.

(b) Consider the two-electron operator $\hat{Q}_2 = \sum_{i>j} \hat{q}_2(i,j)$. The matrix element of the two-electron operator between two Slater determinants becomes

$$\langle \Psi_K | \hat{Q}_2 | \Psi_L \rangle =$$

$$\begin{cases} 0 \\ \qquad \text{if more than two orbitals in } \Psi_K \text{ are different from the} \\ \qquad \text{orbitals in } \Psi_L \\ \\ [\langle u_{K_a} u_{K_b} | \hat{q}_2 | u_{L_c} u_{L_d} \rangle - \langle u_{K_a} u_{K_b} | \hat{q}_2 | u_{L_d} u_{L_c} \rangle](-1)^{p(a)+p(b)+p(c)+p(d)} \\ \qquad \text{if all orbitals of } \Psi_K \text{ and } \Psi_L \text{ are the same except for} \\ \qquad u_{K_a}, u_{K_b}, \text{ of } \Psi_K \text{ and } u_{L_c}, u_{L_d} \text{ of } \Psi_L, \text{ respectively} \\ \\ \sum_{k=1}^{N} [\langle u_{K_a} u_{K_k} | \hat{q}_2 | u_{L_c} u_{L_k} \rangle - \langle u_{K_a} u_{K_k} | \hat{q}_2 | u_{L_k} u_{L_c} \rangle](-1)^{p(a)+p(c)} \\ \qquad \text{if } \Psi_K \text{ and } \Psi_L \text{ differ by only two orbitals } u_{K_a} \text{ and } u_{L_c} \\ \\ \sum_{k<l=1}^{N} [\langle u_{K_k} u_{K_l} | \hat{q}_2 | u_{L_k} u_{L_l} \rangle - \langle u_{K_k} u_{K_l} | \hat{q}_2 | u_{L_l} u_{L_k} \rangle] \\ \qquad \text{if } \Psi_K = \Psi_L \end{cases}$$

$p(a), p(b), \ldots$ are defined as in case (a).

Slater orbital

$$\phi_{nlm}(r,\theta,\phi) = N_{nl}\,r^{n-1}e^{-\zeta_{nl}r}\,Y_{lm}(\theta,\phi)$$

where ζ_{nl} is the orbital exponent, $Y_{lm}(\theta,\phi)$ the normalized spherical harmonics and N_{nl} a normalization constant

$$N_{nl} = \frac{(2\zeta_{nl})^{n+\frac{1}{2}}}{[(2n)!]^{\frac{1}{2}}}$$

The orbital exponents are in most cases taken from standard tables or calculations, and in the problem set ζ_{nl} values will often be given quantities. However, Slater in his original work (Slater 1930) determined ζ_{nl} from the screening constants, S_{nl}, as

$$\zeta_{nl} = \frac{Z - S_{nl}}{n}$$

where Z is the atomic number.

Slater screening constants

The screening constant S_{nl} for the atomic orbital ϕ_{nlm} is evaluated according to the following rules:

1. Divide the orbitals into groups, each having a different screening constant:

$$(1s),\ (2s,2p),\ (3s,3p),\ (3d\,),\ (4s,4p),\ (4d\,),\ (4f\,),\ (5s,5p)$$

2. S_{nl} is the sum of the following contributions:
   a. 0 from any shell outside the one considered;
   b. 0.35 from each electron in the group considered (except 0.30 for the $1s$ orbital);
   c. For s and p orbitals, 0.85 is added for each electron in the next inner shell and 1.00 for all electrons still further in. For d and f orbitals, 1.00 is added for all electrons inside the orbital.

Spectroscopic (or equilibrium) dissociation energy, D_e

$$D_e = E(R = \infty) - E(R = R_e)$$

where $E(R)$ is the electronic energy at the nuclear separation R and R_e is the equilibrium internuclear separation.

Spherical harmonics

$$Y_{lm}(\theta,\phi) = \left[\frac{(2l+1)(l-|m|)!}{4\pi(l+|m|)!}\right]^{\frac{1}{2}} P_l^{|m|}(\cos\theta)e^{im\phi}$$

$P_l^{|m|}$ are the Legendre polynomials:

$$P_l^{|m|}(x) = (1 - x^2)^{|m|/2} \frac{d^{|m|}}{dx^{|m|}} P_l(x)$$

where $P_l(x)$ is given in Eqs. (B.6, 7–10). Explicit expressions for Y_{lm} can be obtained from these equations. Note the phase conventions ("all Y_{lm} have positive phase").

Spin forbidden transition
Transition that is forbidden because the initial and final state has different spin multiplicity.

Term energy
The total electronic energy of LS term, that is, a state with a definite value of the quantum numbers L and S.

Third-order polynomial, roots of
A third-order polynomial

$$E^3 + aE^2 + bE + c = 0$$

has the roots

$$E_\mu = \alpha + 2\beta \cos(\gamma + \frac{2}{3}\pi\mu), \qquad \mu = 1, 2, 3$$

where

$$\alpha = -\frac{a}{3}, \qquad \beta^2 = \alpha^2 - \frac{b}{3}$$

and

$$\cos 3\gamma = \frac{3\alpha\beta^2 - \alpha^3 - c}{2\beta^3}$$

Thomas–Reiche–Kuhn sum rule
See the f-sum rule.

Torsional force constant
If we denote the angle of rotation by θ, the torsional force constant for a particular value of $\theta(= \theta_0)$ is

$$f(\theta_0) = \left(\frac{\partial^2 E(\theta)}{\partial \theta^2}\right)_{\theta = \theta_0}$$

where $E(\theta)$ is the total energy of the system.

Virial theorem

A relation between the average values of the kinetic ($\hat{T}$) and potential ($\hat{V}$) energy that holds for an *exact* eigenstate $|\Psi\rangle$:

$$2\langle\Psi|\hat{T}|\Psi\rangle = \langle\Psi|\mathbf{r}\cdot\boldsymbol{\nabla}\hat{V}|\Psi\rangle$$

For potentials of the form $V(r) = kr^n$ the virial theorem reads

$$2\langle\Psi|\hat{T}|\Psi\rangle = n\langle\Psi|\hat{V}|\Psi\rangle$$

Zero differential overlap (ZDO)

The *differential* overlap is defined as

$$\phi_\mu(i)\phi_\nu(i)$$

where ϕ_μ and ϕ_ν are atomic orbitals. (If we integrate over coordinate i, we have an overlap integral.) The ZDO approximations means that we put

$$\phi_\mu(i)\phi_\nu(i) = \delta_{\mu\nu}$$

whenever it appears in integrals; for example,

$$(\phi_\mu\phi_\nu|\phi_\lambda\phi_\pi) = \delta_{\mu\nu}\,\delta_{\lambda\pi}(\phi_\mu\phi_\mu|\phi_\lambda\phi_\lambda)$$

ATOMIC UNITS

In the atomic unit system the three fundamental units of mass, length, and time are chosen as (1) the rest mass of an electron (m), (2) the radius of the first Bohr orbit for the hydrogen atom (a_0), and (3) the time required for an electron to travel 1 atomic unit (abbreviated a.u.) of length in the first Bohr orbit ($a_0\hbar/e^2$). Thus in atomic units

$$m = 1 \tag{AB. 1}$$

$$a_0 = \frac{\hbar}{me^2} = 1 \tag{AB. 2}$$

$$\frac{a_0\hbar}{e^2} = \frac{\hbar^3}{me^4} = 1 \tag{AB. 3}$$

Equations (AB. 1–3) imply that also $e = \hbar = 1$ in atomic units.

The atomic unit of energy, often called the hartree, is defined as

$$\frac{e^2}{a_0} = \frac{me^4}{\hbar^2} = 27.2107 \text{ eV} \tag{AB. 4}$$

This energy is twice the ionization potential of the hydrogen atom if the reduced mass of the electron is replaced by its rest mass.

Thus, when atomic units are used, one sets $e = \hbar = m = a_0 = 1$ in quantum mechanical expressions and all results will be expressed in atomic units. For example, the kinetic energy operator $-\hbar^2 \nabla^2 / 2m$ becomes $-\nabla^2/2$.

It should be noted that the atomic energy unit defined in Eq. (AB. 4) is twice the atomic unit (Rydberg) often used in physics textbooks, such as that by Slater (1968).

The relation between 1 a.u. of various quantities and the fundamental atomic constants m, e, and $\hbar$ is give below. The relation between other commonly used units and atomic units is also indicated in some cases:

1 a. u. of	In units of m, e, and $\hbar$	Equal to
Angular momentum	$\hbar$	
Area	$a_0^2 = \dfrac{\hbar^4}{m^2 e^4}$	
Charge	e	
Density	$\dfrac{m}{a_0^3} = \dfrac{m^4 e^6}{\hbar^6}$	
Electric moment (dipole moment)	$ea_0 = \dfrac{\hbar^2}{me^2}$	2.54154 debye (D)
Energy (hartree)	$\dfrac{e^2}{a_0} = \dfrac{me^4}{\hbar^2}$	27.2107 eV 627.71 kcal mole^{-1} 2.19476×10^5 cm^{-1} 2 rydberg (Ry)
Length	$a_0 = \dfrac{\hbar^2}{me^2}$	0.529172 Å
Planck's constant	$h = 2\pi\hbar$	
Rest mass of the electron	m	
Time	$t = \dfrac{a_0 \hbar}{e^2} = \dfrac{\hbar^3}{me^4}$	
Velocity	$\dfrac{a_0}{t} = \dfrac{e^2}{\hbar}$	
Volume	$a_0^3 = \dfrac{\hbar^6}{m^3 e^6}$	

REFERENCES

Abramowitz, M., and Stegun, I.A. (1965) *Handbook of Mathematical Functions.* Dover, New York.

Adamantides, V., Neisius, D., and Verhaegen, G. (1980). *Chem. Phys.* **48**, 215.

Ahlrichs, R., and Staemmler, V. (1976). *Chem. Phys. Lett.* **37**, 77.

Aston, J.G., Szasz, G., Woolley, H.W., and Brickwedde, F.G. (1946). *J. Chem. Phys.* **14**, 67.

Atkins, P.W. (1970). *Molecular Quantum Mechanics: An Introduction to Quantum Chemistry.* Oxford (Clarendon Press), London and New York.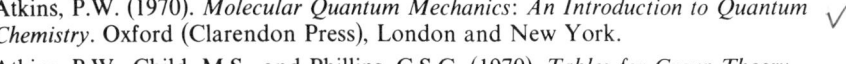

Atkins, P.W., Child, M.S., and Phillips, C.S.G. (1970). *Tables for Group Theory,* Oxford University Press, London and New York.

Avery, J. (1972). *The Quantum Theory of Atoms, Molecules, and Photons.* McGraw-Hill, London.

Ballhausen, C.J. (1962). *Introduction to Ligand Field Theory.* McGraw-Hill, New York.

Ballhausen, C.J., and Gray, H.B. (1965). *Molecular Orbital Theory.* Benjamin, New York.

Ballhausen, C.J. (1979). *Molecular Electronic Structure of Transition Metal Complexes.* McGraw-Hill, New York.

Basch, H., Moskowitz, J.W., Hollister, C., and Hankin, D. (1971). *J. Chem. Phys.* **55**, 1922.

Bellamy, A.J. (1974). *An Introduction to Conservation of Orbital Symmetry: A Programmed Text.* Longmans, London.

Bender, C.F., and Davidson, E.R. (1968). *J. Chem. Phys.* **49**, 4989.

Bernheim, R.A., Bernard, H.W., Wang, P.S., Wood, L.S., and Skell, P.S. (1970). *J. Chem. Phys.* **53**, 1280.

Borden, W.T. (1975). *J. Am. Chem. Soc.* **97**, 5968.

Borden, W.T., Davidson, E.R., and Hart, P. (1978). *J. Am. Chem. Soc.* **100**, 388.

Botterud, I., Lofthus, A., and Veseth, L. (1973) *Physica Scripta* **8**, 218.

Brion, H., and Moser, C. (1960). *J. Chem. Phys.* **32**, 1194.

Brundle, C.R., Jones, G.R., and Basch, H. (1971). *J. Chem. Phys.* **55**, 1098.

Buenker, R.J., and Peyerimhoff, S.D. (1975). *Chem. Phys. Lett.* **34**, 225.

Buenker, R.J., Peyerimhoff, S.D., and Perić, M. (1976). *Chem. Phys. Lett.* **42**, 383.

Cabell-Whiting, P.W., and Hogeveen, H. (1973). *Adv. Phys. Org. Chem.* **10**, 129.

Cade, P.E., Sales, K.D., and Wahl, A.C. (1966). *J. Chem. Phys.* **44**, 1973.

Carr, R.V., Kim, B., McVey, J.K., Yang, N.C., Gerhartz, W., and Michl, J. (1976). *Chem. Phys. Lett.* **39**, 57.

Carré, M., and Dufay, M. (1968). *C.R. Acad. Sci. Paris* **B266**, 1367.

Cederbaum, L.S., and von Niessen, W. (1975). *J. Chem. Phys.* **62**, 3824.

Chandra, A.K. (1974). *Introductory Quantum Chemistry.* Tata McGraw-Hill, New Delhi.

Childs, R.F., Sakai, M., Parrington, B.D., and Winstein, S. (1974). *J. Am. Chem. Soc.* **96**, 6403.

Clementi, E., and Roetti, C. (1974). *At. Data Nucl. Data Tabl.* **14**, 177.

Condon, E.U., and Shortley, G.H. (1935). *The Theory of Atomic Spectra.* Cambridge University Press, New York and London.

Connor, J.A., Hillier, I.H., Saunders, V.R., and Barber, M. (1972). *Mol. Phys.* **23**, 81.

Cook, D.B. (1974). *Ab Initio Valence Calculations in Chemistry.* Butterworths, London.

Cook, D.B. (1978). *Structures and Approximations for Electrons in Molecules.* Ellis Horwood, Chichester.

Cooper, C.D., and Compton, R.N. (1972). *Chem. Phys. Lett.* **14**, 29.

Couch, D.A., Wilkins, C.J., Rossman, G.R., and Gray, H.B. (1970). *J. Am. Chem. Soc.* **92**, 307.

Coulson, C.A. and Rushbrooke, G.S. (1940). *Proc. Cambridge Phil. Soc.* **36**, 193.

DeKock, R.L., and Gray, H.B. (1980). *Chemical Structure and Bonding.* Benjamin/Cummings, Menlo Park, California.

Derrick, P.J., Åsbrink, L., Edqvist, O., Jönsson, B.-Ö., and Lindholm, E. (1971). *Int. J. Mass. Spectrom. Ion. Phys.* **6**, 161.

Dewar, M.J.S. (1969). *The Molecular Orbital Theory of Organic Chemistry.* McGraw-Hill, New York.

Dicke, R.H., and Wittke, J.P. (1960). *Introduction to Quantum Mechanics.* Addison-Wesley, Reading, Massachusetts.

Dressler, K., and Ramsay, D.A. (1959). *Phil. Trans. Roy. Soc. London* **A251**, 553.

England, W.B., Rosenberg, B.J., Fortune, P.J., and Wahl, A.C. (1976). *J. Chem. Phys.* **65**, 684.

Eyring, H., Walter, J., and Kimball, G.E. (1944). *Quantum Chemistry.* Wiley, New York.

Fukui, K., Yonezawa, T., and Shingu, H. (1952). *J. Chem. Phys.* **20**, 722.

George, D.V. (1972). *Principles of Quantum Chemistry.* Pergamon, New York.

Gimarc, B.M. (1979). *Molecular Structure and Bonding: The Qualitative Molecular Orbital Approach.* Academic Press, New York.

Gradshteyn, I.S., and Ryzhik, I.M. (1965). *Tables of Integrals, Series and Products* (4th ed.). Academic Press, New York.

Gray, H.B. (1965). *Electrons and Chemical Bonding.* Benjamin, New York.

Green, S., Bagus, P.S., Liu, B., McLean, A.D., and Yoshimine, M. (1972). *Phys. Rev.* **A5**, 1614.

Griffith, J.S. (1961). *The Theory of Transition-Metal Ions.* Cambridge University Press, New York.

Hall, C.G. (1951). *Proc. Roy. Soc.* (*London*) **A205**, 541.

Haller, I. (1967). *J. Chem. Phys.* **47**, 1117.

Hanna, M.W. (1969). *Quantum Mechanics in Chemistry* (2nd ed.). Benjamin, New York.

Harding, L.B., and Goddard, W.A., III (1977). *J. Chem. Phys.* **67**, 1777.

Harris, F.E., and Michels, H.H. (1967). *Adv. Chem. Phys.* **13**, 205.

Hay, P.J. (1977). *J. Am. Chem. Soc.* **99**, 1003.

Heidt, L.J., Koster, G.F., and Johnson, A.M. (1958). *J. Am. Chem. Soc.* **80**, 6471.

Heilbronner, E., and Straub, P.A. (1966). *Hückel Molecular Orbitals.* Springer-Verlag, Berlin, Heidelberg, New York.

Herzberg, G. (1966). *Molecular Spectra and Molecular Structure, Vol. III: Electronic Spectra and Electronic Structure of Polyatomic Molecules*. Van Nostrand-Reinhold, New York.

Herzberg, G., and Johns, J.W.C. (1971). *J. Chem. Phys.* **54**, 2276.

Hoffmann, R. (1968). *J. Chem. Phys.* **49**, 3739.

Huber, K.P., and Herzberg, G. (1979). *Molecular Spectra and Molecular Structure, Vol. IV: Constants of Diatomic Molecules*. Van Nostrand-Reinhold, New York.

Hyman, H.H., ed. (1963). *Noble Gas Compounds*. University of Chicago Press, Chicago.

Karplus, M., and Porter, R.N. (1970). *Atoms and Molecules*. Benjamin, New York.

Kasha, M. (1961) in *Light and Life* (W.D. McElroy and B. Glass, eds.). Johns Hopkins Press, Baltimore.

Kołos, W., and Wolniewicz, L. (1965). *J. Chem. Phys.* **43**, 2429.

Kołos, W. (1967). *Int. J. Quantum Chem.* **1**, 169.

Kołos, W., and Wolniewicz, L. (1968). *J. Chem. Phys.* **48**, 3672.

Kołos, W., and Wolniewicz, L. (1969). *J. Chem. Phys.* **50**, 3228.

Koopmans, T.A. (1934). *Physica* **1**, 104.

Landau, L.D., and Lifshitz, E.M. (1965). *Quantum Mechanics* (2nd ed.). Addison-Wesley, Reading, Massachusetts.

La Paglia, S.R. (1971). *Introductory Quantum Chemistry*. Harper & Row, New York.

Lathan, W.A., Curtiss, L.A., Hehre, W.J., Lisle, J.B., and Pople, J.A. (1974). *Prog. Phys. Org. Chem.* **11**, 175.

Lemal, D.M., and Lokensgard, J.P. (1966). *J. Am. Chem. Soc.* **88**, 5934.

Levine, I.N. (1974). *Quantum Chemistry* (2nd ed.). Allyn and Bacon, Boston.

Linderberg, J., and Öhrn, Y. (1973). *Propagators in Quantum Chemistry*. Academic Press, New York.

Linderberg, J. Öhrn, Y, and Thulstrup, P. (1976). *Quantum Science Methods and Structure* (J.-L. Calais, O. Goscinski, J. Linderberg, and Y. Öhrn, eds.), p. 93. Plenum,New York.

Lowe, J.P. (1978). *Quantum Chemistry*. Academic Press, New York.

Mataga, N., and Nishimoto, K. (1957). *Z. Physik. Chem.* (*Frankfurt*) **13**, 140.

McConnell, H.M. (1956). *J. Chem. Phys.* **24**, 764.

McLean, A.D., and Yoshimine, M. (1968). "Tables of Linear Molecule Wave Functions," *IBM J. Res. Dev. Suppl.* **12**, 206.

McWeeny, R., and Sutcliffe, B.T. (1969). *Methods of Molecular Quantum Mechanics*. Academic Press, London.

McWeeny, R. (1979). *Coulson's Valence*. Oxford University Press, London and New York.

Merzbacher, E. (1970). *Quantum Mechanics* (2nd ed.). Wiley, New York.

Meyer, W. (1971). *Int. J. Quantum Chem.* S5, 341.

Michl, J. (1977). *Photochem. Photobiol.* **25**, 141.

Mizushima, M. (1970). *Quantum Mechanics of Atomic Spectra and Atomic Structure*. Benjamin, New York.

Moore, C.E. (1949). *Atomic Energy Levels*. (Natl. Bur. Stand. Circular No. 467). U.S. Govt. Printing Office, Washington, D.C.

Moore, J.A., and Barton, T.J. (1978). *Organic Chemistry*: *An Overview*. Saunders, Philadelphia.

Mulliken, R.S., Reike, C.A., Orloff, D., and Orloff, H. (1949). *J. Chem. Phys.* **17**, 1248.

Murrell, J.N. (1963). *The Theory of the Electronic Spectra of Organic Molecules*. Wiley, New York.

Murrell, J.N., Kettle, S.F.A., and Tedder, J.M. (1970). *Valence Theory* (2nd ed. 1978). Wiley, New York.

Murrell, J.N., and Harget, A.J. (1972). *Semi-empirical Self-consistent-Field Molecular Orbital Theory of Molecules*. Wiley-Interscience, London.

Murrell, J.N., Kettle, S.F.A., and Tedder, J.M. (1978). *The Chemical Bond*. Wiley, New York.

Newton, M.D., and Palke, W.E. (1966). *J. Chem. Phys.* **45**, 2329.

Orgel, L.E. (1960). *An Introduction to Transitioh-Metal Chemistry*: *Ligand Field Theory*. Methuen, London.

Osafune, K., and Kimura, K. (1974). *Chem. Phys. Lett.* **25**, 47.

Parr, R.G. (1963). *Quantum Theory of Molecular Electronic Structure*. Benjamin, New York.

Pauling, L., and Wilson, E.B. (1935). *Introduction to Quantum Mechanics*. McGraw-Hill, New York.

Pearson, R.G. (1976). *Symmetry Rules for Chemical Reactions*: *Orbital Topology and Elementary Processes*. Wiley-Interscience, New York.

Pilar, F.L. (1968). *Elementary Quantum Chemistry*. McGraw-Hill, New York.

Platt, J.R., Klevens, H.B., and Price, W.C. (1949). *J. Chem. Phys.* **17**, 466.

Pople, J.A., and Beveridge, D.L. (1970). *Approximate Molecular Orbital Theory*. McGraw-Hill, New York.

Purvis, G.D., and Öhrn, Y. (1974). *J. Chem. Phys.* **60**, 4063.

Roothaan, C.C.J. (1951). *Rev. Mod. Phys.* **23**, 69.

Rossi, A.R., and Bartram, R.H. (1979). *J. Chem. Phys.* **70**, 532.

Ruedenberg, K. (1977). *J. Chem. Phys.* **66**, 375.

Salem, L. (1966). *The Molecular Orbital Theory of Conjugated Systems*. Benjamin, New York.

Schiff, L. (1968). *Quantum Mechanics* (*3rd ed.*). McGraw-Hill, New York.

Schläfer, H.L., and Gliemann, G. (1967). *Einführung in die Ligandenfeldtheorie*. Akademische Verlagsgesellschaft Frankfurt am Main, Frankfurt am Main.

Shoppee, C.W., and Cook, B.J.A. (1972). *J. Chem. Soc. Perkin Trans.* I, 2271.

Siegbahn, K., Nordling, C., Johansson, G., Hedman, J., Hedén, P.F., Hamrin, K., Gelius, U., Bergmark, T., Werme, L.O., Manne, R., and Baer, Y. (1969). *ESCA Applied to Free Molecules*. North-Holland, Amsterdam.

✳ Simons, J. (1983). *Symmetry in Chemical Reactions*. Science Books International, Portola Valley, California.

Slater, J.C. (1930). *Phys. Rev.* **36**, 57.

Slater, J.C. (1960). *Quantum Theory of Atomic Structure*, Vols. I and II. McGraw-Hill, New York.

Slater, J.C. (1968). *Quantum Theory of Matter* (2nd ed.). McGraw-Hill, New York.

Slegeir, W., Case, R., McKennies, J.S., and Pettit, R. (1974). *J. Am. Chem. Soc.* **96**, 287.

Sobelman, I.I. (1979). *Atomic Spectra and Radiative Transitions*. Springer-Verlag, Berlin.

Streitwieser, A., Jr. (1961). *Molecular Orbital Theory for Organic Chemists*. Wiley, New York.

Sullivan, J.H. (1967). *J. Chem. Phys.* **46**, 73.

Takahashi, M., Kaya, K., and Ito, M. (1978). *Chem. Phys.* **35**, 293.

Tinkham, M. (1964). *Group Theory and Quantum Mechanics*. McGraw-Hill, New York.

Turner, D.W., Baker, C., Baker, A.D., and Brundle, C.R. (1970). *Molecular Photoelectron Spectroscopy*. Wiley-Interscience, New York.

von Niesen, W., Kraemer, W.P., and Cederbaum, L.S. (1976). *J. Electr. Spectrosc. Rel. Phen.* **8**, 179.

Wasserman, E., Kuck, V.J., Hutton, R.S., and Yager, W.A. (1970). *J. Am. Chem. Soc.* **92**, 7491.

Weast, R.C., ed. (1977). *Handbook of Chemistry and Physics*. (57th ed.). The Chemical Rubber Co., Cleveland, Ohio.

Wind, H. (1965). *J. Chem. Phys.* **42**, 2371.

Wolniewicz, L. (1969). *J. Chem. Phys.* **51**, 5002.

Woodward, R.B., and Hoffmann, R. (1970). *The Conservation of Orbital Symmetry*. Verlag Chemie, GMBH, Weinheim/Bergstrasse.

Zelikoff, M., and Watanabe, K. (1953). *J. Opt. Soc. Am.* **43**, 756.

INDEX